P9-DEE-676

Carpentry Framing and Finishing

second edition

Byron W. Maguire

Prentice Hall, *Englewood Cliffs, New Jersey 07632*

Library of Congress Cataloging-in-Publication Data

MAGUIRE, BYRON W. (date)
Carpentry : framing and finishing / Byron W. Maguire.—2nd ed.
p. cm.
Bibliography: p.
Includes index.
ISBN 0-13-115494-X
1. Carpentry. 2. House construction. I. Title.
TH5606.M22 1989
694—dc19 88-27545
CIP

Editorial/production supervision
and interior design: *Jean Lapidus*
Cover design: *Ben Santora*
Manufacturing buyer: *Mary Ann Gloriande*

This book can be made available to businesses and organizations at a special discount when ordered in large quantities. For more information contact:

Prentice-Hall, Inc.
Special Sales and Markets
College Division
Englewood Cliffs, N.J. 07632

© 1989, 1979 by Prentice-Hall, Inc.
A Division of Simon & Schuster
Englewood Cliffs, New Jersey 07632

All rights reserved. No part of this book may be reproduced, in any form or by any means, without permission in writing from the publisher.

Printed in the United States of America

10 9 8 7 6 5 4 3 2 1

ISBN 0-13-115494-X

PRENTICE-HALL INTERNATIONAL (UK) LIMITED, *London*
PRENTICE-HALL OF AUSTRALIA PTY. LIMITED, *Sydney*
PRENTICE-HALL CANADA INC., *Toronto*
PRENTICE-HALL HISPANOAMERICANA, S.A., *Mexico*
PRENTICE-HALL OF INDIA PRIVATE LIMITED, *New Delhi*
PRENTICE-HALL OF JAPAN, INC., *Tokyo*
SIMON & SCHUSTER ASIA PTE. LTD., *Singapore*
EDITORA PRENTICE-HALL DO BRASIL, LTDA., *Rio de Janeiro*

Contents

Preface

This book is about residential carpentry knowledge and related skills that advance the understanding of this subject. It can be used in two different ways. If you want to increase your understanding of residential carpentry's framing and finishing principles, you would study the book beginning with the first chapter, complete Chapter 1's project, and then proceed through Chapters 2, 3, and so on to the end. If, on the other hand, you had a particular framing project to solve, you could reference the chapter covering the subject and use the tables, illustrations, and text to gain the know-how for your project. These two reasons would probably be sufficient reason to add this book to your library; however, there are many more.

In this book I have attempted to convey the professional carpenter's viewpoint in the hope that a strong language bond will be established between us and that you become more conversant in the more advanced principles of carpentry associated with framing and finishing. For example, in the first two chapters, information and projects are provided for you that should build your ability to place the various framing members mentally and lead you to the correct choice as to their order of installation or functional arrangement. Most professionals can visualize these fundamentals and therefore understand the order of construction. You may have to complete each project at chapter's end to test the depth of your new understanding. Let me assure you that these projects will cause you to apply the fundamentals of the chapter in a very real way. They will also build your planning, estimating, and organizing knowledge.

I have included some rudimentary data merely to set a starting point. From that point, the text, figures, and tables are used to build understanding about framing and finishing. In many cases I have used proven engineering data but have avoided using the engineering approach so as to maintain a simpler reading style. Still you may have to devote more time to study than, say, to reading a newspaper.

With this book you could build a room onto your house, screen in your porch with professional style and quality, build decks and stairs, redesign a roof, change the character and style of your house, and even build a formal set of stairs from the first to second floor if, for example, you expanded your attic.

Now let's go over the contents of the book briefly so that you have a better grasp of the book than that you may have gotten by scanning the table of contents. Chapters 1 and 2 will cause you to look at houses from several viewpoints. One is for style: whether a ranch, split-level, Cape Cod, southern colonial, or other. The other viewpoint is to mentally place framing and finishing pieces in their proper place and understand the complexities of integrating them. For example, when you see a wide, overhanging cornice, you should have a mental image of the rafter tail, rafter tail cut at the roof's pitch, and lookouts to support the soffit, and actually see fascias, soffits, and friezes. When you look at a valley in a roof, you should mentally see a valley rafter with a *drop,* the vertical cut made with a rise of 17 inches versus a rise of 12 inches; the valley jack rafters that anchor to the ridge with a common cut; and valley cripple rafters if there is a hip roof section closeby.

You know that studs, joists, and rafters are used for framing. But in Chapter 3 you will learn which woods are capable of handling sustained (dead) loads and transient (live) loads. You will learn to select the MSR (machine stress rated) lumber or select structural or common lumber best suited to carry snow and ice or to accommodate a 40-psf live load. The project at the chapter's end and Appendix A, B, C, and D data should focus the learning effort in a practical way. Plywood is also covered since many grades and species are used in construction.

Chapter 4 is important, as it covers the insulation of living spaces. Although there are some technical descriptions, there are also some very useful reference data. For example, the R values for many building materials are provided. This is especially useful if you are planning a new addition or improving your home's thermal protection.

In Chapter 5 we discuss the principles of floor systems and provide a variety of technical information on design and construction. You will appreciate the figures showing girder construction, distributing loads, and the grade of lumber best suited for beams of various lengths. Table 5–6 provides, at a glance, the proper thickness and type of plywood for flooring. Floor assembly plans and the project at the end of the chapter will round out your understanding of the design and construction of floor assembly.

Proceeding in logical building order, Chapter 6 describes walls and wall systems. In this chapter you will discover useful information on all types of wall materials, beginning with studs. You will learn about the force placed on studs and how each grade is capable of handling the load. Figure 6–10 is especially useful because at a glance you will know what materials to use for headers of different lengths, and then we discuss why the different types are used. You will learn where common, trimmer, cripple, and jack studs are used, as well as plates and sills, and what sheathing does to add strength and other properties to a wall.

Two other parts of this chapter will benefit you in building. They are the section on exterior wall systems, which is well illustrated and described, and that on interior wall systems of both load- and non-load-bearing types. Providing nailing surfaces for drywall will be easy if you use one or more of the suggestions shown in Figure 6–21. Preparing window and door openings is also made easy with the use of Figures 6–23 and 6–24. Finally, the project will be a true planning–learning experience.

The next four chapters deal with the roof. If you know and understand all this information, you will be ready to tackle any roof project. I have not counted them, but there must be 50 to 75 terms introduced which are related directly to roofs; they are all defined and used in applications and illustrations and included in tables. Chapter 7 is a study of the common rafter and things related to it, such as pitch, rise, run, rafter tail, bird's mouth, layout, defining length using the step-off method and others, and more. The figures and Table 7–1 will really help you understand what you are about to do. I get a little technical in selecting the proper-size rafter so that you will realize that not just any piece of lumber will do. In Chapter 8 the study gets more complex since the subject is hip and valley rafters for equal-pitch roofs. One thing that you will learn quickly is to use rise and 17-in. versus rise and 12 in. for stepping-off rafter length. The figures and tables support and make the reading understandable—if you study. Here you will learn what "backing" or "drop" mean. You should make good use of Table 8–4, 8–5, or 8–6 if you cut a hip or valley with an overhanging rafter tail. Chapter 9 deals with the jack and cripple rafter. This information is technical, so I have added quite a few illustrations, close-ups of cuts, and cutaways to help visualize what is being done. As before, tables and charts show a lot of handy reference materials, and I stress interfacing with the framing square. You will find a practical solution for defining jack and cripple rafter length in this chapter. If you do the chapter-end project, you will learn quite a bit more and should add to your planning and estimating knowledge. There are a variety of roof-related subjects in Chapter 10, including the building of dormers, gable ends, and chimney saddles; fly rafter installation; and sheathing.

In Chapter 11 we study the various ways to keep a house dry. You will learn about using flashing, caulking, tar paper, and grout. The ample illustrations clearly show the techniques used to solve each type of situation.

In Chapter 12 I show you how to build on-site columns for one- and two-story porches. The work is not difficult but is exacting; there are several technical points in the chapter that need careful study. Figures 12–5 and 12–6 help explain these.

Should you consider screening in a porch as I suggest in Chapter 13, you will gain cabinetmaking-joinery skills in the process. Making a screened-in-porch is challenging for several reasons: because a great deal of preplanning is required, and because mortise-and-tenon or dowel jointing must be used. However, beginning on page 271 an estimating layout is given to you and on the next page a suggested construction schedule is provided in three parts. Following the construction is a section on panel installation which completes the building job.

Chapters 14 through 16 deal with stairs and staircases. In Chapter 14 you will gain an understanding of the principles of staircase construction and design. Many new terms are used and shown in illustrations. Some principles include the rule of 17, stairwell sizing, angle of incline, and anchoring of stringers. In Chapter 15 you will study the practical aspects of building interior stairs. The process is well illustrated in step-by-step fashion. In Chapter 16 the focus is on staircases from decks to the ground, and I cover the unique situations frequently encountered. I also explain the need to prevent decay by using treated lumber and natural conditions. The many illustrations should answer special applications frequently faced when designing such stairs.

I wish you personal success with each project you undertake.

ACKNOWLEDGMENTS

Many associations and suppliers of building materials have contributed to this book. Special thanks are due the following: American Plywood Association, Boise Cascade Corporation, Celotex Corporation, Kingsbury Homes Company, National Forest Products Association, Southern Forest Products Association, and Western Wood Products Association. A special thanks goes to my wife, Betty, for her work and support on this book, and to the people of Prentice-Hall for making the publication possible.

Biloxi, Mississippi *Byron W. Maguire*

Chapter 1

Functional Properties in House Construction

The novice carpenter is most often thrown without training into the construction of houses by the contractor, takes the instructions on each task and applies them to that task. In this environment there is little, if any, time to reflect on why the task is done in a certain way or to notice that a building principle is involved. When the occasion arises for a repetition of the same task, it is repeated as often as necessary. Naturally, if a variety of tasks are performed over and over, a definite skill is obtained. But has any learning of construction principles taken place? All too often the answer is "no," because when this "trained" carpenter is confronted with a slight variation of the principle applied, he or she cannot cope, and thus makes mistakes that result in a weakness in the structure of the house.

When weekend carpenters build or modify their homes, they frequently use the same approach. They too make improper decisions that result in added costs and structurally unsound improvements.

Not everyone fits into these categories. Many novice carpenters and weekend carpenters are successful with most jobs and seek guidance from books like this one on the most difficult and technical parts of the construction project.

To be effective in applying the principles advanced in this book, the functional properties of house construction need to be defined to set a common ground for meaningful study.

ESTABLISH PARAMETERS

The first step in establishing the parameters of a house is to obtain a total concept of the exterior proportions of the house to be built. From what data can such an idea be developed, and what should be included? It could be that the concept to be formed is the *character* of the house's exterior developed from its parameters. Parameters include the shape and a kind of roof or cornice, proportions of the house using straight lengths of walls and offsets, doors of common and special character, window placement as well as kind of window, and exterior wall materials. This is an ideal way of approaching the problem. If there is an architect's rendering of the building, much of its character can be gleaned from careful study.

Parameters from the Architect's Rendering

Windows provide an example of what can be discovered from the architect's rendering. The style of the windows and their relative positions can be seen. They could be single or double hung, casement, bay, or picture windows. If they are single hung, double hung, or casement, installation should be simple and rather commonplace. However, suppose that one is a bay window; then many questions need to be answered. Are there to be special framing requirements? Probably. The framing is not usual because of the many angles involved with bay windows. There may also be some roof framing as well. The angles will be difficult for an apprentice to cope with. Some journeymen also have problems framing the roof over a bay window. So you can see, windows and their shapes give a house some of its character and immediately bring forth mental images of the structural needs for proper anchorage and inclusion.

What else may be gleaned from an architect's rendering? Doors often play a significant role in a house's character. The door or doors may be any of a variety of designs, usually solid and quite heavy. More significant, though, is how they are framed, and this leads directly to thinking about how they will be supported. Perhaps a colonial-style door assembly is used as in Figure 1-1. The wide moldings around the outside of the frame require special care and handling. The trim also poses problems involving flashing and siding, and some thought must be given to the best method for installing the unit so that it is plumb and level, well anchored, securely braced, and waterproofed for many years.

Other features of the house that can be obtained from the rendering are the type of siding and cornice, the porch and its features, the carport, and the breezeway, to name a few. As each part of the house is thought of, the person viewing the rendering should consider the normal structural requirements, as well as the usual tasks, and, if possible, the sequence of construction. If a part of the house is observed but the underlying construction can be visualized only piecemeal if at all, a need for learning exists.

Figure 1–1 Colonial door assembly.

Parameters from the Blueprints and Specifications

To continue, suppose that the house is to be framed and have a brick veneer. How would the frieze boards be installed? Figure 1–2 illustrates a detail drawing from a

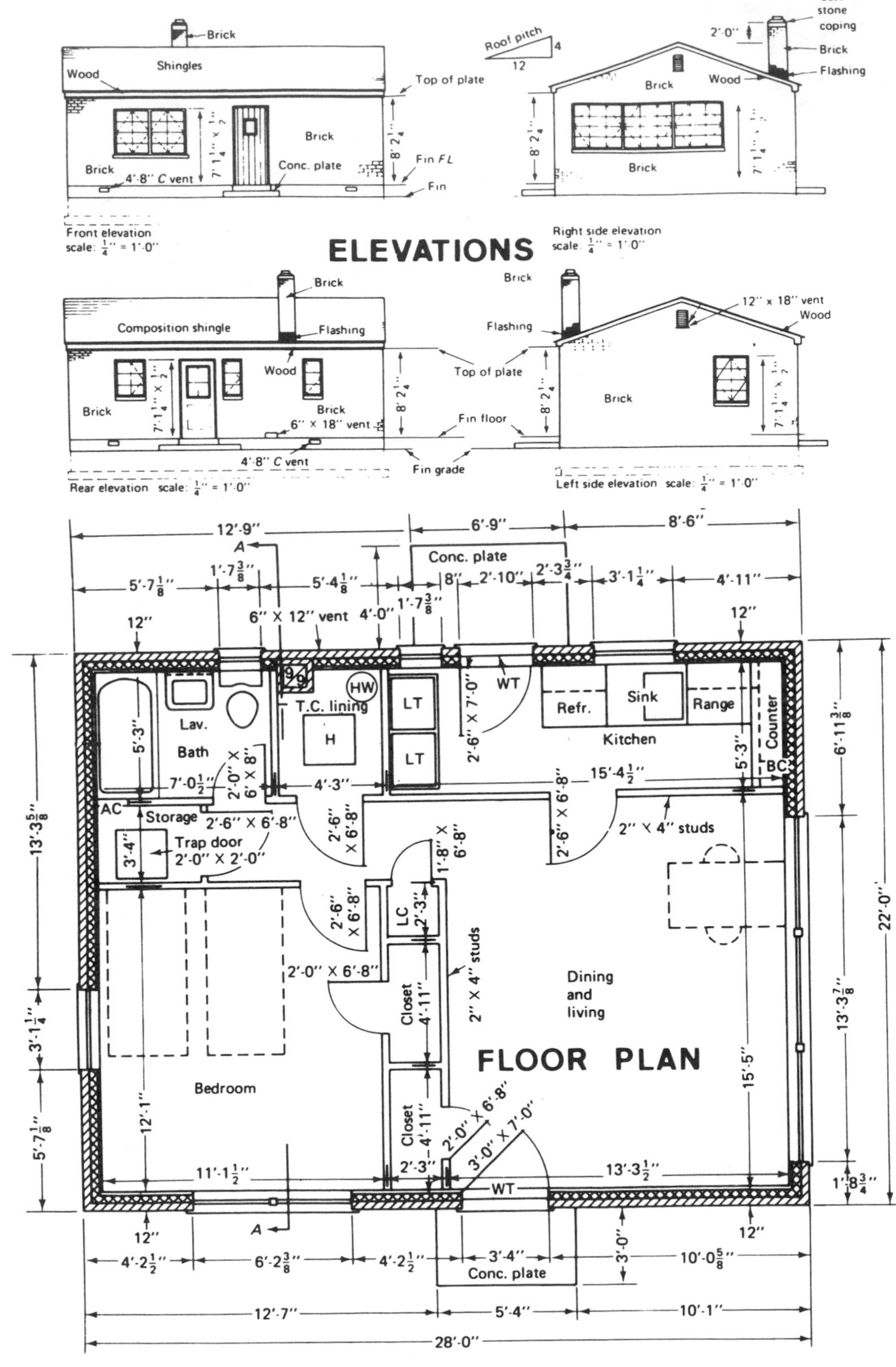

ELEVATIONS
Brick
Wood
Shingles
Top of plate
Brick
Conc. plate
Brick
4'-8" C vent
Fin FL
Fin
Front elevation
scale: 1/4" = 1'-0"
Roof pitch
4
12
Cast stone coping
2'-0"
Brick
Flashing
Brick
Wood
Brick
Right side elevation
scale: 1/4" = 1'-0"
Brick
Composition shingle
Flashing
Wood
Brick
Brick
6" × 18" vent
4'-8" C vent
Rear elevation scale: 1/4" = 1'-0"
Top of plate
Fin floor
Fin grade
Brick
Flashing
12" x 18" vent
Wood
Brick
Left side elevation scale: 1/4" = 1'-0"
FLOOR PLAN
Conc. plate
6" X 12" vent
Lav.
Bath
T.C. lining
HW
H
LT
LT
WT
Refr.
Sink
Range
Kitchen
Counter
BC
AC
Storage
Trap door 2'-0" X 2'-0"
2" X 4" studs
LC
Closet
Closet
2" X 4" studs
Dining and living
Bedroom
WT
Conc. plate
12'-9"
6'-9"
8'-6"
28'-0"
22'-0"
13'-3 5/8"
12'-7"
5'-4"
10'-1"

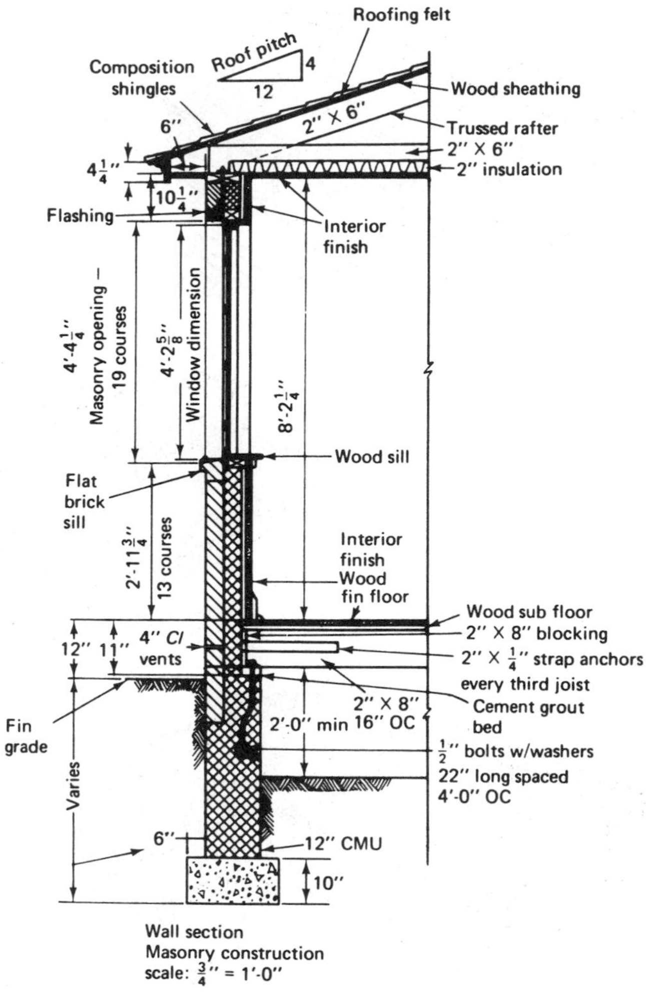

Figure 1–2 Details of blueprints reveal much of the final look and of the details of fabrication (see also facing page).

blueprint that gives these data. This leads to the second method of developing the character of the exterior of the building: the use of blueprints and specifications.

Much can be learned from a detailed study of these plans. They can be used to develop a variety of images of the character within the house. For example, the shape of the perimeter is readily seen from the floor plans. This should convey an image of breaks in wall lines; interior and exterior corners; window placement; type of wall to be constructed; parts to be attached, such as porches, breezeways, and carports; and entranceway details.

From the elevation drawings each of the house's sides can be seen, studied, understood in terms of how common, selected, or special principles of construction

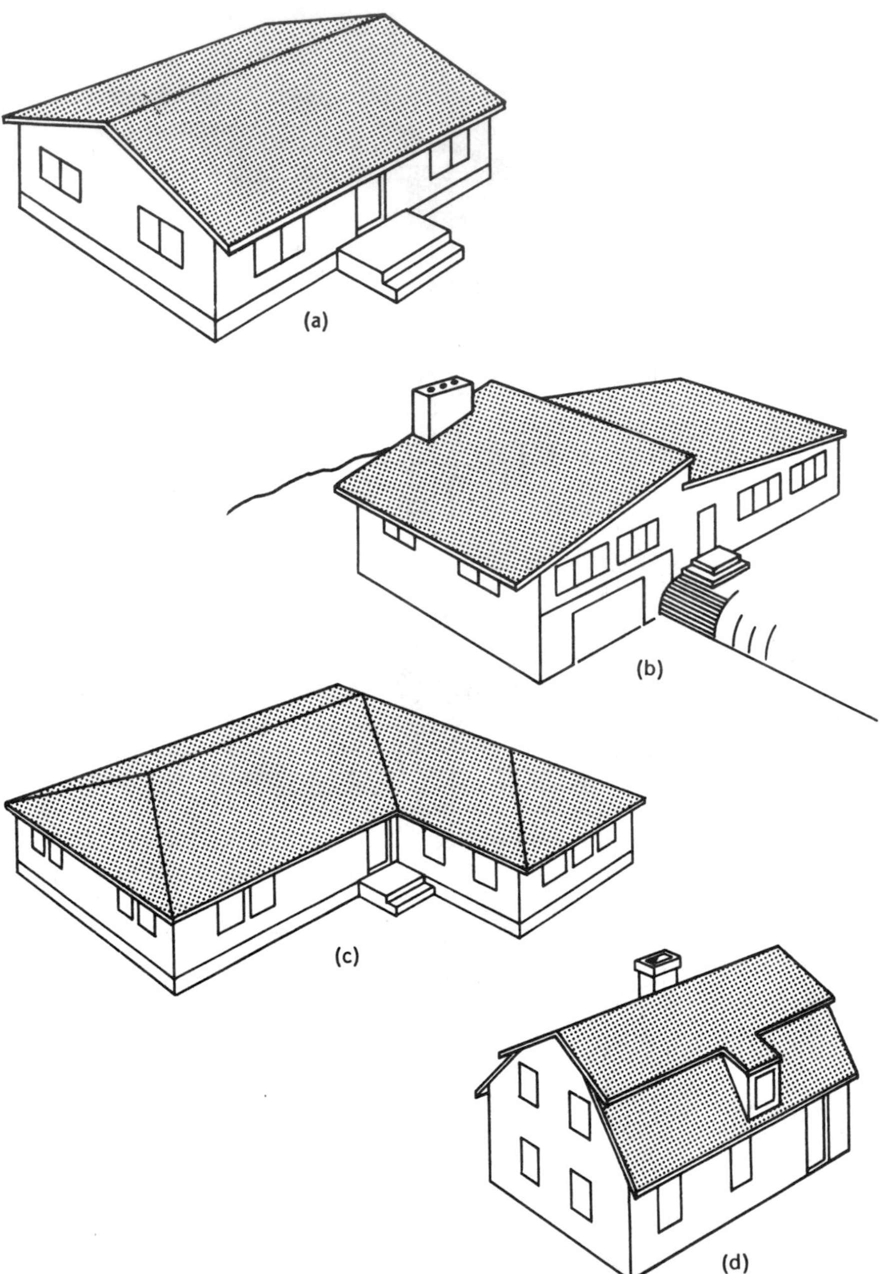

Figure 1–3 Determine how rafters and bracing are included by mentally constructing the roof.

are applied. In addition, detail drawings can also provide a vast amount of data that relate to the character of the house. From these drawings and the elevation drawings, a mental picture of the roofline can be developed. The roof design can be determined to be a simple *gable* or *A-framed, hip* or *hip and valley, flat roof, gambrel,* or *mansard* type, as Figure 1–3 illustrates. Also, determine if the roof has dormers. If it does, are they gable or shed style? These designs mean a great deal in terms of the principles of construction. Chapter 2 covers the designs of roofs in more detail, but for now the idea of developing a mental image of the construction involved should clearly point to using trusses or building the roof with individual rafters and braces. A look at the plan will quickly define the type of construction.

From the outside view, a mental image should be developed of where, for example, hip and valley rafters are needed, where the jack rafters are to be installed, where common rafters should be used, and finally where cripple rafters might be needed. The roof design and shape will tell all of these things. The roof design should also show the type of cornice used. This feature influences the layout of the rafter. If the cornice overhangs the house, the rafter length must be increased to allow for construction of the cornice. If the cornice is the *close* variety, as shown in Figure 1–4, no rafter overhang is needed. However, proper backing for the cornice must be used and this information is seldom given in plans or specifications.

Where the roof is attached to the wall, as in a breezeway, the intersection of roof and wall requires precision installation of roofing shingles, flashing, and wall siding to ensure a sound, waterproof job. After visualizing this picture, you should develop mental images of the construction sequence. For that matter, thinking could go into more detail on such matters as the method of tying the rafters to wall framing, sound application of sheathing, and preparation of the flashing materials.

Considerable information on the character of a house can be obtained from a close study of the exterior views. Some of the most prominent features such as windows, doors, siding, roof, and cornice allow the carpenter's mind to delve into

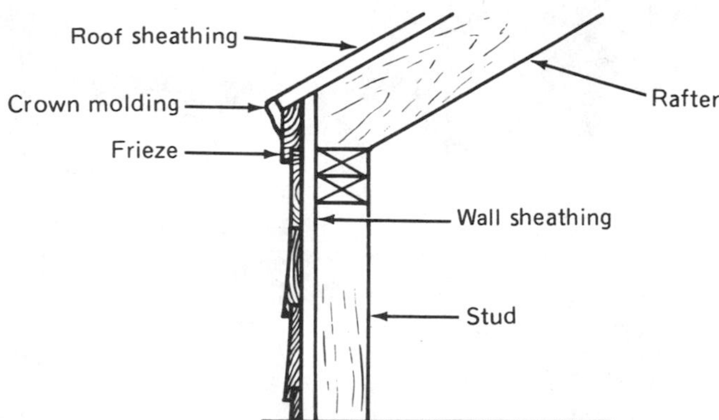

Figure 1–4 A mental picture should be made of the factors that will affect cornice installation.

the structural needs that make these features possible. Obviously, not all parameters have been developed in these few pages. More information can be obtained by study of the plans, but there is another avenue open to the carpenter that can develop a mental image of the structure.

Parameters from the Styles of Houses

Although the definitions of the specific character of a house are fundamental to the builder, there is some regularity of construction if the style of a house can be identified by its overall appearance. Figure 1–5 shows four different styles of houses. Each can be built with standard building materials; yet certain characteristics are more predominant in one than in others. For example, the Cape Cod style (Figure 1–5a) seldom has much overhanging cornice. This is understandable, because in the Cape Cod region of the northeastern United States, the sun helps heat the house during the long cold winters. In contrast, the ranch house (Figure 1–5b) has a large overhang so that the sun is blocked from the walls and windows during the long hot summers of the South and Southwest.

The split-level style of house (Figure 1–5c) came into being after World War II and found wide acceptance on sloping property. It may be styled for the North with small overhangs or the South with large overhangs. The two-story house (Figure 1–5d) has always been popular in America and is the most economical house to build and maintain on a per-square-foot basis. Heat from the lower floor drifts upstairs, rooflines are usually simple, and dormers are frequently included.

With these particularly distinctive characteristics, the carpenter can create mental images of the structural needs for floors, walls, window and door installation, as well as rafter and cornice needs. These are important objectives that develop much understanding of the principles used in constructing the exterior of a house.

Figure 1–5 Styles of houses tell the story of their construction. (Photos courtesy of Kingsberry Homes, Inc.; formerly a part of the Boise Cascade Corporation.)

Figure 1–5 (continued)

DEFINE INTERIOR PROPORTIONS AND SPECIFIC CHARACTER

Several concepts can be developed about the interior character of a house from the plans and specifications. Room sizes and arrangement are the most obvious. Why, though, are these rooms placed and sized as they are? The life-style of the family, number of members, and family income probably dictate much of the size and location of the rooms. However, these values frequently are secondary to the architect who designs the house. He or she is concerned with other, and to him or her more important, considerations. He or she knows that more people can afford homes if they are built economically, with a minimum of waste materials, yet meet FHA and other national building codes. The architect therefore selects a group of rectangularly proportioned rooms and arranges them into a house. Using this method maximizes the use of all materials, providing a general-purpose house to meet the general needs of the average family. Once the basics are established, he or she applies trim and other individual features much as auto manufacturers do when dressing up cars. Some rooms may be paneled, others papered, and some with chair rails for individuality and sales appeal.

Parameters for Bearing Walls and Proper Corner Construction

The carpenter, on the other hand, must view the drawing and specifications for other purposes. Namely, are there load-bearing or non-load-bearing partitions? If they are load-bearing, it means that ceiling joists will rest on top of them. It also means that if the bearing partition runs parallel to the floor joists, trimmer joists or double joists need to be installed in the floor. It could also mean that a girder needs to be installed beneath or within the floor joist assembly. A non-load-bearing partition does not need reinforcement to the extent that a load-bearing partition does. Some reinforcement, such as solid bridging, should be included in both floor and ceiling joist assemblies.

Another aspect of interior character of a house is that all wall materials meet at corners and along the ceiling. So the need for proper backing to nail or bond the wall materials must be provided. A study of the plans should reveal that the partitions that run perpendicular (at right angles) to the ceiling joists need no additional ceiling-corner backing. Those that run parallel to the ceiling joists do. Interior sidewall corners need backing for both types of walls. Assemblies similar to those shown in Figure 1–6 must be built and installed to provide the adequate, sound backing needed.

Parameters for Sound Control

Television, radio, stereo, and a need for quiet are all important parts of the average family's daily living. To provide these for the individual, and yet allow others in the same household privacy for their needs, rooms are increasingly being soundproofed. The specifications could call for sound control at STC 50, for example. This level of sound control means that regular levels of noise will not penetrate the

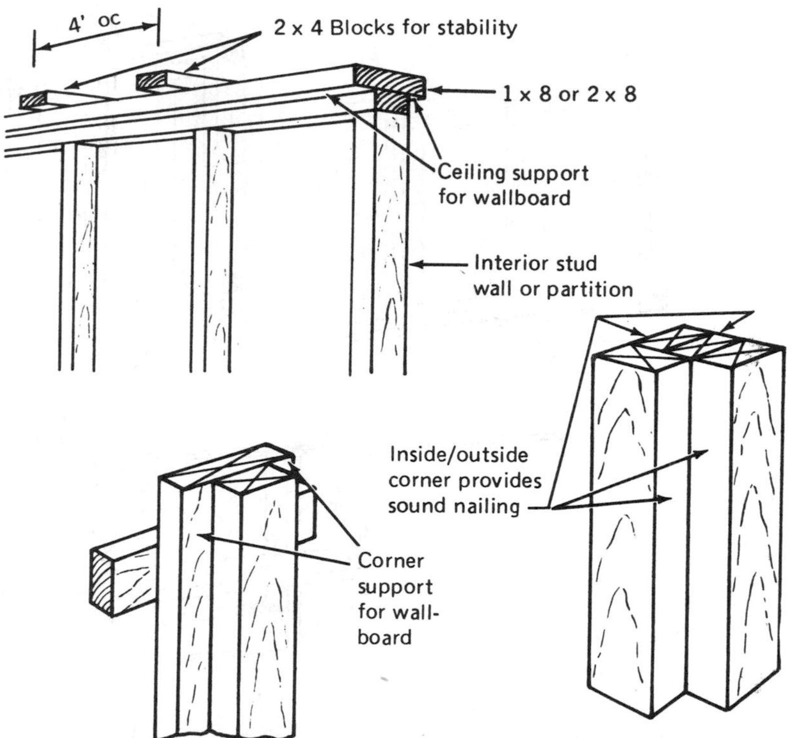

Figure 1-6 Sound anchorage for walls and wall coverings is not found in plans, so it must be thought of by the carpenter.

wall or radiate from room to room. However, sharp sounds, such as a chair falling over or a book falling to the floor, may be audible. Soundproofing is accomplished in many ways; two that can be seen from plans are the *double-stud double-plate* (Figure 1–7a) *and double-stud single-plate* (Figure 1–7b) construction techniques. Other data taken from specifications can call attention to other techniques. As Figure 1–7c and d show, resilient channels, sound deadening board, wool or glass fiber insulating batts, and carpeting may be required to achieve desired results.* Other sources of information on sound control must be studied to learn what is needed for a thorough and effective job.

Parameters for Modular Wall Panels

Finally, the character of the interior may be set by modular wall panels. These are wall panels varying in thickness from 1½ inches to several inches thick and are finished on both surfaces. They are usually set into channels that are fastened to

*For more on sound control, see Chapter 6 of *Carpentry in Commercial Construction* (Craftsman Book Co., Carlsbad, CA).

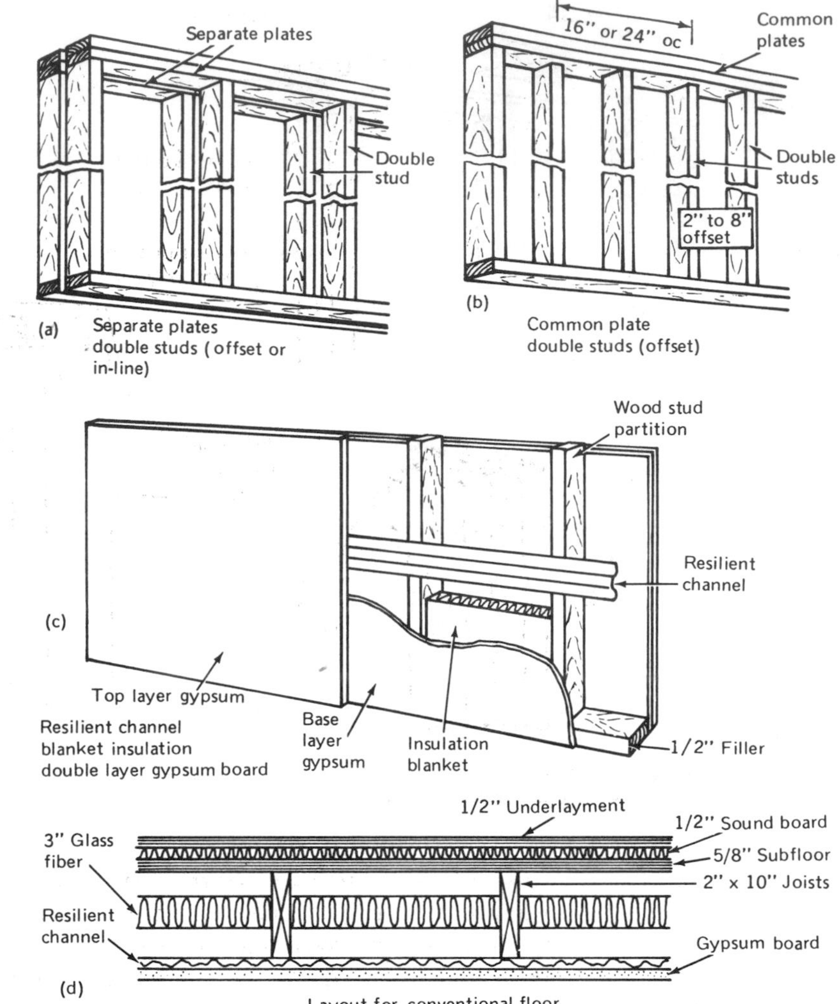

Figure 1-7 Understanding the principles of sound control through alternative construction principles and methods.

the floor and ceiling. There installation and removal is simple and quick. These walls could be installed in residential houses where truss roofing is used. Since the truss is self-supporting, the lower chord (same place as the ceiling joist) acts as a ceiling joist, and partitions may be installed anywhere and at any time. All the interior partitions are non-load-bearing.

Several constraints are placed on moving walls with wiring and plumbing. A study of the plans and specifications should reveal if modular walls are included. If they are, a procedural sequence of construction can be developed. For example, blocking between trusses can be installed where partitions will initially be set; then the entire ceiling of the house may be covered with wall board. All perimeter walls should be wired and have the plumbing installed, then insulated and wallboarded. Finally, modular wall units as partitions can be installed.

There may be other methods of developing the character of the house's interior, but all should ultimately lead the carpenter to the application of construction principles to meet any and all situations.

DETERMINE SUPPORT REQUIREMENTS

Thus far this chapter has illustrated the need for the carpenter to study closely the pictorials, plans, and specifications of a house to develop mental images of each completed segment. To this end, examples were given to help you understand how to relate what you see to what is needed during construction. There are several additional and fundamental requirements that must also be made a part of the image. These are the support requirements for floor and ceiling openings, the alignment of joists, studs, and rafters, and the slope of the roof.

Openings in Joist Assemblies

A structural deficiency exists wherever an opening is made in a floor or ceiling joist assembly, whether it be for a chimney, stair, or fan. Additional materials and strengthening methods must be used to overcome the deficiency. As a rule, headers and trimmer joists as shown in Figure 1–8 are employed to reinforce the structure. The headers support the free end of the joists. The double header provides needed strength. The trimmer joist provides support in a parallel direction to the joist. It also provides sound anchoring for partitions or other construction forms that may be attached to the opening.

Pressure and Force from the Roof

Figure 1–9 shows the direction and points of force that are present in any house. Notice that the rafter force is directed to the plate. Through the plate and into the stud is a downward force that continues through the floor joist to the foundation. It is therefore vital that the rafter and wall stud be aligned one on top of the other and that the studs align over the floor joists. Should the rafter be offset from the stud, its force and weight may be sufficient to shear the plate members. Some types of lumber have extremely low shear values.

The support requirements of the roof, that is, the ability of a roof to carry its normal dead load of plywood and shingles as well as a snow load, depends to a large extent on its slope. The greater the slope, the more the force is directed to the rafter–wall–plate junction. Therefore, less direct weight is applied to the rafters.

However, in a low-sloping roof, say a 4- or 5-in.* pitch, the weight of snow would cause a more direct downward force, and thus possibly cause a structural

*Please note: Throughout this book in. refers to inch or inches; ft. refers to foot or feet.

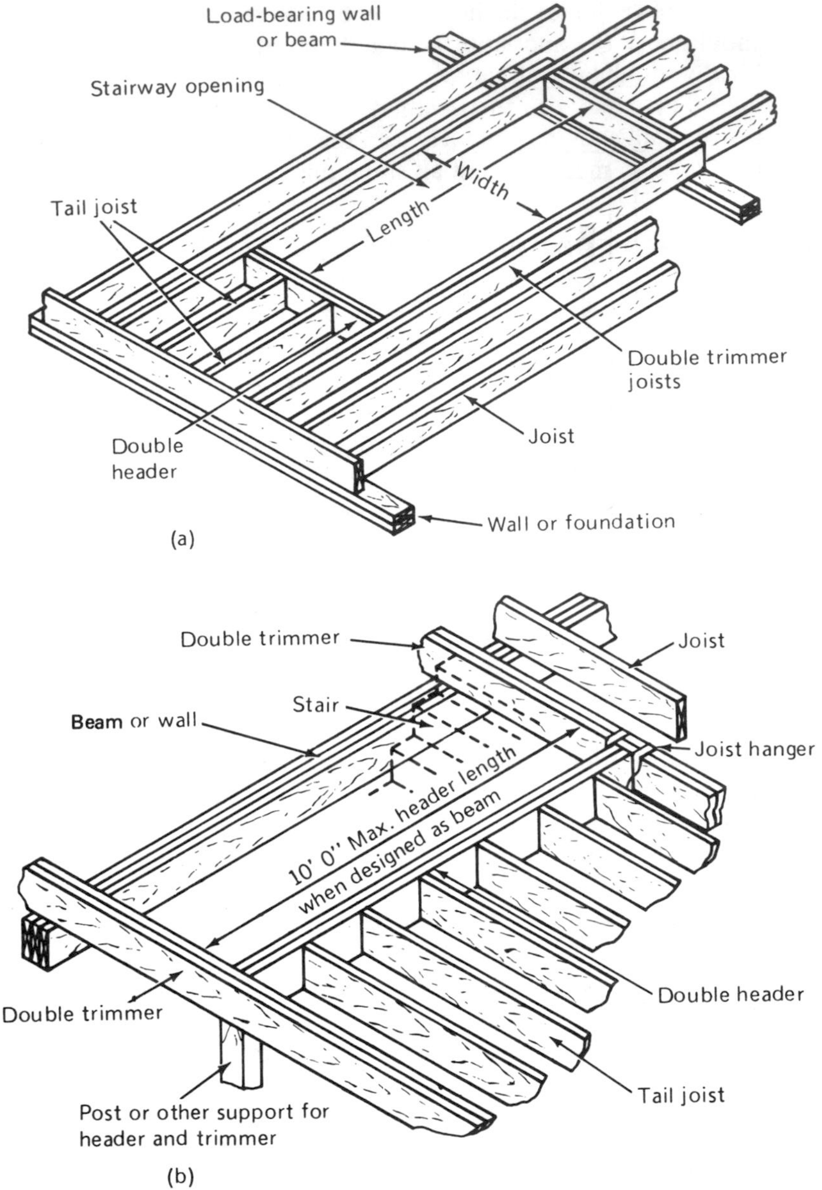

Figure 1–8 Understanding the principles necessary to overcome structurally weak areas.

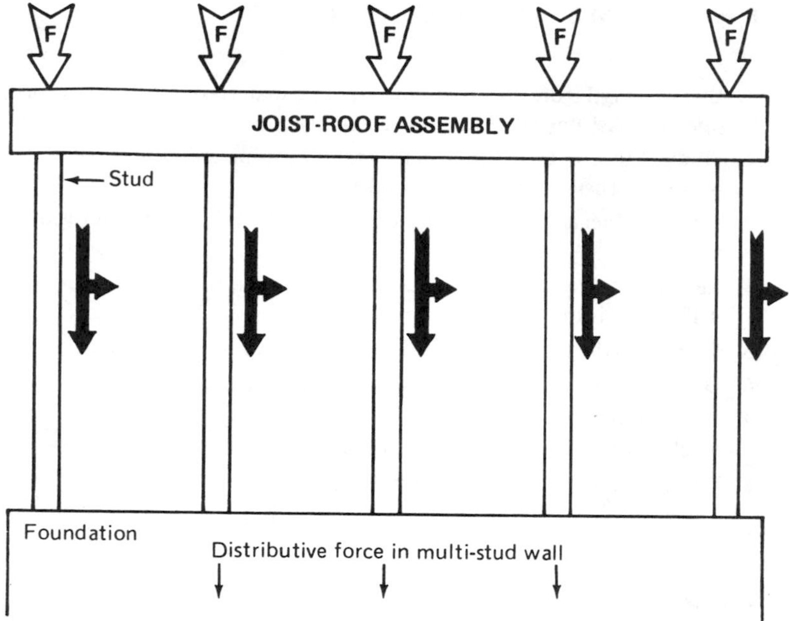

Figure 1–9 Forces always exist, but how do we cope with those developed by the roof?

collapse. So what must be done? The members selected for rafters must meet all the requirements to support the dead load, a live load, and a snow load, where applicable. To this end, certain species of fir or pine are used. They must be properly cured and seasoned and must be of appropriate size. Tables giving these parameters are provided in the appendix.

From this brief discussion, it is clear that a study of the related factors, including physical layout, geographical location, snowfall predictions, slope of roof, and materials of sufficient quality and strength is imperative to the carpenter to ensure that the house is soundly constructed.

In conclusion, the ideas put forth in this introductory chapter were developed from the experience of many master carpenters. The best of the trade can and do mentally develop a sequence of construction long before the first board is selected. The best men and women know the exact size of each piece of lumber, where it goes, and when to install it. They know where the points of force are greatest and what steps are necessary to ensure a sound, reliable structure. It therefore should be the goal of every carpenter to become a master craftsman. If the goal is ever achieved that carpenter will have developed the ability to mentally construct each segment of a house as well as the whole house. He or she will then translate those mental images into practical tasks in proper sequence, and thus create a beautiful work.

PROBLEMS AND QUESTIONS

1. By studying Figure 1–1, state the framing materials you would expect to use to frame the opening. Assume that the door assembly is 5 ft. 6 in. wide.
2. Study Figure 1–2, the floor plan, and determine the number of 2 × 4s you would use for outside corners on the exterior walls.
3. What problems would you expect to have to solve if the elevation plan called for a bay window?
4. Referring to Figure 1–2, the elevation drawings, what is the exterior covering of the building? What is the pitch of the roof?
5. Why does the frieze board in Figure 1–4 have to be rabbeted?
6. Describe two methods of framing a soundproof wall.
7. Which style of house uses
   a. close cornices **b.** large overhangs
8. Define a load-bearing wall or partition.

PROJECT

The next several projects will increase your knowledge on recognizing structure and form. (Several questions have more than one correct option.)

A. Using Figure 1–5, make the following observations.
1. Which houses have close cornices or the gable ends?
   a. Cape Cod b. Ranch
   c. Split-level d. Two-story
2. Which houses use corner boards with the siding?
   a. Cape Cod b. Ranch
   c. Split-level d. Two-story
3. Which houses use a balanced window placement design?
   a. Cape Cod b. Ranch
   c. Split-level d. Two-story
4. Match the type or style of door to its house.
   a. Cape Cod (1) Six-panel colonial
   b. Ranch (2) Colonial assembly with sidelights
   c. Split-level (3) Cross buck or Dutch
   d. Two-story
5. Which houses need valley rafters and valley jack rafters?
   a. Cape Cod b. Ranch
   c. Split-level d. Two-story

B. Using Figure 1–2 as a reference, what details provide structural form?
1. Is the rear wall line broken by the chimney?
2. Is the house square?
3. Is either the front or rear door recessed?

4. The glass area on the right side appears to be either doors or windows in the floor plan. Which are they?
5. Does the specification call for a truss roof?
6. What is the interior finished ceiling height?
7. Assume that the floor and ceiling joists run parallel to the front of the house. Where would the bearing wall be located?
   a. The living and dining area left wall
   b. The bedroom and heating closet right walls
   c. The bedroom rear wall; includes a header across the living and dining area

C. Again using Figure 1–5 as a reference, find the following.

1. Where is the force of the roof directed on the:
   a. Cape Code house? Front, left side, back, right side
   b. Ranch house? Front, ends, back
   c. Split-level house?
      (1) Top level Front, back, sides
      (2) Lower level Front, back, sides
   d. Two-story house?
      (1) Top level Front, back, sides
      (2) Pediment Front, back, sides
      (3) Breezeway Front, back, sides
      (4) Garage Front, back, sides

Answers:

Part A: **1.** a,d
2. a,c,d
3. a,c,d
4. a1, b3, c2, d2
5. b,d.

Part B: **1.** No
2. no, rectangular
3. no, rectangular
4. windows
5. yes
6. 8 ft. 2¼ in.
8. b.

Part C: **1.** a. front, back
b. front, back
c. (1) sides
(2) front, back
d. (1) front, back
(2) sides
(3) front, back
(4) sides.

Chapter 2

Designs of Residential Houses

Chapter 1 was devoted to explaining how the experienced carpenter or master craftsman visualizes all of the different tasks that will be performed. Subjects were generalized. Examples of portions of homes were selected to show how and why trained carpenters become so. In this chapter the approach is more selective. The focus is on discrete designs of homes found throughout the country. Each subject represents a particular style or design of home. Most frequently, the design resulted from environmental conditions, social attitudes, and availability of resources.

In the development of the colonial home design, most included a porch or veranda that was used socially during the cool mornings and evenings. The porch ceiling and its roof also provided shade for the sidewalls of the home, thereby acting as an insulator between outdoors and indoors. Large windows and high ceilings also contributed to comfortable living. Consider also that the houses were not insulated and had no central heat or air-conditioning.

Contrast the other designs with the colonial. There are also sound reasons for their design. The two-story without colonial styling is beautiful, not at all commonplace, but is a very excellent design. The ranch style is a rambling, spread-out design. Its uniquely large overhanging cornices provide protection from the hot summer sun. The split-level is extremely popular in hilly country because of the added rooms that may be obtained without altering the terrain or resorting to extensive landfill or excavation. The saltbox and Cape Cod styles are simple in design. They were originally designed to make use of the sun year round. Finally, the contemporary or ultramodern style evolved along the coastline and in special mountain areas where views are exceptionally beautiful. Its design maximizes the use of natural light and attempts to integrate the indoors with nature.

This brief introduction to housing designs narrows the selections that follow and focuses attention not only on style but also on the simplicity or complexity of design that translates to specific tasks that must be done when constructing any of the houses.

CONSTRUCTION FEATURES OF HOUSING STYLES

The Southern Colonial Style

One of the main characteristics of the southern colonial design is the decorative columns that support a porch or pediment. Frequently, these are round hollow assemblies with bases and crowns, but they may also be square. Figure 2–1 shows four round columns. When columns like these are used, they are usually purchased from a wood mill; the carpenter merely installs them. Square columns, on the other hand, are frequently made on-site.

The roof design in Figure 2–1 is the simplest of all designs, the gable. The main roof over the two stories and porch or pediment extends beyond the columns and all sides to create wide overhangs. Since the roofline is unbroken, it may be assumed that ceiling joists extend out over the porch and are used as part of the roof assembly and porch ceiling. On the other hand, simple truss assemblies may be used quite easily for the entire structure.

Trusses may also be used for the lower gable roofs, or they may be made from properly designed and braced rafters. The lower roofs also carry the wide overhang design.

All cornices are *closed* or *boxed* in. The *rakes* and *soffits* are all enclosed with plywood. Although not shown, ventilating panels or screens are installed in the soffits to allow proper airflow in the attic.

The house is wood framed with the probability that *platform framing* is used rather than *balloon framing*. The reason is the unbalanced window and door layout of the front wall. Notice that there are different window widths on the second floor compared with the first. When balloon framing is used, windows on each floor are usually the same width and are aligned vertically. Another reason for using the platform framing technique is that the upper floor extends slightly out over the lower wall.

Lap siding is used to cover the outer walls except where brick veneer is installed. This creates a special problem for the carpenter. He or she must make sure that the siding brick joints are well sealed with caulking and overlap in the direction of rainfall.

Figure 2-1 Southern colonial style. (Photo courtesy of Kingsberry Homes, Inc.; formerly a part of the Boise Cascade Corporation.)

The Dutch Colonial Style

The Dutch colonial style employs the gambrel roof design seen in both homes and barns. The home shown in Figure 2-2 makes use of the small section of gambrel roof to add the style. This roof is very complex in design. The main roof is a gable frame design. Where the gambrel section ties in, valley rafters and valley jacks are needed for framing the upper level of the gambrel and main roof. The lower, steeper sections of the gambrel are framed and attached to extended floor assemblies and sidewall. Overhangs are large but straightforward in design.

Framing the front wall is somewhat complicated because of the extension under the gambrel roof (right) and the projected upstairs wall that contains the bay window (left). In addition, the entrance door assembly is recessed 2 ft. from the outer wall line.

Siding is of the lap variety. Rather than using self-lap on corners, vertical corner boards are used. This makes all joints butt. It also means that tar paper or vinyl should be installed behind these corners and joints to prevent leaks.

Figure 2–2 Dutch colonial style. (Photo courtesy of Kingsberry Homes, Inc.; formerly a part of the Boise Cascade Corporation.)

Aside from the special considerations listed above, the house is easily built using the western or platform framing technique.

The Simpler Colonial Style

Not all colonial designs are elaborate or ornate, as Figure 2–3 shows. Here is one of simpler lines, yet it retains a colonial character. The porch across the front protects the entrance and provides a place to socialize. The four columns support the porch roof and its ceiling. The porch ceiling joists extend from ceiling joists of the house.

The roof is a simple gable design with ridges at two different levels. The center section of roof uses longer rafters to allow for coverage of the porch roof; yet the rafters along the entire rear of the house are uniform in size and are in alignment. Overhangs are generous and finished with *soffits* supported by *fly rafters, ledgers,* and *lookouts.*

The frame of the house is rectangular with no breaks in any walls. This simpli-

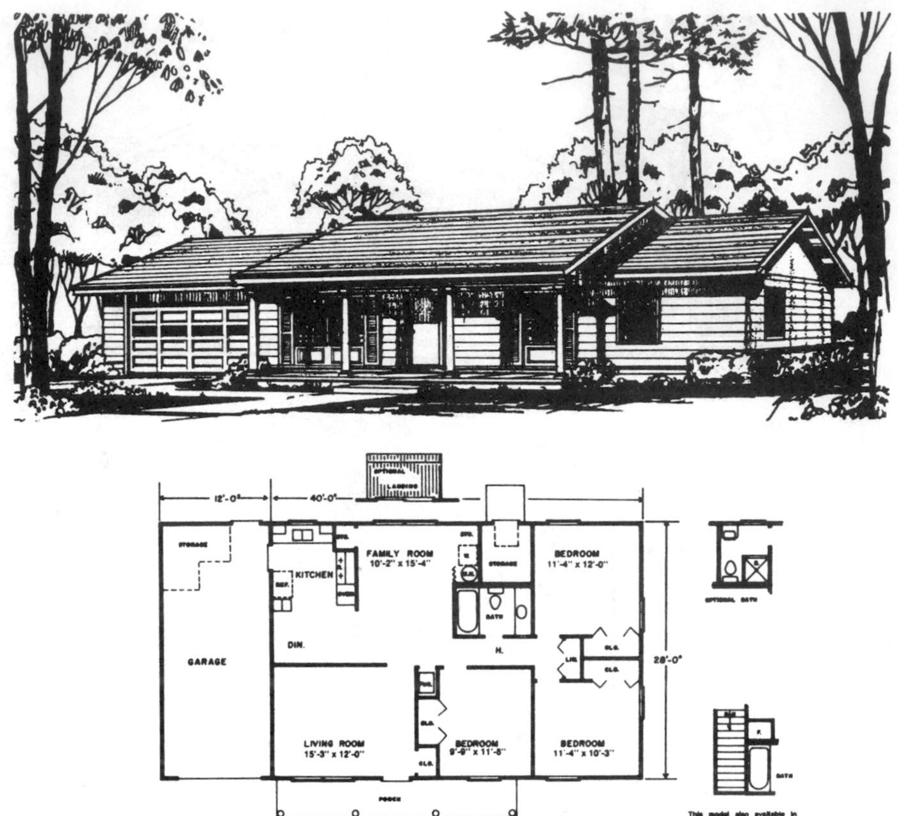

Figure 2–3 Simple colonial styling. (Photo courtesy of Kingsberry Homes, Inc.; formerly a part of the Boise Cascade Corporation.)

fies construction. Yet, because of the roofline it appears as if there are breaks in the line of the front wall. Siding is lap with self-corners. Windows are used in unbalanced design.

The Country Classic

The country classic design shown in Figure 2–4 is very complex. It has many interior and exterior corners. Its roof is especially complex. A portion of the house is two-story. The remainder is one-story.

The roof, for example, employs a gable with valley design. The main roof over the second floor employs two sideless dormers. The lower roof extends along the porch and develops a valley with the roof over the short extension to the right. This arrangement complicates construction. The garage roof is a gable that joins the main roof. It is probable that equal pitches are used for both roofs; but because the spans are different, the ridges do not align. Or it could be that an unequal valley

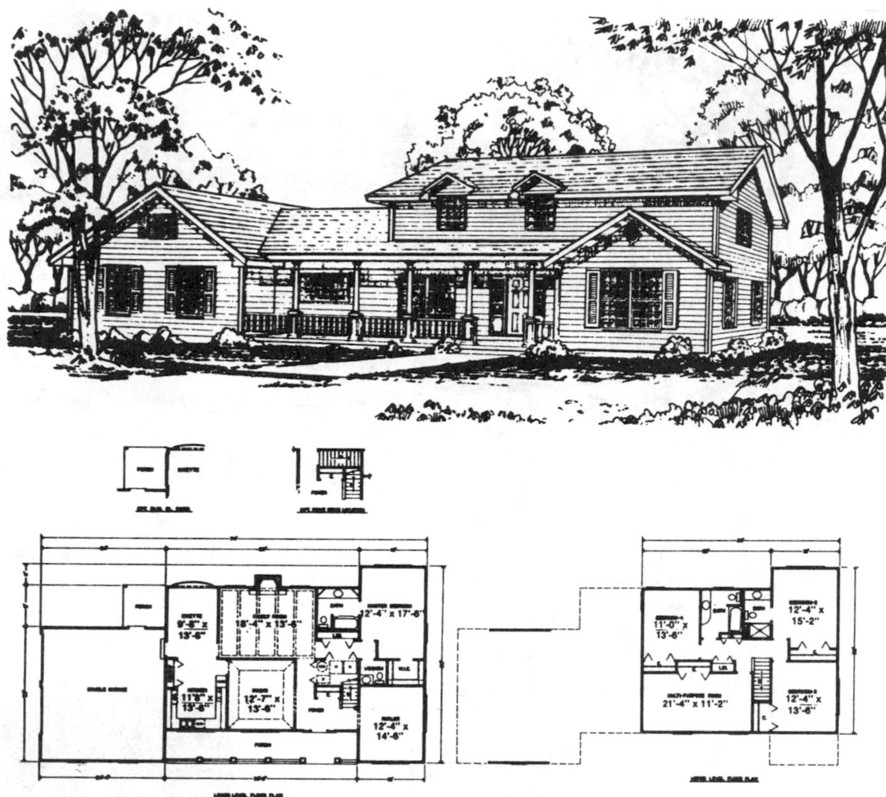

Figure 2–4 Country classic style. (Photo courtesy of Kingsberry Homes, Inc.; formerly a part of the Boise Cascade Corporation.)

pitch is created. Elevation plans would need to be used to determine the pitches of the different roofs. The fireplace in the rear wall will require special carpentry work in the roof as well as wall and floor framing.

The columns may be built on-site; if so, they should challenge the craftsman carpenter. Railings with spindles complete the porch. Siding is lap with corner boards used throughout. Again, special attention must be paid to corners and siding installed this way.

The Two-Story Using Balanced Design

The two-story shown if Figure 2–5 is one of balanced design insofar as the windows on each floor are identical to those on each side of the entrance and the entrance is a balanced design. Its roof is a simple unbroken line of gable design and may easily employ trusses. Western or platform framing should be used because of the cantilever projection of the upstairs rooms.

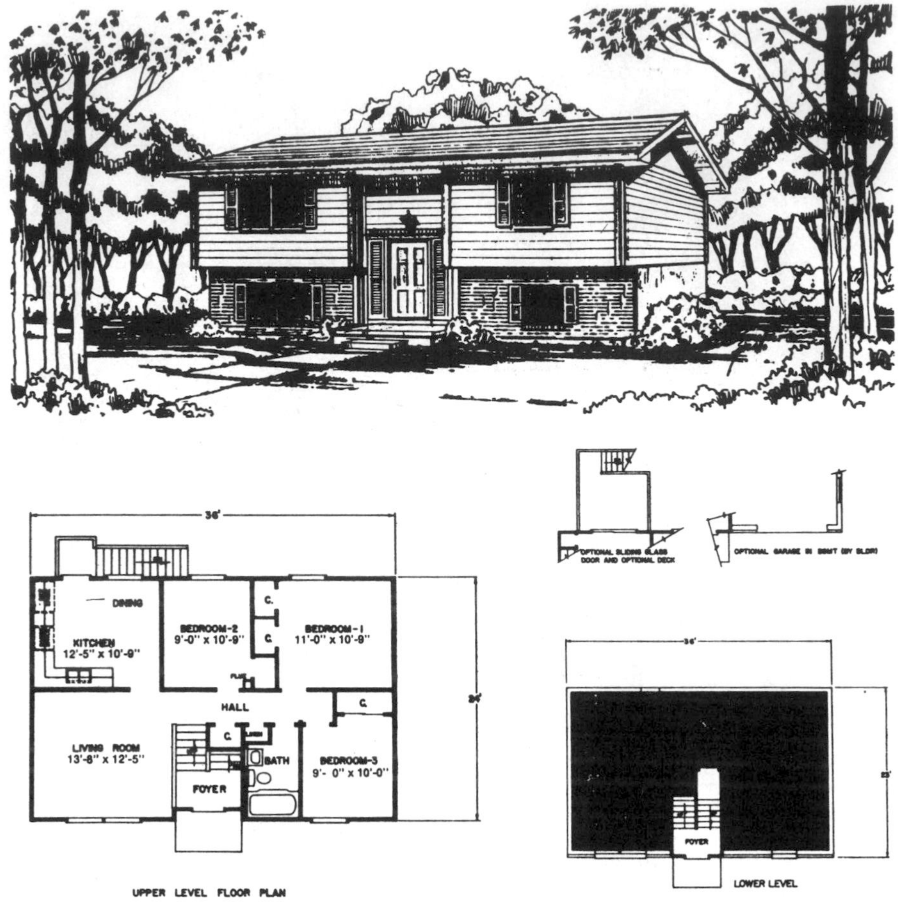

Figure 2–5 Two-story house with balanced design. (Photo courtesy of Kingsberry Homes, Inc.; formerly a part of the Boise Cascade Corporation.)

Brick veneer on the lower walls requires the carpenter to extend the siding over the brick. Again, siding is lap with corner boards used. A simple framing technique prepares the opening for the front-door assembly. Notice that it is recessed slightly and that the recessed line carries through to the roof. As on several of the other homes, staircases need to be constructed inside.

The Two-Story Using Balanced Design

The two-story design shown in Figure 2–6 truly has a balanced design on the main section. The windows are the same size and are equidistant either side of the entrance. The roof is a simply designed gable with finished overhangs. There is a

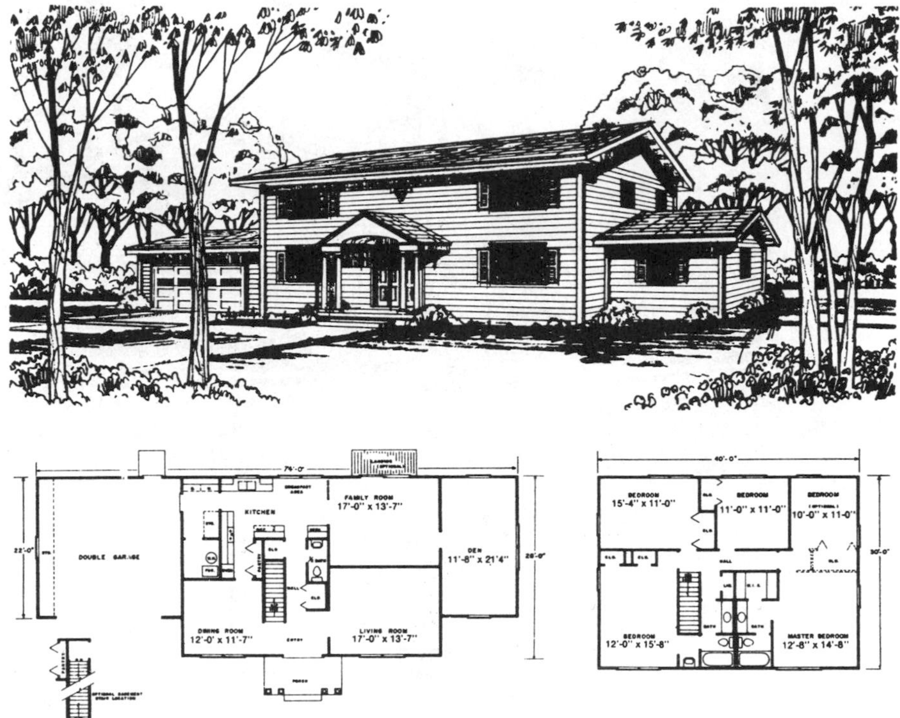

Figure 2–6 Another two-story house with balanced design. (Photo courtesy of Kingsberry Homes, Inc.; formerly a part of the Boise Cascade Corporation.)

fireplace at the rear, so the roof will need framing around it, as will the walls on either side.

The porch is also a simple gable, but its ceiling is not as simple to construct. It is elliptical. This means that special ceiling joists need to be cut. The ceiling materials will then be anchored to these specially prepared ceiling joists. In all probability ¼-in. plywood or hardboard will be used. The porch roof is supported by four columns that may be fabricated on site. Lapsiding is butted to corner boards. Gable ends are finished in plywood, which breaks up the line created by the lap siding.

Inside the house a staircase is built that requires a special stairwell framing in the second-floor frame. The staircase is not too difficult to construct since it has no turns or curves.

The Split-Level

Designed for sloping land, the split-level shown in Figure 2–7 is constructed with the platform framing technique. It has a simple gable roof with wide finished over-

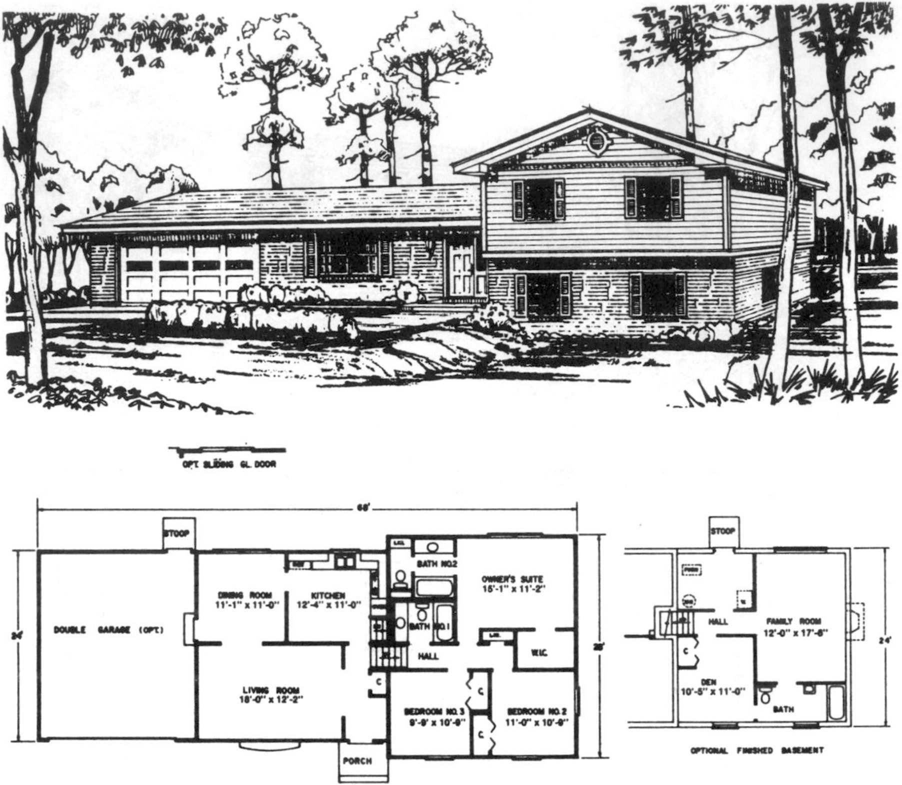

Figure 2–7 Split-level style. (Photo courtesy of Kingsberry Homes, Inc.; formerly a part of the Boise Cascade Corporation.)

hangs. The lower levels are brick veneered. This means that the carpenters must fashion the *frieze boards* to overlap the top course of brick. This also means that the upper story must project out over the lower walls to accommodate the brick veneer.

Siding used on the upper level reduces the weight on the footings and also provides a contrast in the lines of the design. A partially balanced design is used on the house where the windows of the two stories are of the same sizes and are in vertical and horizontal alignment. Two short sections of stairs are needed to allow passage from each level.

The Split-Level with Spanish Accent

There are several differences between the split-level shown in Figures 2–7 and 2–8. They are the porch, which requires a roof, and the vertical siding on the upper level; both are shown in Figure 2–8. Aside from these changes, the houses are alike.

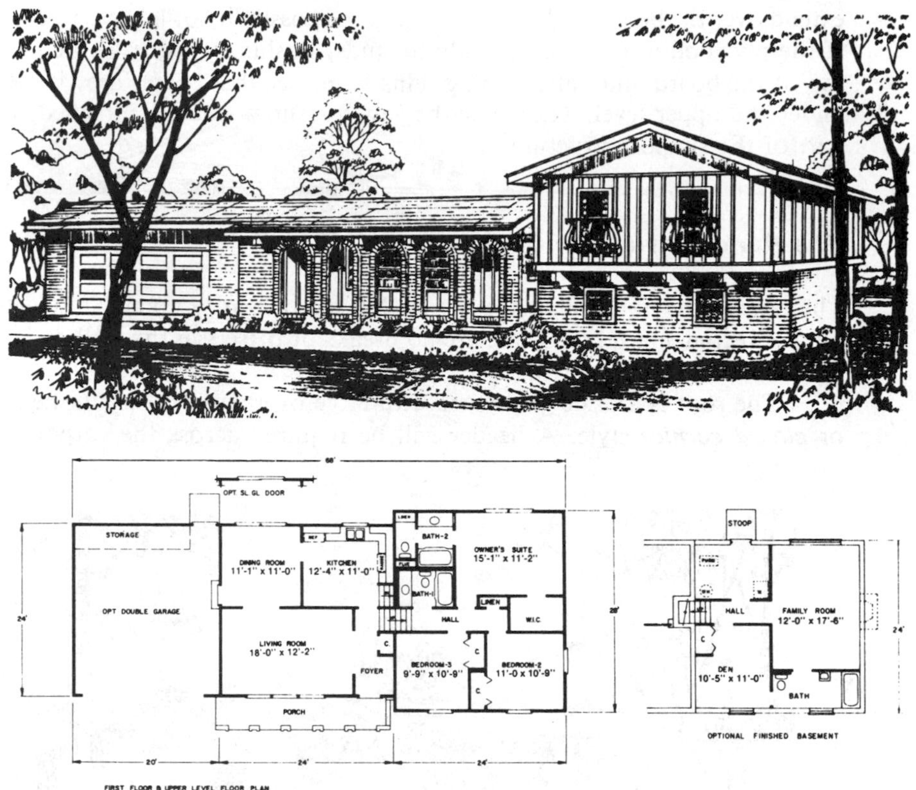

Figure 2–8 Split-level style with Spanish accent. (Photo courtesy of Kingsberry Homes, Inc.; formerly a part of the Boise Cascade Corporation.)

Therefore the comments about structure made before apply to this house except for the additional roofing required to cover the porch and vertical siding and plywood gable end covering.

The rafters are extended to cover the porch, but notice that the line remains unbroken. That is, all rafters are in alignment and a single ridge level is used. There are two building requirements to note though. First, the framing above the garage door is higher than normally expected because of the shortened rafters. Second, there is a small valley in the porch roof over the entrance side of the porch. This valley sheds water from the upper roof away from the wall of the upper level, thereby reducing the risk of leaks. It also causes the water to flow from the lower roof to the shed and into the garden, not on the entrance sidewalk.

The vertical siding is board and batten. Notice that it does not extend to the rakes of the roof. Nor does it extend to the bottom of the wall. Based on these observations, flashing is placed on top and behind the lower ends of the lower board band and vertical siding. The vertical plywood panels used for gable ends must extend over the board-and-batten siding if the possibility of leaks is to be eliminated. If the plywood were nailed directly to studs, flashing would also be needed at the top of the board-and-batten siding. Finally, notice that beams are visible extending under the upper level. These may be just for show—simply attached, rather than part of the support structure.

The Ranch-Style with Carport

The ranch-style house shown in Figure 2–9 is a one-story home constructed around a basic rectangular plan. There are no breaks or offsets in the wall designs, so construction is very straightforward.

The roof is a basic gable type with large overhang that is finished in a *boxed* or *closed cornice* style. A header will be required across the carport opening to

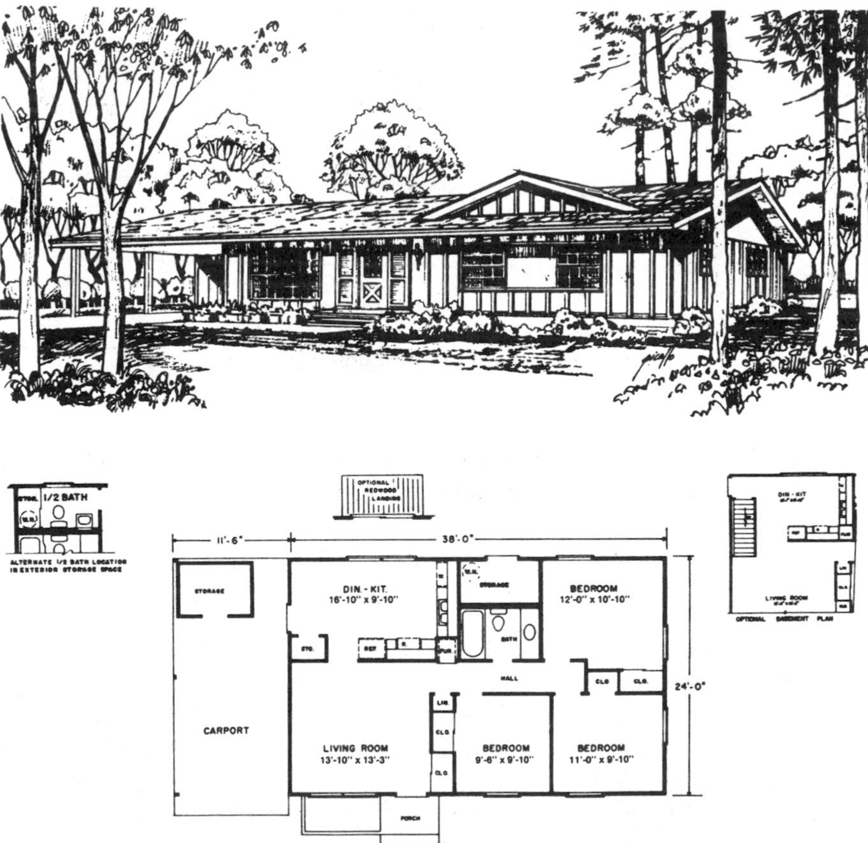

Figure 2–9 Ranch-style house. (Photo courtesy of Kingsberry Homes, Inc.; formerly a part of the Boise Cascade Corporation.)

support the roof. Truss roof assemblies could easily be employed for the roof's construction. The dormer is of the sideless variety and adds a dimension of style.

Vertical board-and-batten siding covers the exterior walls except on the gable ends under the main roof. Here exterior plywood panels are used.

The columns on the carport that support the roof are probably square and made on site. They may be 4 × 4s or doubled 2 × 4s cased with 1-in. stock, trimmed at the base and crown.

The Ranch-Style with Single-Car Garage

The basic structure of the ranch-style house shown in Figure 2–10 is rectangular on a single level. No offsets or breaks in the walls are employed, which simplifies construction. The garage is framed along with the house and a header is used across the garage door opening.

Figure 2–10 Modest ranch style with garage. (Photo courtesy of Kingsberry Homes, Inc.; formerly a part of the Boise Cascade Corporation.)

The roof design is a basic gable style with a few of the rafters extended to cover the porch. There is a fairly large overhang in *closed cornice style* that adds a bit of formality. The porch roof is supported by three simply designed posts. The severity of straight lines and plainness is broken by the inclusion of curved frieze pieces between posts. Notice that the porch ceiling slants along the roof line. This is necessary to provide headroom for the windows and the door.

Brick veneer is used as the exterior wall covering for the front of the house. This means that the frieze board under the cornice soffit must be fabricated away from the wall about 4½ to 5 in.Vertical siding covers the remainder of the house, including the gable ends. Board and battens are used, and on the gable ends the battens must line up with those on the wall if the line is to look neat and proper.

The Two-Story Saltbox

The two-story saltbox-style house shown in Figure 2–11 is of balanced design. Windows are aligned over each other and are of the same style. The door is centered in the front wall. A breezeway connects the two-car garage to the house.

The roof is somewhat complex because of the breezeway and elongated rafters on one half of both the main house roof and garage roof. This is typical of this style of house, however. In many earlier homes of this style, rooms were added as the family increased in size. The roofline was retained to cover the new additions at the rear of the house.

The main roof is a gable type with overhangs to the front and back. A *closed* cornice design is used on the gable ends. Here facia boards cover the lap siding. The breezeway roof ties into the side of the house and garage. Where it ties into the house flashing must be used for leakproof installation. Where it ties into the garage roof, valley rafters and valley jack rafters are used.

The garage roof uses rafters of two different lengths. Those on the breezeway side are shorter than those opposite. The amount of overhang at rafter ends equals those of the main house, as do the *closed* cornices on gable ends.

A cantilever projection of second-floor joists is made along the front wall, where a projection of several feet is shown. The extended ends of these joists must carry the load of the wall and roof, so special care needs to be taken during construction. Sound design as well as good-quality strong lumber needs to be used. Lapsiding is used on all the walls, with corner boards installed first.

The breezeway uses four simply designed columns. They are incorporated in the design of the porch. The frieze board along the ceiling level of the porch is fashioned in a manner similar to the tops of the garage door openings.

The Contemporary

The contemporary-style house shown in Figure 2–12 is very complex in design. Many breaks in the exterior wall line result in interior and exterior corners. Many different roof segments are used. Some walls are single-story, while others are two-story.

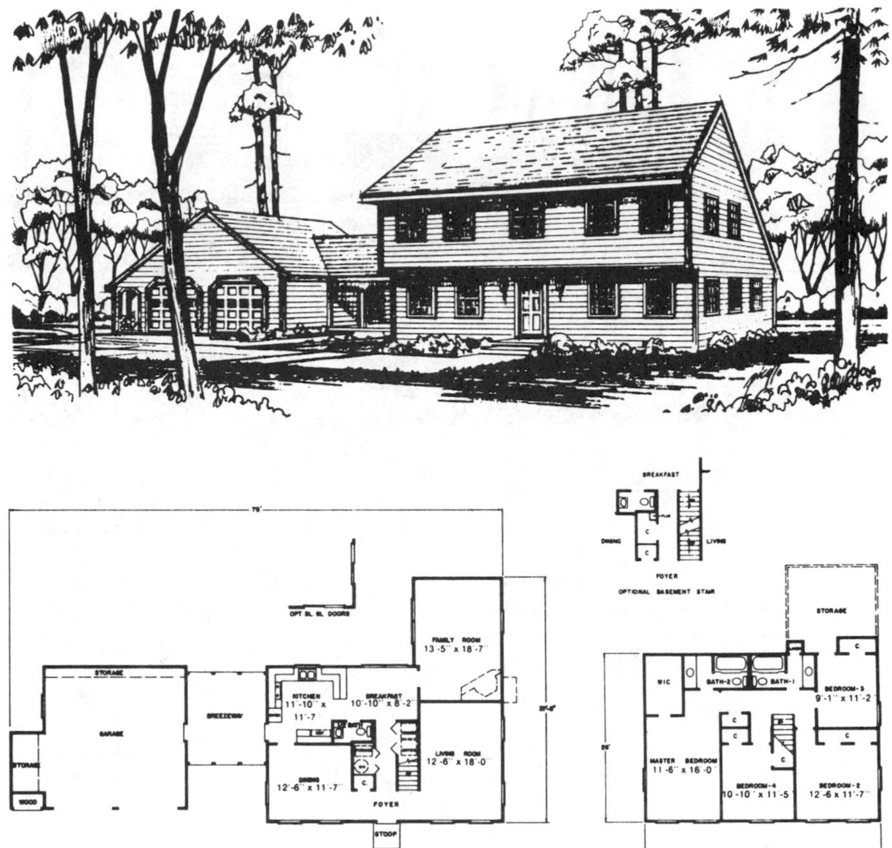

Figure 2–11 Two-story saltbox-style house. (Photo courtesy of Kingsberry Homes, Inc.; formerly a part of the Boise Cascade Corporation.)

The pitches of the roofs are quite steep; however, this does not pose a construction problem. The fact that they are at different levels and lengths does. Rafters will be custom cut to a variety of lengths. The various lengths do not alter the pitch, nor will they change the ridge, bird's mouth, or overhang rafter tail cuts.

Much of the house is two-story, yet because of the roof design, exterior walls will extend above ceiling levels. Cathedral ceilings are used in some rooms.

The siding is installed in two styles, diagonally across the front and horizontally on the remainder of the house. Corner boards are used to simplify corner joints where the two directions of siding would be difficult to join. Cornices are *close* on sidewalls and boxed on overhangs in front and rear.

Framing for the clerestory windows between the front roof and higher rear roof needs to be sound to avoid structural problems and to provide a solid base for windows. Windows installed in this area need to be of very high quality and well insulated so that no possibility of leaks is present.

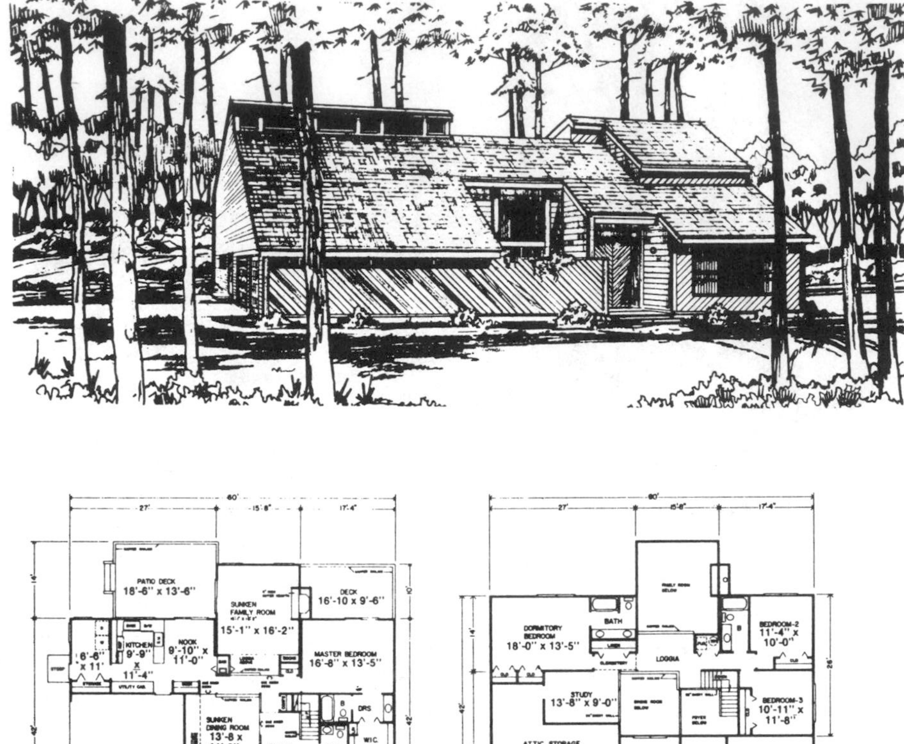

Figure 2–12 Contemporary-style house. (Photo courtesy of Kingsberry Homes, Inc.; formerly a part of the Boise Cascade Corporation.)

The inside staircase includes a landing. This means that two sections of stairs are to be built and installed. Another short section of staircase needs to be built onto the rear deck. Both sets of stairs are different in design and structure. The interior one is more formal with risers and treads fit into side stringers, all made of fine softwoods or hardwoods. In contrast, the deck stairs are made with step treads and stringers under the treads. No riser boards will probably be used. Further, these stairs will probably be made from treated pine or fir.

LEARNING MORE ABOUT PRINCIPLES OF CONSTRUCTION

The pictorials of houses can provide much valuable information to the carpenter. Sometimes this information is readily translated into tasks that are already understood. Other times the information poses problems of two kinds: What is the task or tasks that must be performed, and what is the way or ways to perform the task?

If the task is readily recognizable, its completion during the construction phase should pose no problem. Further, if a thorough understanding of the principles of construction and design are known, a full grasp of it is appreciated. Should any slight variations from the commonplace be seen, they too are understood.

On the other hand, if the pictorial does not communicate a task or series of tasks that must be performed, the carpenter lacks knowledge of the principles of construction and design. A clear need for learning exists. Studies can be conducted in formal classrooms and at home with books.

Most of the designs of houses identified in this chapter are simple in principle and construction. A large percentage of them are rectangular and employ the simple gable roof. Many are built on concrete slabs and are single-story. Therefore, the western or platform framing technique is used. Modular construction is also possible. Balloon framing is seldom used. Even though cornices are boxed, they are plain and few if any problems are expected during their construction.

Several of the houses do, however, employ complex roof and wall design. These features relate to design and construction principles that require considerable study. Understanding the hip-and-valley roof design requires a knowledge of force, loads, principles of design, and adaptation of the tools to be used. Several chapters to follow devote considerable attention to roof design and planning. Walls must also be constructed soundly so that forces from the roof, joists, winds, earthquakes, and so on, are predictable and accounted for. For instance, how does a header carry its load and distribute it to the studs? Or, what size header must be used to carry a load of a certain size? Questions like these are a part of learning and understanding the principles of construction and design.

Another example where adaptations of tasks must be used to employ a single principle is installing flashing. The principle requires that a metal strip be formed with a bend and installed so that the natural fall of rain sheds down its surfaces. If, for example, the tops of windows or door frames need flashing, these pieces extend up under the siding and out and over the window or door top. On the other hand, if flashing is used where a roof intersects with a sidewall as shown in Figures 2–1, 2–2, 2–3, and others, small pieces of flashing are installed on top of each row of shingles. Afterward the siding is installed on top of the pieces of flashing.

There are literally hundreds of different tasks that have many subtle or large variations on both simple and complex principles. To name or list them would be useless. They cannot be found in a set of plans or specifications either. They must be learned on the job and from study. Many of the chapters that follow are developed for the single purpose of providing a clear and complete understanding of the principles employed in various tasks so that the carpenter knows what he or she is doing is correct.

PROBLEMS AND QUESTIONS

1. What is a distinctive feature of a southern colonial house?
2. Look at Figure 2-1 and determine the type of roof each section uses. What type of rafter is used?
   a. common **b.** hip **c.** cripple **d.** valley
3. Why would western framing probably be used for the house shown in Figure 2-1?
4. Determine the different types of rafters used in the house shown in Figure 2-2.
5. Does the house shown in Figure 2-2 use self-corners on lap siding or corner boards?
6. What is the purpose of the four columns on the house in Figure 2-3?
7. Cite at least four details of the country classic house (Figure 2-4) that complicate its construction.
8. Does the two-story house (Figure 2-5) have a balanced or unbalanced design? The two-story in Figure 2-6?
9. Why would platform framing be used for the split-level house (Figure 2-8)?
10. What would be the difficulty in building the contemporary house (Figure 2-12)? List at least three factors.
11. Take a pad and pencil and observe a neighbor's house. Write down as much design detail and construction detail as you can. Have your instructor or a fellow carpenter verify your data.

PROJECT

The project for this chapter will make you more knowledgeable about styles and structures that create style.

A. Using Figure 2-1 as a reference, which characteristics of the house make it a southern colonial? What other characteristics can we identify about this house?

1. Which of the following are characteristics of the house? (Circle your options)
   a. Large columns in the front
   b. Pediment
   c. Large overhangs
   d. Six-panel door
   e. Overhead garage door
   f. Brick veneer
   g. Lap siding

2. How many outside corner assemblies would the carpenter make for the first level/second level?
   a. 4/4 b. 5/4
   c. 6/5 d. 6/6
3. What type or name(s) can be used to describe the cornice used on the house?
   a. Close b. Closed
   c. Boxed d. Open
4. What type of lap siding installation technique is used on the upper floor?
   a. Corner boards with butt joints
   b. Self-returns with bevel cuts

B. Using Figure 2–2 as a reference, which characteristics create specific construction techniques?
1. What construction technique is used to install the siding?
   a. Self-returned corners
   b. Corner boards and butt joints are used
2. On which part of the house would the specific need for structural members with high shear ratings be required?
   a. Lower floor joists
   b. Upper floor joists on the front wall
   c. Upper ceiling joist on the front wall

C. Refer to Figure 2–4 and decide what characteristics relate to different construction techniques.
1. How many individual valley rafters will need to be cut for the roof?
   a. Two for each dormer
   b. Four for the dormers and two for the house garage connection
2. What do we know about the porch ceiling joists?
   a. They will be installed perpendicular to the front of the house.
   b. They will be installed parallel to the front of the house.
   c. They will be at the same ceiling height as the first-floor ceiling joists.
   d. They will probably be spaced the same as the main house joist separation.
3. Let's see if we can identify every place that flashing would be used. Which of the following options are correct?
   a. One piece is required over each dormer valley rafter.
   b. One piece is required over each valley rafter that joins the house to garage.
   c. One piece will be required across the front at the intersection where the second floor wall extends above the roofline.
   d. One piece will be required where the right side of the front bedroom roof joins the front wall of the second floor.
   e. One piece will be required over the short valley rafter on the left side of the front bedroom roof.
   f. One piece of siding will be required for the right side of the front bedroom wall where the roof joins the second floor wall.
   g. One piece of flashing will be required on each side of the roof that connects to the rear of the second floor wall at the rear. This roof covers part of the master bedroom.
   h. Flashing will be required around the chimney that extends through the roof on the rear wall.
   i. Flashing will be required over every window.

Answers:

Part A: **1.** a. yes, b. yes, c. yes, d. yes, e. no, f. sometimes, g. yes
2. a
3. b,c
4. a.

Part B: **1.** b
2. b

Part C: **1.** b
2. a. yes
b. no
c. yes
d. yes
3. all options are yes.

Chapter 3

Lumber—A Key Building Material

In its various forms, lumber is no stranger to the carpenter. It has almost limitless uses in building construction. To name a few, it is used in making forms for concrete, framing, trim, and wall coverings.

Because this material is so versatile and common, many studies have been conducted to determine its characteristics. The results of these experiments are beneficial to every builder. In this chapter we define and examine several of the most important characteristics of lumber to the builder. First we identify the two types of woods and some species of each type. Grading names and definitions are presented next. Wood properties, which play an important role in sound construction, are then defined and several examples of their use are provided.

TYPES OF WOOD

Each species of wood is placed in either the *softwood* or *hardwood* group, depending on the type of tree. By definition, the softwood tree is cone bearing and has needles or scalelike leaves. In contrast, the hardwood trees usually drop their leaves in autumn.

Softwoods

Many properties of softwoods are shown in Figure 3–1. Notice that the area of *early wood* (5) is much larger than that of late wood (6). Also notice where the resin duct is located. With some imagination you can visualize the grain of a board as well as the end cut. The most common softwood species used in house construction are the firs, pines, cedars, hemlocks, larches, and spruces. Table 3–1 lists many of the different kinds of each species.

Many of the species listed are used for framing. Some, such as the Douglas fir and southern pine, are very good. Others that are used but do not have the strength of these two are the hemlocks, spruces, cedars, and some pines. We discuss the individual qualities of each species later in the chapter.

Hardwoods

Figure 3–2 shows several properties of hardwoods. For example, the sizes of early and late wood cells are both small and almost equal in area. This characteristic accounts for the density of hardwood. The closeness in annual rings also accounts for the fine grains usually found in hardwoods and explains why fine end cuts can be made.

The number of hardwood species used in construction is limited to birch, elm, maple, oak, mahogany, and walnut. These woods are seldom used for framing. However, they are used for paneling, cabinets, doors, trim, and molding wherever stained woods are desired for their beautiful grain and color.

GRADING OF LUMBER

Every builder needs to know and understand how lumber is graded and what the grades mean. The grading rules are established and modified by the lumber-

TABLE 3–1 SPECIES OF SOFTWOODS

Fir	Pine	Larch
Douglas fir	White	Western
Silver (Pacific)	Yellow	Alpine
White	Ponderosa	American or black
Noble	Sugar	Tamarack or hackmatack
Red (California)	Loblolly	
	Shortleaf	
	Slash	
Hemlock	**Cedar**	**Spruce**
Western	Port Arford	Sitka
Eastern	Alaska	Engelmann
	Incense	White
	Western red	Black

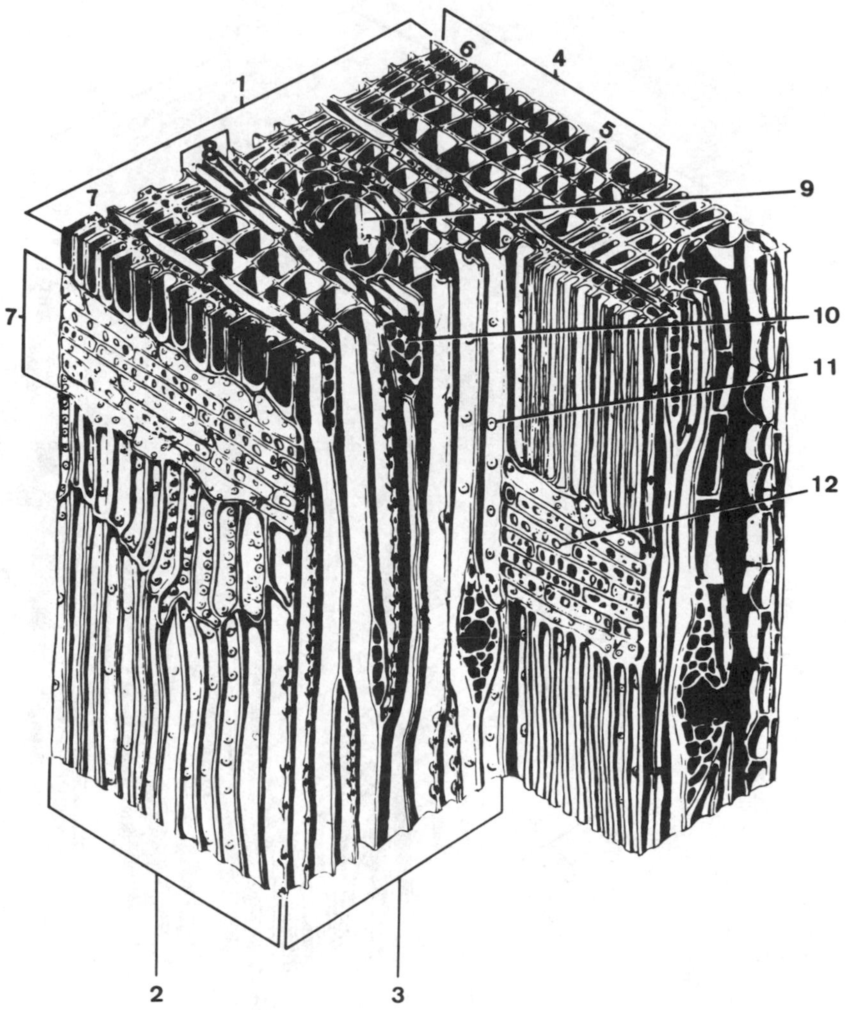

Softwood key

1. Cross sectional face
2. Radial face
3. Tangential face
4. Annual ring
5. Earlywood
6. Latewood
7. Wood ray
8. Fusiform ray
9. Vertical resin duct
10. Horizontal resin
11. Bordered pit
12. Simple pit

Figure 3–1 Wood structure of a softwood. (From U.S. Department of Agriculture, PA 900.)

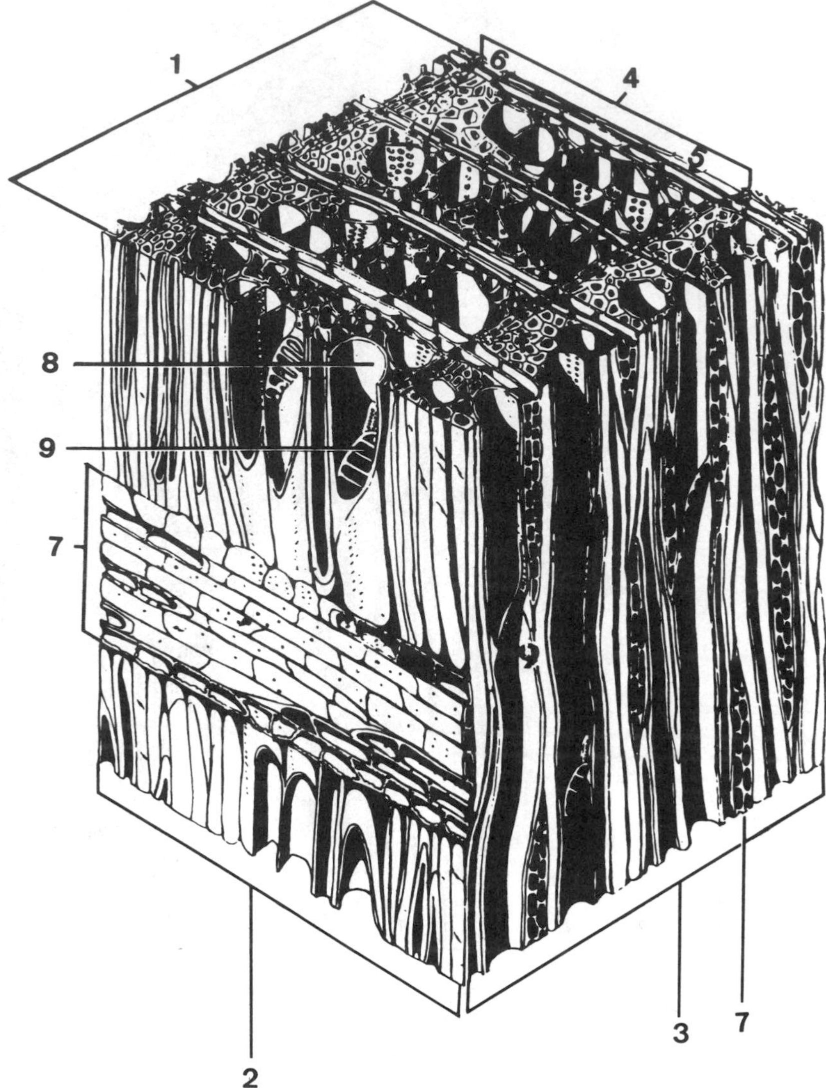

Hardwood key

1. Cross sectional face
2. Radial face
3. Tangential face
4. Annual ring
5. Earlywood
6. Latewood
7. Wood ray
8. Vessel
9. Perforation plate

Figure 3–2 Wood structure of a hardwood. (From U.S. Department of Agriculture, PA 900.)

producing companies, and associations of these companies, in coordination with the U.S. Departments of Commerce and Agriculture.

Through the efforts of committees from industry and government, a fair set of standards has evolved. Naturally, the grading must be equitable so that the user can rely on minimum quality and structural capabilities of the lumber, yet provide a source of income and profit for producers.

The Various Grades

There are grades established for all types of lumber. Some of the more common that are important to the house builder are the select, finish, and structural grades. Table 3–2 provides a brief identification of some of the most commonly used grades.

The Standard of a Grade

Associated with each grade is a standard that sets the types of defects that are acceptable and the limitations of these defects. The inspector-grader first selects the best-looking surface and then applies the rules for grading. Since human judgment is applied, the highest grade is selected rather than the lowest. This is a rather difficult task for an inexperienced person. Consider the need to grade a 2 × 6 joist as *select structural.*

> Dimensional lumber of this quality is limited in characteristics that affect strength and stiffness values to provide a fiber stress in bending value of 65% of that allowed for clear, straight-grain wood and to provide a recommended design value for modulus of elasticity of 100% of that allowed for the clear wood average. This grade is recommended for use in applications where both high strength and stiffness values and good appearance may be required.

Characteristics permitted and limiting provisions shall be:

Checks—Surface seasoning checks, not limited. Through checks at ends are limited as splits.

Grain—Medium.

Knots—Sound, firm, encased, and pith knots, if tight and well spaced, are permitted in sizes not to exceed the following, or equivalent displacement:

Nominal width	At edge wide face	Centerline wide face	Unsound or loose knots and holes (any cause) (in.)	
5	1	1½	⅞	One hole or equivalent smaller per 4 lin. ft
6	1⅛	1⅞	1	
8	1½	2¼	1¼	
10	1⅞	2⅝	1¼	
12	2¼	3	1¼	
14	2⅜	3¼	1¼	
16	2⅜	3⅜	1¼	
18	2½	3½	1¼	

TABLE 3-2 GRADES OF LUMBER[a]

A. *Select grades* B & BTR & No. 1 and 2 clear C select D select	Relatively clear lumber used where finishes allow some imperfections; *uses:* trims and cornices, flooring casing, base
B. *Finish grades* Superior finish Prime finish E finish	The highest quality of finish lumber, often clear and excellent for staining
C. *Common board grades* No. 1 common No. 2 common No. 3 common (standard) No. 4 common (utility) No. 5 common (industrial)	Available in most sizes, but not always in all grades; frequently used in house construction and often finished with transparent coatings
D. *Light framing and studs grades (dimension lumber)* Construction Standard Utility Studs	Recommended for general framing purposes; grades are listed in order of best to poorest
E. *Structural light framing grades* Select structural No. 1 No. 2 No. 3 Economy	These grades are limited in characteristics; thus they can be used where high strength and stiffness values are required; appearance and quality decrease from top to bottom of the list; these apply to joists and planks, beams and stringers, posts and timbers
F. *Dimension pullout grades* C & BTR dimension D dimension	Stock is frequently used for remanufacture of moldings, stairways, etc.
G. *Machine stress-related grade* MSR	Lumber grades with extreme fiber stress in bending, modulus of elasticity; tension parallel to grain, and compression parallel to grain
H. *Dense graded* Dense Douglas fir and larch	

[a]Grade titles and listing are those generally used by Western Woods Products Association and many other inspection bureaus.

Manufacture—Standard E.

Pitch and pitch streaks—Not limited.

Pockets—pitch or bark—Not limited.

Shake—On ends, limited to ½ the thickness. Away from ends, several heart shakes up to 2 ft long, none through.

Skips—Hit and miss skips in 10% of the pieces.

Slope of grain— 1 in 12.

Splits—Equal in length to the width of the piece.

Stain—Stained sapwood. Firm heart stain or firm red heart limited to 10% of the piece.

Wane—¼ the thickness, ¼ the width. Five percent of the pieces may have wane up to ½ the thickness and ⅓ the width for ¼ the length.

Warp—½ of medium.[1]

Clearly, all the different characteristics of trees are included in establishing a standard for a single grade of lumber. Each grade has a unique set of standards, including most if not all those listed for select structural. Of course, the allowable defects are stated differently.

WOOD PROPERTIES IMPORTANT TO THE BUILDER

Lumber cut from any tree contains many individual and common properties. For example, a 2 × 4 may have been cut from sapwood or heartwood, may be strong or weak, may have internal compression or tension, and may be green or dry. The importance of these properties is developed and is fundamental to future studies.

Sapwood versus Heartwood

Sapwood is made up of the outermost growth rings of the tree. It is usually the lighter-colored wood and contains some resins and gums. Heartwood, by contrast, is the dead tissue surrounded by the sapwood. Its only function is to support the tree. The heartwood is frequently referred to as *pith.*

Heartwood is generally more durable than sapwood because it resists decay. On the other hand, it is usually brittle and thus tends to have higher shear values. Sapwood, although less durable, can be chemically treated more easily. Sapwood also has greater tensile strength because of its superior flexibility.

Specific Gravity

The specific gravity of each species of wood is the measure of the amount of cell wall material in the wood. It thus is a useful factor for determining the strength of the wood, which then aids in selecting it for various purposes. To cite several examples, consider that Douglas fir, southern pine, and oak all have high specific gravities. This indicates that they are all suitable for framing materials where many forces, stresses, and weights must be dealt with. On the other hand, ponderosa pine,

[1]*Source:* "Grading Rules for Western Lumber," 1988 ed., p. 122, courtesy of Western Wood Products Association, Portland, OR.

spruce, and other lighter woods with lower specific gravities are not desirable where load bearing is expected. They are desirable, however, for trim and other non-load-bearing uses.

The method for determining the specific gravity of a species of wood is as follows:

1. A block of wood 1 × 1 × 10 in. is cut and dried in an oven at 105°C until a constant weight is determined.
2. A volume of water equal in weight to the dried wood specimen is poured into a tall glass cylinder (the shape of the cylinder forces the block of wood to float upright).
3. The dried wood block is lowered into the water to the point where it floats, then removed.
4. The specific gravity is found by measuring the length of wood under water (which is the weight of the wood) by the weight of equal volume of water.

Let us assume that the block sinks 3 in. into the water. This means that the weight of the wood is supported by the force of 1 × 1 × 3 in. = 3 in.^{3} of water. *Therefore:*

$$\text{specific gravity} = \frac{\text{weight of object}}{\text{weight of equal volume of water}}$$

$$= \frac{3\text{ in.}^3}{10\text{ in.}^3} = 0.3$$

Table 3–3 lists some of the softwoods and hardwoods and their specific gravities so that a comparison may be made of their relative strengths.

Table 3–3 shows considerable range in the specific gravities of both softwoods and hardwoods. It would appear that more hardwoods are suitable for construction

TABLE 3–3 WOODS AND SPECIFIC GRAVITY

Hardwoods	Specific gravity	Softwoods	Specific gravity
White oak	0.57	Western larch	0.51
Elm	0.57	Douglas fir	0.48
Beech	0.56	Shortleaf pine	0.46
Birch	0.55	Western hemlock	0.38
Black walnut	0.51	Ponderosa pine	0.37
Basswood	0.32	Sitka spruce	0.37
		Western white pine	0.36
		White fir	0.35
		Incense cedar	0.35
		Englemann spruce	0.32

of houses. This is true, but fortunately several varieties of softwoods are equally strong and suitable for this purpose. It is also fortunate that softwood trees outnumber hardwood trees by millions to one, thus making lower housing costs possible.

Modulus of Elasticity

By definition, the *modulus of elasticity,* symbolized as E is the relationship between the amount a piece of lumber deflects and the size of the load causing the deflection. In simpler terms, the modulus of elasticity is the measure of a board's stiffness. Inherent in the definition is the understanding that the member (board) must recover its original shape and size after first being strained and then having the force removed. Although each species of lumber has an assigned modulus of elasticity, the values have been assigned to the various grades of lumber. This simplifies use of the information when selecting joists, rafters, and other structural members.

Table 3–4 provides a distinct relationship of the modulus of elasticity to the grade of lumber. As shown, a 2-in. piece of lumber 4 in. or wider graded as *dense select structural kiln dried* has a modulus of elasticity of *1.9 million pounds per square inch.* In contrast, a utility stud (2 × 4) has a modulus of elasticity of only 1.4 million psi. Since modulus of elasticity is related to stiffness, the clear interpretation is that a utility stud does not have the quality of stiffness of a stud graded Dense Sel Str KD.

Tables containing modulus of elasticity data by species and grade will be presented in later chapters. In these tables the values of modulus of elasticity vary from those in Table 3–4 and may in some cases be less than 1 million psi. Species with a modulus of elasticity of less than 1 million psi are not recommended for use in load-bearing structures because it is doubtful that they could recover their original shape after being placed under heavy stress.

Determination of the allowable stress in lumber cannot be made from the E value alone; its F_b or extreme fiber in bending rating must also be considered.

Extreme Fiber in Bending

Extreme fiber in bending, F_b, means that lumber is assessed according to its fiber bending and resisting properties. F_b occurs when loads are applied producing tension and compression as illustrated in Figure 3–3. The tension occurs along the bottom of the member, or the point farthest from the applied load. Compression is induced in fibers along the face nearest to the applied load. The F_b values used in tables such as Table 3–4 have been measured in laboratory tests.

Although each grade of lumber has a specific F_b when three or more joists are used to distribute the load in a floor assembly, allowable variations are acceptable. Figure 3–3 shows that the force on a single member must be borne by that member; however, when three (or more) joists are used, the load is distributed over a wide surface. In effect, this reduces the load on each member and allows for an accept-

TABLE 3-4 ALLOWABLE STRESS BY GRADE OF LUMBER

Grade	Extreme Fiber In Bending "F_b"* 2–4″ thick, 5″ & wider psi	2–6″ thick, 2–4″ wide psi	Modulus of elasticity "E" psi
Dense Sel Str KD****	2200	2500	1,900,000
Dense Sel Str	2050	2350	1,800,000
Sel Str KD	1850	2150	1,800,000
Sel Str	1750	2000	1,700,000
No. 1 Dense KD	1850	2150	1,900,000
No. 1 Dense	1700	2000	1,800,000
No. 1 KD	1600	1850	1,800,000
No. 1	1450	1700	1,700,000
No. 2 Dense KD	1550	1800	1,700,000
No. 2 Dense	1400	1650	1,600,000
No. 2 KD	1300	1550	1,600,000
No. 2	1200	1400	1,600,000
No. 3 Dense KD	875	1000	1,500,000
No. 3 Dense	825	925	1,500,000
No. 3 KD	750	850	1,500,000
No. 3	700	775	1,400,000
Construction KD		1100	1,500,000
Construction		1000	1,400,000
Standard KD		625	1,500,000
Standard		575	1,400,000
Utility KD		275	1,500,000
Utility		275	1,400,000
Stud KD	800**	850	1,500,000
Stud	725**	775	1,400,000

Source: Southern Forest Products Association, New Orleans, LA.
Note: Terms and abbreviations: Sel. Str. means select structural; KD means kiln dried to a moisture content 15% or less; MG means medium grain; and where KD is not shown the material is dried to a moisture content 19% or less.

able variation in the F_b value. Table 3-4 lists both F_b for a single member and F_b with a variation of 1.15 F_b.

Deflection

The two previous topics illustrated certain properties of lumber when under stress. When any load is placed on a member, deflection must be considered as well. By definition, deflection is the degree to which a member may safely deflect or bend

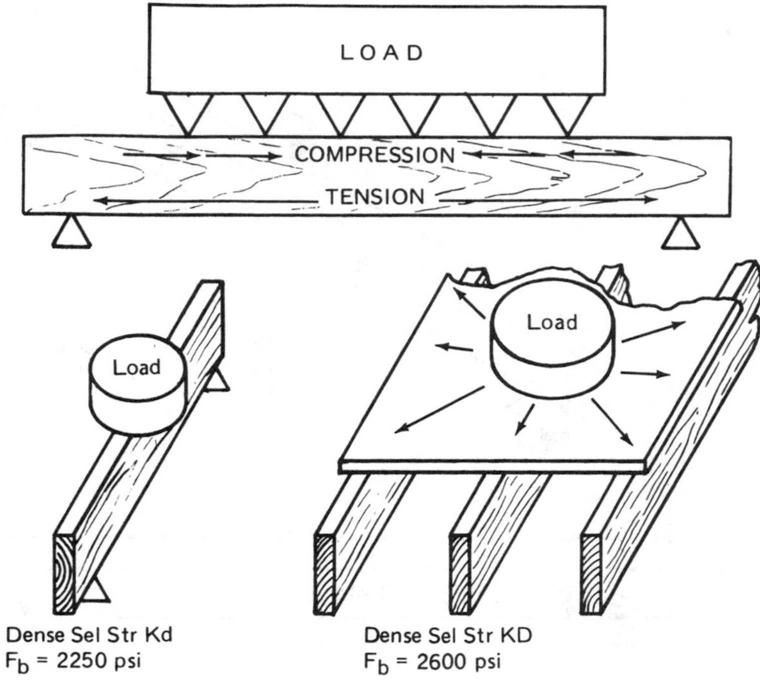

Figure 3–3 Demonstrating compression and tension with load applied and distributed.

under load. For appearances' sake the deflection must be limited so that sag is not visible. In joist and rafter construction the deflection factor may often be the controlling one in the determination of the size, grade, and species of lumber to be used. For structural joists and rafters it is usually limited to one of three proportions: $S/360$, $S/240$, and $S/180$, where S is the variable of span stated in inches.

Figure 3–4 illustrates deflection. Notice that deflection is the amount of bending a member has when spanning a relatively long distance. This is determined as shown. Where a beam of 16 ft spans an area for its full length, convert 16 ft to 192 in. to obtain the variable S. If a joist calls for a $S/360$ deflection, the total allowable deflection is ½ in. The other two ratios, $S/240$ and $S/180$, result in considerably more deflection.

Based on such properties, the building industry has set down the following general guide.

- $S/360$ Floor joists in living and sleeping areas. Ceiling joists with attic and plastered ceiling joists with no attic if the live load is predicted 10 psf. Low- and high-slope rafters, plastered ceilings.
- $S/240$ Ceiling joists with limited attic, drywall, and a live load of 20 psf. Ceiling joists with no attic, drywall, and a live load of 10 psf. Low-

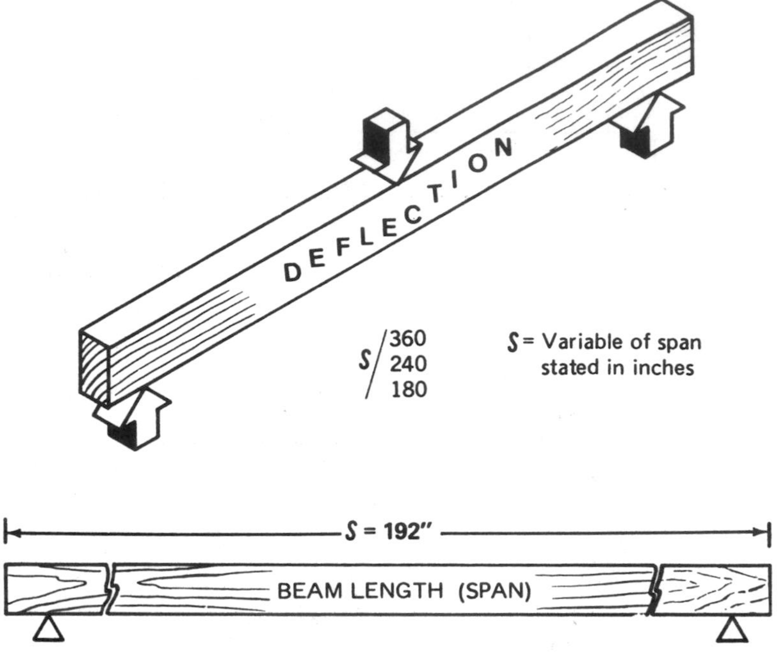

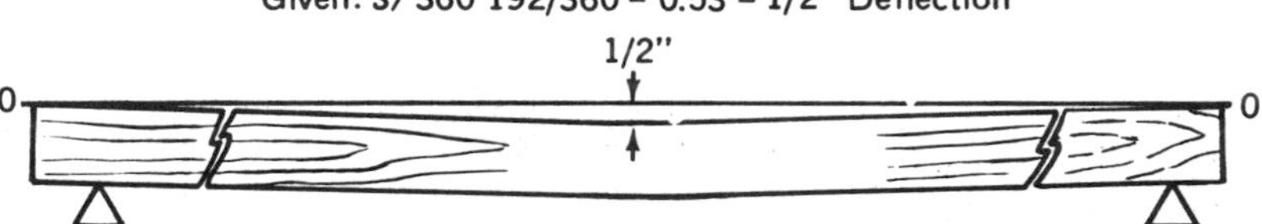

Given: S/ 360 192/360 = 0.53 = 1/2" Deflection

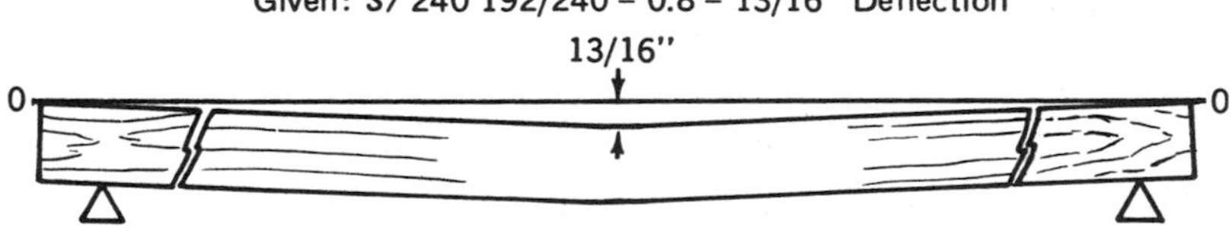

Given: S/ 240 192/240 = 0.8 = 13/16" Deflection

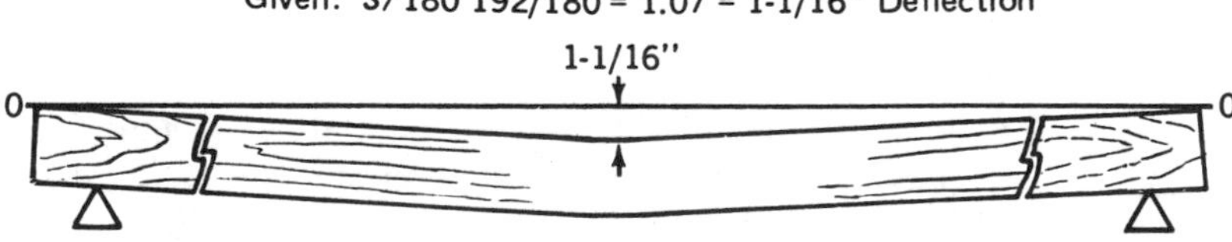

Given: S/180 192/180 = 1.07 = 1-1/16" Deflection

1-1/16"

0 0

Figure 3-4 Demonstrating deflection and the effects of deflection with a variable *S* and different constants 360, 240, and 180.

or high-slope rafters, drywall ceiling. Low-slope rafters (3 in. to 12 ft or less), no finished ceilings.

S/180 High slope rafters, no finished ceilings, heavy roof covering, loads from 35 to 55 psf. High-slope rafters, no finished ceilings, light roof covering, loads 27 to 47 psf.

In summary, floor joists require the greatest stiffness, which means the least amount of deflection. Therefore, *S*/360 must be observed. Rafters and ceiling joists may be selected from tables where deflection of *S*/240 is indicated, provided that other qualifications as stated before are met. Notice that floor joists are not listed under *S*/240. Finally, *S*/180 may be used for rafter construction only.

Comparative Strength Values

Sometimes it is not easy to choose lumber by its various properties, especially when trying to match species for species. Figure 3-5 provides in graph form a list of many of the common species of lumber used in construction. From the data presented, longleaf pine (which is in relatively short supply) is the best overall species in every category. It would have the best F_b value, the greatest modulus of elasticity, and the least deflection. It would be used for floor joists as well as rafters. Douglas fir ranks only slightly less high in all categories, making it an excellent building material.

In contrast, ponderosa pine is low in bending strength, stiffness, toughness, and hardness. Because of this, it is not suited for use as structural members. However, it makes fine moldings and trim and it does have a relatively high nail-holding strength.

Dry Lumber—Types and Percentages of Moisture Content

What do the symbols in Figure 3-6 mean? The various symbols KD, MC 15, MC 19, S Dry, S Grn, and A Dry indicate the moisture content of a piece of lumber. They all denote the degree of drying that has taken place (see p. 51).

Kiln dried (KD) means that the lumber has been dried in an oven. Stacks of properly spaced lumber are heated in the oven until the moisture content is reduced to 10% or less for finish lumber and 19% for framing lumber.

Moisture content (MC) identifies simply the moisture content of the lumber, not the method of drying that was used. MC 15 indicates a 15% moisture content and MC 19, a 19% moisture content.

Surface dry (S Dry) and *Surface green (S Grn)* indicate a 19% moisture content and usually indicate air-dried lumber.

Air dried (A Dry) is self-defining. It is a process where lumber is properly stacked in a dry climate for periods of 3 to 4 months.

Freshly cut lumber may contain as little as 30% moisture to as much as 115% or more. Once the fiber becomes saturated, the hollow spaces or cells fill with water.

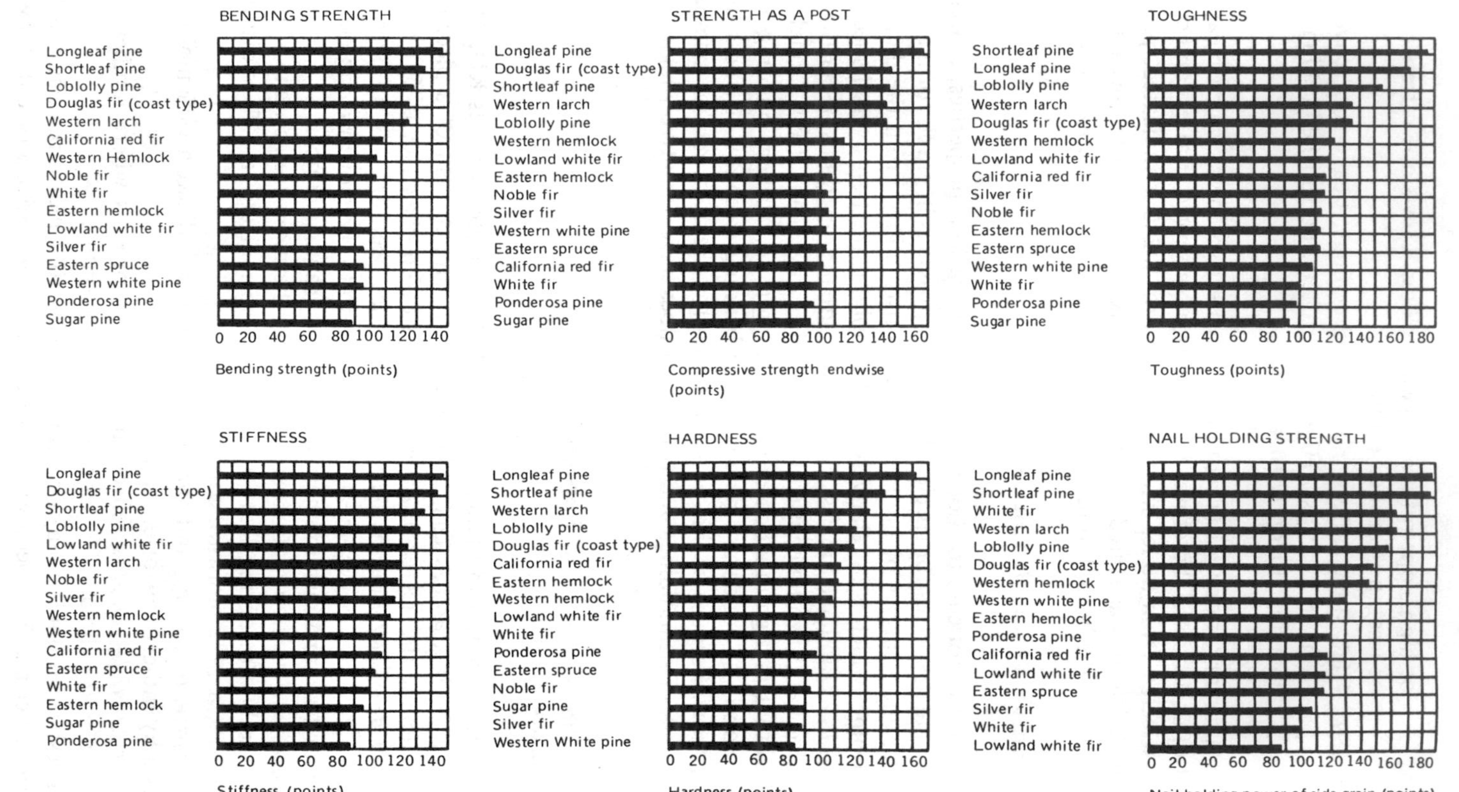

Figure 3–5 Comparative strength values of lumber by species. (From U.S. Department of Agriculture, Technical Bulletin 408.)

Figure 3–6 Six stamps on lumber that reveal its moisture content.

Drying the lumber allows the excess water in the cells to evaporate; however, this only reduces the moisture content to an average of 30%. Further time or effort must be used to reduce the moisture content in the woody cell walls. This reduction causes fibers to shrink. Thus, unless shrinkage occurs lumber is not considered dried. Naturally, the drier the lumber, the more shrinkage has taken place.

This chapter has been developed with a single purpose, to make the reader aware of the many factors that are used in the selection of a piece of lumber for a particular purpose. Many factors were brought out that defined the best lumber suited to trim and non-load-bearing uses, as well as lumber that could be used where stress is a factor as in rafters and joists. In the following section we examine in problem form the use of the previously developed factors in the selection of a single member in the hopes of bringing into focus their importance in construction.

SELECTING THE PROPER MEMBER

The examples that are to be presented involve all the properties of wood defined in this chapter. Fortunately, the government and many private associations have compiled much of the needed data in tabular form. This simplifies the selection problem. However, two selection methods are presented. Example 1 uses stated requirements that must be refined from Table 3–4 and tables of species and grades in Appendix A. In contrast, Example 2, which achieves the same results as Example 1, uses prepared tables (Appendix B) rather than calculations to satisfy the requirement needs. Finally, Example 3 shows how another table (Appendix C) may be formed that includes all the requirements of the first two problems and determines in one step the proper structural member to use.

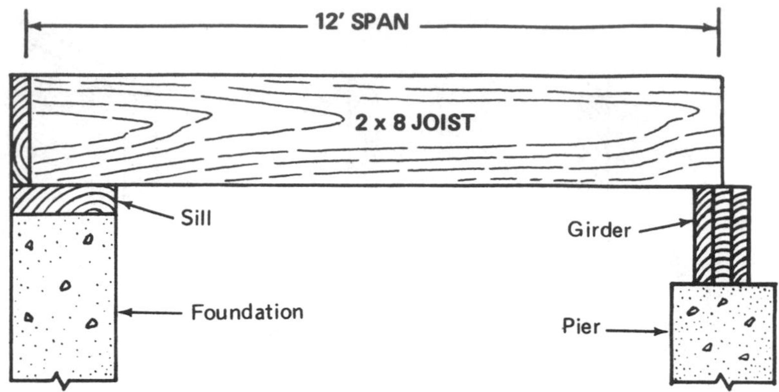

Figure 3–7 Detail of a joist assembly defining span and joist size, but not grade or species.

All three problems are faced with the same basic need: Select a species of lumber with the proper grade to support a floor in the living quarters of a residential house. Figure 3–7 illustrates the problem we are faced with. It shows a joist assembly made from 2 × 8 joists that is to span slightly less than 12 ft.

Example 1

Requirements:
$S/360$ = ½ in.; given S = 144 in./360, $S/360$ = 0.4
Modulus of elasticity = 1.9 million psi
F_b including variable factor 1.15 = 1897.5 psi
o.c. spacing 24-in. multiple joist assembly
Live load (LL) = 40 psf; dead load (DL) = 10 psf

Grades available (from Table 3–4)	F_b (psi)	E (psi)
Dense Sel Str KD	2200	1.9 million
Dense Sel Str	2050	1.8 million
Sel Str KD	1850	1.8 million
Sel Str	1750	1.7 million
No. 1 Dense KD	1850	1.9 million
No. 1 Dense	1700	1.8 million
No. 1 KD	1600	1.8 million

Note: All of the grades above exceed the strength (F_b) needs for the example floor joist when used in the assembly.

Species based on grade determined above can be found in Appendix A.

Douglas fir–larch (S Dry, S Grn, 19% max, moisture content)

Douglas fir–larch (north) (S Dry, S Grn, 19% max. moisture content)

Southern pine (S Dry, 19% max. moisture content)

Southern pine (KD, 15% max. moisture content)

The choices available for selection are Douglas fir, larch, or southern pine. No others have the stiffness required. Since both are acceptable, price and availability probably will dictate the selection. Now let's see if the same results can be obtained using different tables.

Example 2

Requirements:
$S/360$ = ½ in. (same as Example 1)
2 × 8 joists spaced 24-in. o.c. in multijoist assembly
LL = 40 psf; DL = 10 psf
Modulus of elasticity = 1.9 million psi; F_b = 1610
Floor joists 40 psf (LL) and 10 psf (DL); 2 × 8 on 24-in. o.c.
(from Table J1 (Figure 3–8 is an extract of Table J1, Appendix B))
Span slightly less than 12 ft

Tracing across from left in Figure 3–8 at 2 × 8 and 24.0 in. to length of member (11 ft 11 in.), we find F_b = 1610. Trace up from 11–11 to top of the chart and find that E = 1.9 million psi. Therefore, the *grades and species* available are the same as in Example 1 and are obtained from the tables in Appendix A.

Example 3

Requirements; The same as Examples 1 and 2.

Refer to Table No. 2 in Appendix C and obtain the grade from the table as follows:

1. Trace down the left column to 2 × 8, and 24.0 in. o.c.
2. Trace across to 11–11 (11 ft 11 in.) and then up to top.
3. Applicable grades are all those listed in "Grade" columns (1, 2, and 3) and notice that they are the same as those listed in Example 1.

Therefore, because the grades are the same as in Example 1, Douglas fir, larch, and southern pine are the species needed for joists. Further tables in Appendix A provide these data.

Summary

What has been proven by the exercise these problems have presented? First, several methods may be used by the builder in selecting the proper species and grade of lumber for a particular job. Next, an understanding of what is included in a table is extremely important if it is to be a working tool. Third, many of the properties of lumber are included in the selection process. Recall that grades are assigned according to limitations and defects. Various drying ratings may be assigned, but in general 15 to 19% moisture content is acceptable. Predicted loads must be selected, and the function of the member (joist, rafter, stud) must be known if the proper variable ($S/360$, $S/240$, $S/180$) is to be used. Finally, E and F_b values are needed.

Joist size (in.)	spacing (in.)	Modulus of elasticity, E (1,000,000 psi) 0.4	0.5	0.6	0.7	0.8	0.9	1.0	1.1	1.2	1.3	1.4	1.5	1.6	1.7	1.8	1.9	2.0	2.2	2.4
2 × 6	12.0	6-9 450	7-3 520	7-9 590	8-2 660	8-6 720	8-10 780	9-2 830	9-6 890	9-9 940	10-0 990	10-3 1040	10-6 1090	10-9 1140	10-11 1190	11-2 1230	11-4 1280	11-7 1320	11-11 1410	12-3 1490
	13.7	6-6 470	7-0 550	7-5 620	7-9 690	8-2 750	8-6 810	8-9 870	9-1 930	9-4 980	9-7 1040	9-10 1090	10-0 1140	10-3 1190	10-6 1240	10-8 1290	10-10 1340	11-1 1380	11-5 1470	11-9 1560
	16.0	6-2 500	6-7 580	7-0 650	7-5 720	7-9 790	8-0 860	8-4 920	8-7 980	8-10 1040	9-1 1090	9-4 1150	9-6 1200	9-9 1250	9-11 1310	10-2 1360	10-4 1410	10-6 1460	10-10 1550	11-2 1640
	19.2	5-9 530	6-3 610	6-7 690	7-0 770	7-3 840	7-7 910	7-10 970	8-1 1040	8-4 1100	8-7 1160	8-9 1220	9-0 1280	9-2 1330	9-4 1390	9-6 1440	9-8 1500	9-10 1550	10-2 1650	10-6 1750
	24.0	5-4 570	5-9 660	6-2 750	6-6 830	6-9 900	7-0 980	7-3 1050	7-6 1120	7-9 1190	7-11 1250	8-2 1310	8-4 1380	8-6 1440	8-8 1500	8-10 1550	9-0 1610	9-2 1670	9-6 1780	9-9 1880
	32.0					6-2 1010	6-5 1090	6-7 1150	6-10 1230	7-0 1300	7-3 1390	7-5 1450	7-7 1520	7-9 1590	7-11 1660	8-0 1690	8-2 1760	8-4 1840	8-7 1950	8-10 2060
2 × 8	12.0	8-11 450	9-7 520	10-2 590	10-9 660	11-3 720	11-8 780	12-1 830	12-6 890	12-10 940	13-2 990	13-6 1040	13-10 1090	14-2 1140	14-5 1190	14-8 1230	15-0 1280	15-3 1320	15-9 1410	16-2 1490
	13.7	8-6 470	9-2 550	9-9 620	10-3 690	10-9 750	11-2 810	11-7 870	11-11 930	12-3 980	12-7 1040	12-11 1090	13-3 1140	13-6 1190	13-10 1240	14-1 1290	14-4 1340	14-7 1380	15-0 1470	15-6 1560
	16.0	8-1 500	8-9 580	9-3 650	9-9 720	10-2 790	10-7 850	11-0 920	11-4 980	11-8 1040	12-0 1090	12-3 1150	12-7 1200	12-10 1250	13-1 1310	13-4 1360	13-7 1410	13-10 1460	14-3 1550	14-8 1640
	19.2	7-7 530	8-2 610	8-9 690	9-2 770	9-7 840	10-0 910	10-4 970	10-8 1040	11-0 1100	11-3 1160	11-7 1220	11-10 1280	12-1 1330	12-4 1390	12-7 1440	12-10 1500	13-0 1550	13-5 1650	13-10 1750
	24.0	7-1 570	7-7 660	8-1 750	8-6 830	8-11 900	9-3 980	9-7 1050	9-11 1120	10-2 1190	10-6 1250	10-9 1310	11-0 1380	11-3 1440	11-5 1500	11-8 1550	11-11 1610	12-1 1670	12-6 1780	12-10 1880
	32.0					8-1 990	8-5 1080	8-9 1170	9-0 1230	9-3 1300	9-6 1370	9-9 1450	10-0 1520	10-2 1570	10-5 1650	10-7 1700	10-10 1790	11-0 1840	11-4 1950	11-8 2070

Figure 3–8 Table J-1 from Appendix B, determining the floor joist to be used by modulus of elasticity, size and length, and F_b value.

PROBLEMS AND QUESTIONS

1. Define the type of tree that softwood is obtained from. List three types of softwood commonly used in construction.
2. Go to your local lumberyard and make a study of two grades of lumber in stock. Try to list the differences in grain, knots, stains, splits, and so on, between them.
3. What are the three select grades of lumber?
4. Of the framing grades, which is the best (worst) for framing?
5. What is the allowable difference in fiber-in-bending percentage for *select structure* graded lumber compared with clear straight grain? What is the percentage of modulus of elasticity?
6. Which wood is best for framing?
   a. sapwood **b.** heartwood
7. Define *specific gravity.*
8. Perform the test explained on pages 43 to 44 with a piece of pine and compare your results to those in Table 3–3.
9. To what does the *modulus of elasticity* of a board refer?
   a. its strength **b.** its stiffness
10. To what does a board's *extreme fiber in bending* refer?
    a. its stiffness **b.** its strength
11. If a joist is 14 ft long and is to have a deflection limitation of $S/240$, what would the maximum deflection be?
    a. ½ in. **b.** 11⁄16 in. **c.** 15⁄16 in.
12. What does each symbol mean?
    a. KD **b.** MC 15 **c.** S Grn **d.** S Dry
13. Redo Example 1 (page 52), but change $S/360$ = ½ in. to $S/240$ = ¾ in. and define the lowest grade acceptable as a joist.

PROJECT

Complete the following project so that you are absolutely certain that the lumber used in a modification to an existing house or new house will carry the loads. The project is to build a porch roof over an existing patio whose measurements are 12 ft 6 in. × 20 ft. The concrete slab projects 12 ft 6 in. from the exterior wall. The builder wants to use a *shed roof design* as shown in Figure P3–1. The ceiling will be flat, so it will have *ceiling joist* and be *finished with plywood.* The house is in a northern climate where *snow falls and frequently lasts for 7 days or more.*

A. Your first problem, should you decide to complete this project, is to be convinced that the proper lumber is used for the rafter framing.

1. We will decide which size and type of rafters are needed to do the job. The first consid-

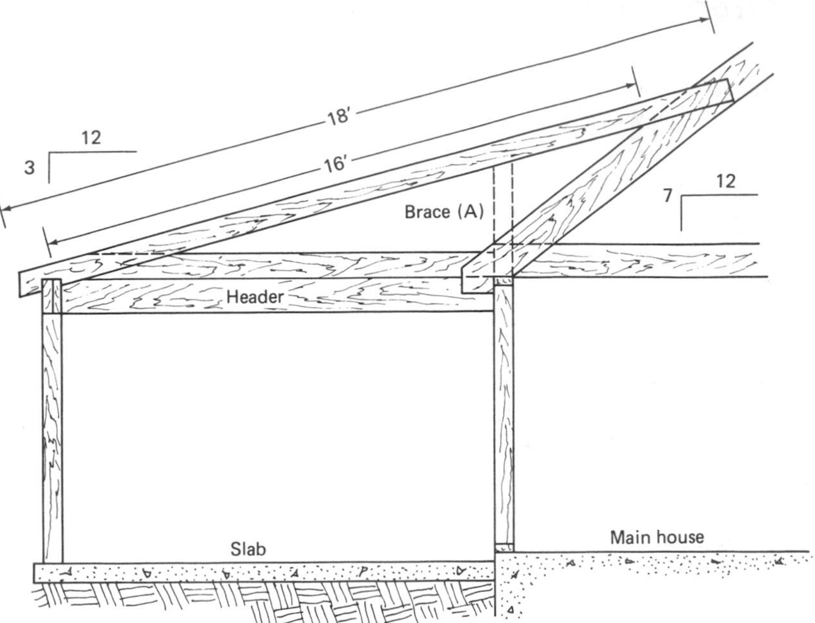

Figure P3–1

eration is to define the slope. As shown, it is "3 in 12." This is a low slope. The length of the rafter is shown from two viewing angles.

a. The rafter length, including overhang, is ____________ ft.
b. The clear slope length is ____________ ft. (Check your answers before continuing.)

2. Let's extract data from Appendix B (Table B–3) first. Then we'll use Appendix A data for southern pine and fill in the missing data below.

Data	Size of stock	o.c. spacing	Data from drawing (Appendixes B and A)
Free slope span	2 × 6	16 in.	____________
F_b			____________
Modulus of elasticity, E			____________
Lowest acceptable grade of lumber			____________
Free slope span	2 × 8	16 in.	____________
F_b			____________
E			____________

Data	Size of stock	o.c. spacing	Data from drawing (Appendixes B and A)
Lowest acceptable grade			________
Free slope span with brace (A)	2 × 6	16 in.	________
F_b			________
E			________
Lowest acceptable grade			________

What would it do for us if we were to use a brace (A) under each 2 × 6 straight up over the outside bearing wall? Did you consider snow loading for 7 days or more? You should have used the column "7-day loading" for F_b from the table in Appendix A.

3. So you now have the data on the rafters. Let's interpret your findings.
   a. True or false? There is no southern pine 2 × 6 18 ft long that can safely withstand the weight from a modulus of elasticity viewpoint.
   b. Which grade of southern pine is the lowest for use with the 2 × 8?
      (1) Sel structural (3) No. 1
      (2) No. 2 (4) No. 3
   c. Which grade of southern pine is the lowest grade for use with the 2 × 6 with brace (A)?
      (1) No. 1 dense (3) No. 2
      (2) No. 1 (4) No. 3 dense
   d. Given that 2 × 8s cost more per board foot than 2 × 6s of the same grade, which of the solutions points to the least costly option with no reduction in structural strength?
      (1) 2 × 6 18 ft
      (2) 2 × 6 18 ft with brace
      (3) 2 × 8 18 ft
      (Check your answers for accuracy.)

B. Now let's select the size and grade of southern pine best suited for the ceiling joists. From the drawing we see that the span is about 12 ft. The o.c. spacing will be 16 in., just like the rafters. There will be no live load since there will not even be room to crawl. Dead weight will be the ceiling joists and plywood ceiling covering (¼-in. plywood weights about 20 lb per 4 × 8 sheet or about 0.6 lb per square foot).

1. From Appendix B, Table B–2:

Stock size	E	F_b
2 × 4	________	________
2 × 6	________	________
2 × 8	________	________

2. From Appendix A, Southern Pine, KD

Size/grade	Acceptable/not acceptable	First preference
2 × 4 construction	______	______
2 × 6 No. 3	______	______
2 × 6 No. 2	______	______
2 × 6 No. 1	______	______
2 × 8 No. 3	______	______
2 × 8 No. 2	______	______
2 × 8 No. 1	______	______

In summary, we would definitely opt for the use of No. 2 grade southern pine for both the rafters and ceiling joists. But to use 2 × 6s for rafters we would need to brace each one from the exterior wall to the underside or alongside the rafter. I hope that you enjoyed this project and learned from it.

Answers:

Part A: **1.** a. Approximately 18 ft,
b. approximately 16 ft;

2. Free slope span 16 ft, F_b = 2100/7-day load 2690; E = 2,400,000 psi; lowest acceptable grade, none
Free slope span 16 ft, F_b = 1200/7-day load 1880; E = 1,200,000 psi; lowest acceptable grade, No. 2
Free slope span 12 ft, F_b = 1200/7-day load 1880; E =1,200,000 psi; lowest acceptable grade, No. 2

3. a. true, b. (2), c. (3), d. (2).

Part B:

1.	Stock size	E	F_b
	2 × 4	None	None
	2 × 6	1,000,000	1140
	2 × 8	500,000	720

2.	Size/grade	Acceptable/not acceptable	First preference
	2 × 4 construction	Not acceptable	
	2 × 6 No. 3	Not acceptable	
	2 × 6 No. 2	Acceptable	Yes
	2 × 6 No. 1	Acceptable	No
	2 × 8 No. 3	Acceptable	Second choice
	2 × 8 No. 2	Acceptable	Too costly
	2 × 8 No. 1	Acceptable	Too costly

Chapter 4

Energy Efficiency and the Residential Carpenter

For many years the carpenter has been the person installing insulation. Very early in this century, insulation consisted mainly of dead air, then paper board usually ½ in. thick was installed in walls. Later, thin batts of insulation were manufactured and used in walls and ceilings. Only recently has emphasis been placed on full insulation of the residential house.

The most compelling reason for this interest in insulation is the cost incurred in operating heating and cooling systems. Since high utility costs have such a significant impact on the homeowner's outlay, there is a common demand for energy-efficient houses throughout the country. In the northern climate, fully insulated houses reduce heating costs for seven, eight, or more months per year; in contrast, fully insulated houses in southern climates reduce air-conditioning costs. These factors, plus increased legislation and improvements in regulations at the federal level, have made insulation mandatory.

A variety of insulating products are available and the carpenter should be able to choose among them. Later in this chapter they are examined more closely. But to generalize, there are insulating panels, batts, loose fill, and foam. Insulating also includes proper design and installation of windows and doors, which should include insulation features. The carpenter may also be required to install storm windows and storm doors that trap air between two surfaces and thus have insulating value.

Radiation of thermal energy (measured in Btu, British thermal units) can occur in either direction through a wall or ceiling, door, or window, and all too often through the floor. Therefore, special protective measures need to be applied to eliminate the losses or gains of radiation from the ground as well as the air. The carpenter must include various procedures during the construction of the floor assemblies that ensure a control of Btu loss or gain to the house.

Installing insulation is not a very difficult task, even though cramped quarters may be encountered. More significant, though, is the need to understand how insulating boards are currently being used and what factors the use of these boards have on the structural soundness of the building. To illustrate in a general way what could be involved, consider that a 2-in.-thick foam insulation panel is being installed over a stud wall. Are let-in corner braces needed to provide adequate strength? It is currently acceptable to use bituminous impregnated fiberboard without let-in corners in some areas of the country. Answers to other such questions are provided later in the chapter.

Finally, solar energy is beginning to play a significant role in house construction in many parts of the country. Solar collectors, which include metal, plastic, and water, are placed into roof assemblies. What problems do these units pose for the carpenter? He or she probably needs to learn more about how these systems work, but also needs to know and prepare for the additional weight the roof must carry. It is likely, for instance, that the *S*/180 criteria allowable for rafters cannot be used. Such deflection may prove unsatisfactory, necessitating specialized bracing or trussing.

It should be evident after reading this brief introduction that the carpenter is involved in many ways with establishing an energy-efficient house that has sound structural qualities. It is obvious that he or she needs to have some basic knowledge of thermal control and be aware of the variety of materials available for use.

PRODUCTS AND TERMS

Various products defined in this section are used in residential and commercial construction to provide insulating factors for the control of thermal radiation. Each is assigned an "R" value, which is a numerical value indicating the relative resistance of the product to the penetration or loss of Btu per hour. All insulations should have a high resistance to crushing and sagging. Some may retain moisture, so care must be used to avoid these problems when using products with limitations.

Cellular-glass insulation board. Boards are available in slabs 2, 3, 4, and 5 in. thick. The assigned R value per inch thickness is 1.8 to 2.2. Thus a 2-in. board would be rated as R-3.6 to R-4.4. This type is classified as rigid insulation, is easily cut to fit, but must be protected from moisture.

Glass fiber with plastic binder. This rigid board is available in thicknesses of ¾, 1, 1½, and 2 in. and has a resistance value of R-3.3 to R-3.9 per inch of thickness. This board is easily cut. If coated with waterproofing, it may withstand considerable moisture.

Foamed plastic. Insulation sheets made from polystyrene, polyurethane, and other plastics are usually available in thicknesses of ½, 1, 1½, and 2 in. The resistance value may vary from R-3.7 for polystyrenes to over R-6.0 for polyurethane per inch of thickness. These panels are currently being used as sheathing in some energy-efficient houses.

Drywall. Drywall made from paper, asbestos, and other minerals has some ability to control Btu radiation. The R value for ½-in.-thick sheets is approximately R-0.45. Although relatively low, it does contribute to the overall sum of a wall's or ceiling's R value.

Plywood sheathing. All lumber has thermal protection properties. The species of lumber, because of its structure, may have more or less ability to control Btu penetration. However, an approximation of R = 0.156 per ⅛ in. of thickness or R = 1.25 per inch is customarily applicable.

Batts. The insulation batt, commonly called a *blanket,* is available in rolls of varying widths and thicknesses. The width does not have any bearing on the R value, only the thickness and type of material used. Batts are available in 1½-, 2-, 3-, 4-, and 6-in. thicknesses and are assigned R values ranging from R-3.5 to R-19.

Loose fill insulation. Loose fill insulation is customarily blown into place with a machine, although it can be placed manually. It affords a unique insulating quality in that the wool and trapped air combine to form an effective thermal barrier. However, to be effective it usually must be installed quite thickly, 6 in. or more. Its R value ranges from R-19 to R-36.

Trapped air. Trapped dry air, usually called dead air space, has an insulating quality if and only if it is stagnate. Such air spaces may be found between sheathing and insulating batts such as stud walls. It may have an R value of R-4 per inch of thickness.

Glass and glazine materials.

1. *Single strength:* R value 1 or below
2. *Double strength:* R value 1 or below
3. *Patio doors:* R value 1 or slightly more than 1
4. *Thermal (vacuum) glass assemblies:* R value 1.6 or slightly better

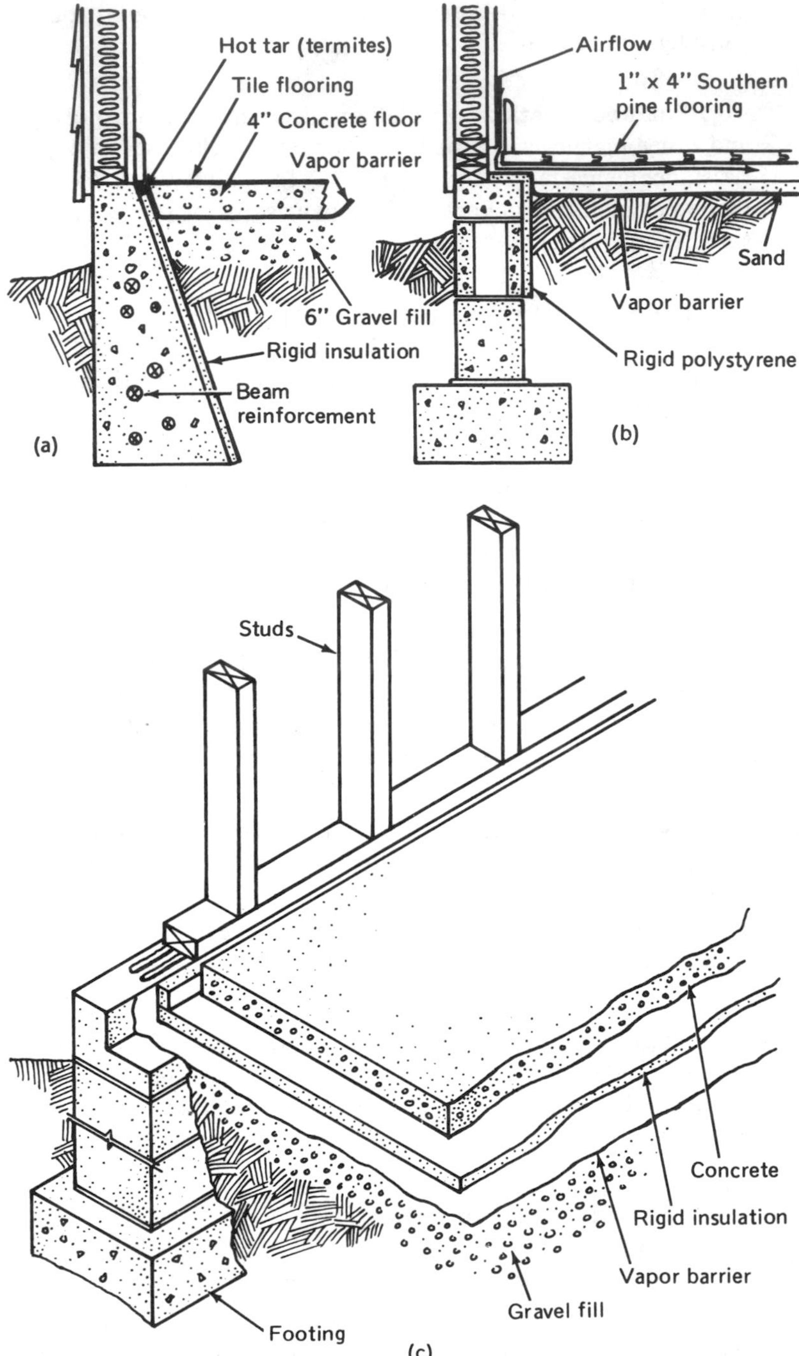

Figure 4–1 Rigid insulation makes good, nondeteriorating thermal protection for foundations, basements, and under slabs.

INSULATING THE FLOORS AND FOUNDATION

In the foundation area, plans and specifications may call for rigid insulation board to be installed. Several typical applications are illustrated in Figure 4–1. In Figure 4–1a and b the insulation is installed vertically along the inside of the foundation wall. In Figure 4–1c, insulation of the same variety is laid horizontally over the vapor barrier and a short way up the wall, at least to the top of the concrete pour. The use of this material in this manner provides good thermal protection for slabs poured on the ground.

Basement inside walls may also be insulated with rigid insulation or batts. Customarily, 2-in. insulation is used because 2 × 2s are nailed to the concrete block walls on 24-in. centers. Insulation is inserted between the studs. If there has been any previous indication of dampness, the walls should first be coated with tar or bituminous material and the rigid board should be used.

The third area that must have insulation is under a joist assembly that is built over a crawl space. There are two methods that may be used to accomplish this objective. Figure 4–2 shows that 1-in. batts are laid on top of the joists and then plywood sheathing is laid on top of it. Figure 4–3 illustrates the second method of insulating a floor joist assembly: batts are installed between the floor joists and are supported by spring-in wood strips, asphalt cement, or friction.

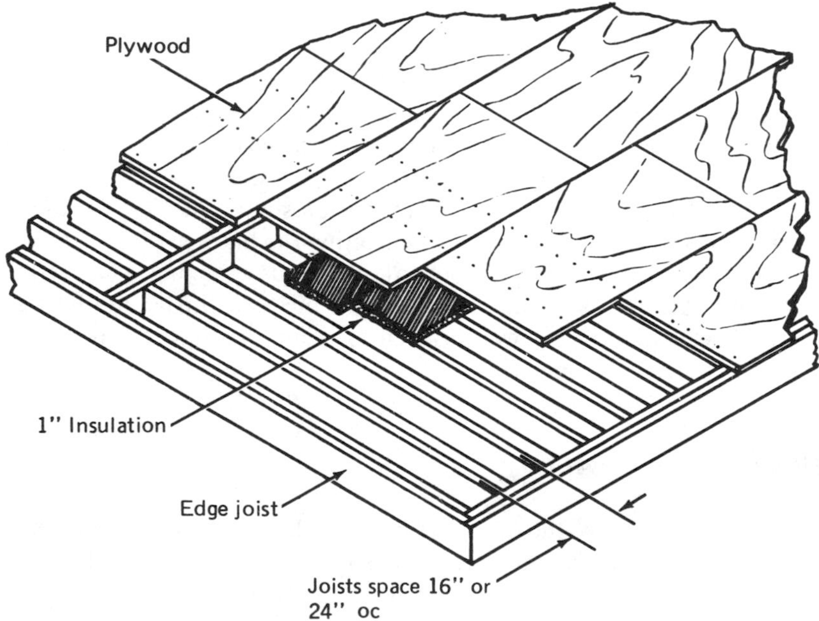

Figure 4–2 Rolls of insulation beneath subfloors provide minimal thermal protection.

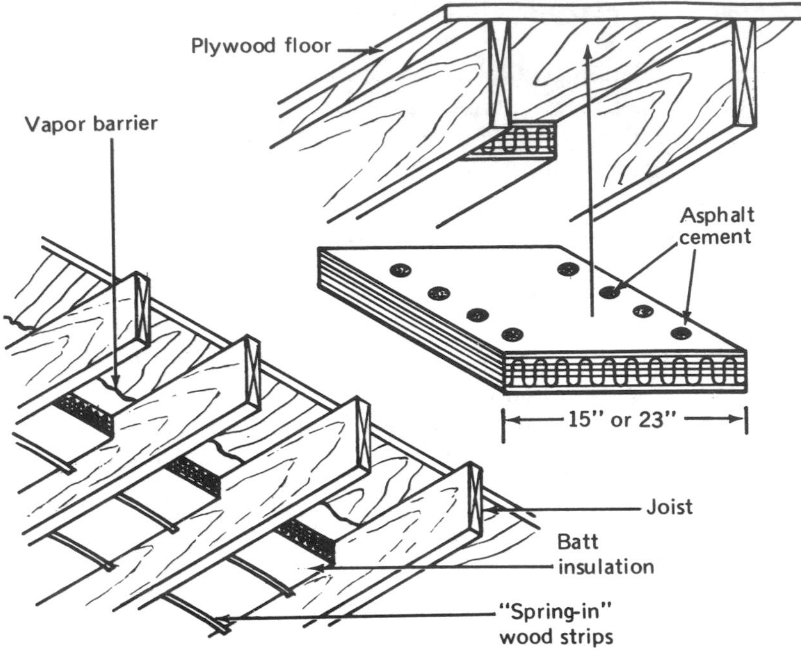

Figure 4–3 Batt or blankets cemented or compressed between joists provide good thermal protection.

In the first method, flexible blankets of insulation may be used. The strips of wood should be cut approximately ½ in. thick and ⅜ in. longer than the space between joists. These strips should be spaced close so that the insulation remains close to the subfloor.

A more rigid batt type of insulation should be used for the cementing or friction method of installation. If cementing the batts in place, an adequate amount of cement must be applied; too little can cause sagging and eventual falling of the insulation. It is also important to stuff pieces of insulation into all areas along the perimeter of the joist wall, as well as around cutouts for plumbing, heating ducts, and electrical wiring.

INSULATING THE WALL

Achieving an R value of R-14 for an insulated wall may be accomplished in several ways. It is currently possible and desirable to increase the value beyond R-14 to R-24 by the use of rigid polyurethane as sheathing. In the preceding section, the average R values were defined for sheathing, air, siding, and batts on a per-inch-thickness basis. The three examples following illustrate how these products work together to accomplish thermal protection.

Single-Wall Construction with Entrapped Air

Figure 4–4 depicts the character of a single-wall system where siding is applied directly to the studs. In this example, an air space of ¾ in. is planned for and included between insulation and wallboard. This means that if a normal 3½-in. stud is used, the insulation thickness is 2 to 3 in. thick. Notice from the legend that each product used in the wall adds to the overall R value of the wall. There is a problem using this insulation method; it is difficult to seal each area between the studs so that air is truly trapped. If a leak is present, the air temperature, moisture content, and drafts eliminate the insulating quality. It therefore may be assumed that a better insulated wall may be achieved by substituting R-13 3½-in. batts and eliminating the air space.

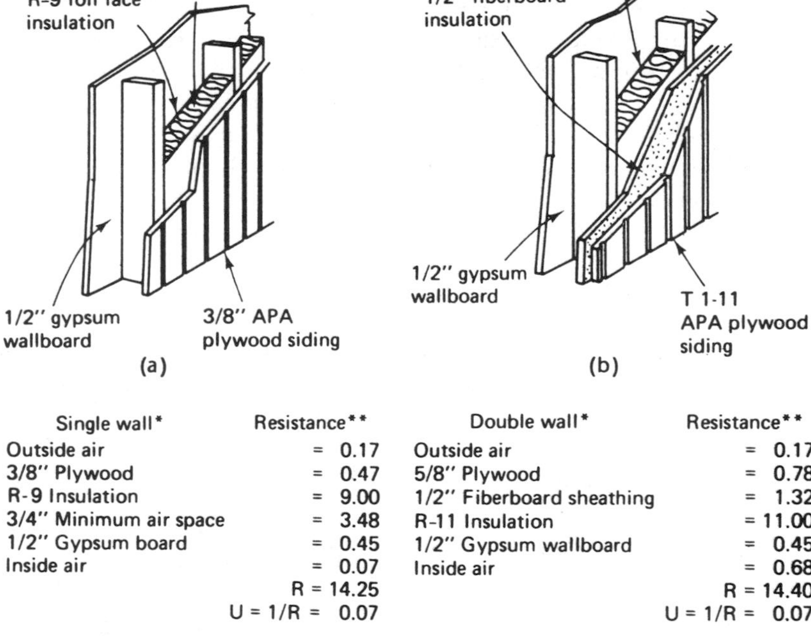

Single wall*	Resistance**
Outside air	= 0.17
3/8" Plywood	= 0.47
R-9 Insulation	= 9.00
3/4" Minimum air space	= 3.48
1/2" Gypsum board	= 0.45
Inside air	= 0.07
	R = 14.25
	U = 1/R = 0.07

Double wall*	Resistance**
Outside air	= 0.17
5/8" Plywood	= 0.78
1/2" Fiberboard sheathing	= 1.32
R-11 Insulation	= 11.00
1/2" Gypsum wallboard	= 0.45
Inside air	= 0.68
	R = 14.40
	U = 1/R = 0.07

‡R-13 insulation (3-5/8") with or without foil may be substituted.

* Meets or exceeds FHA-MPS insulation requirements for multifamily construction, all regions.

**R values from ASHRAE Guide, Actual thickness and species groups affect plywood insulation values.

Figure 4–4 All materials add to the R value of walls, whether single- or double-wall systems.

Double-Wall Construction with No Entrapped Air

In the construction of the double wall as Figure 4–4b shows, sheathing is installed on the outside of studs before siding is applied. In addition, full-thickness (3½-in.) insulation is used between sheathing and wallboard. A slightly better total R value is obtained using this method, and the problem of completely sealing off each section to entrap air is eliminated by use of full-thickness insulation.

The 6-in. Wall

The third method of insulating a wall is relatively new, even though the materials have been available for some time. Figure 4–5 illustrates its construction. Sheathing consisting of 2-in. rigid polyurethane sheets are nailed to the outside of the studs, creating a 6-in. thermal wall. The addition of this insulation can provide an R value as high as R-24 to R-26, which is very desirable. The carpenter who builds such a wall must, of course, understand that these rigid panels have little or no strength, so they cannot keep the walls in alignment. It therefore becomes absolutely essential that let-in corner braces be installed. Only by inclusion of these braces can proper alignment of walls be established and maintained. In addition to the condition above, this rigid insulation has no holding power, so siding must be installed such that nails or wall ties in brick veneer are securely driven into the studs behind the rigid boards.

INSULATING THE CEILING AND ATTIC

The latest trend in insulating ceilings is the "more the better" concept. This is a distinct change from the past, where the theory was that 2 to 4 in. of insulation was

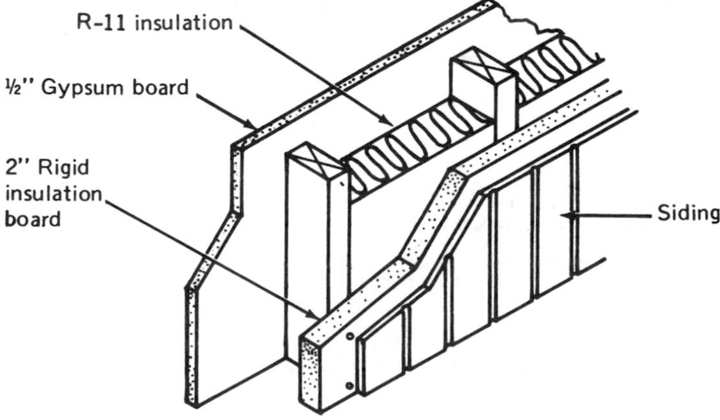

Figure 4–5 Six-inch wall.

sufficient and that the addition of more insulation resulted in very little thermal protection. R values as high as R-36 can be achieved with 12-in.-thick insulation. Generally, though, the depth of ceiling insulation is set at 6 to 6½ in. (R-19) for two reasons: (1) the insulation provides considerable protection in all but the most extreme climates, and (2) ceiling joists are usually 2 × 6 or wider except where trusses are used. Therefore, the installation of insulation is relatively simple.

There are several problems that the carpenter may encounter during the construction of attics as shown in Figure 4–6. One is that the attic may be partially finished with rooms framed. In this event insulation is usually installed between ceiling joists as though the attic were unfinished. Another type of problem exists when there is a finished attic. In this event, the ceiling insulation must be installed between the outer wall and partial partition that is the attic room wall, as well as the partition. Third, insulation must be installed between rafters and dormer studs and rafters before dry wall is fastened to these structures.

A word of caution: Almost all cornices have some sort of ventilating screen so that proper air circulation is obtained in the attic area. Improperly placed insulation that blocks the vents and stops the flow of air in the attic could cause serious problems from two points. (1) Snow and ice will tend to remain on the roof long after they should have melted away; this puts added strain on all roof components, especially the rafters. Also, there is the continual radiation of the temperatures produced by the frozen ice and snow into the attic. (2) In the summer the attic temperature may easily rise to 160 to 180°. This heat soon penetrates the insulation and ceiling causing higher cooling costs.

All these problems can and should be avoided by properly installing the ceiling and rafter insulation so that air can circulate. The temperature and moisture content of air driven by the wind varies throughout the day and night, alternately cooling and heating the attic.

OPENINGS—WINDOWS AND DOORS

Almost without exception, aluminum and wood window units are manufactured with insulation as a part of the design and construction process. Many years ago a special material, weatherstripping, was used to insulate windows. It was very effective in blocking the passage of air. The insulation included in today's windows is quite effective in controlling the passage of air. Overall, however, a window consists of glass, and the radiation of heat and cold through the glass is considerable, as the section on *products* and *terms* illustrated.

The carpenter can do little to change these factors; however, he or she may be required to install storm windows. If the job is well done, a dead air space ranging from 2 to 4 in. is established and this dead air space provides considerable R-value thermal protection.

Of course, windows may be purchased with thermal panes of glass. Again, this does not affect the carpenter's role. There is, of course, one job the carpenter

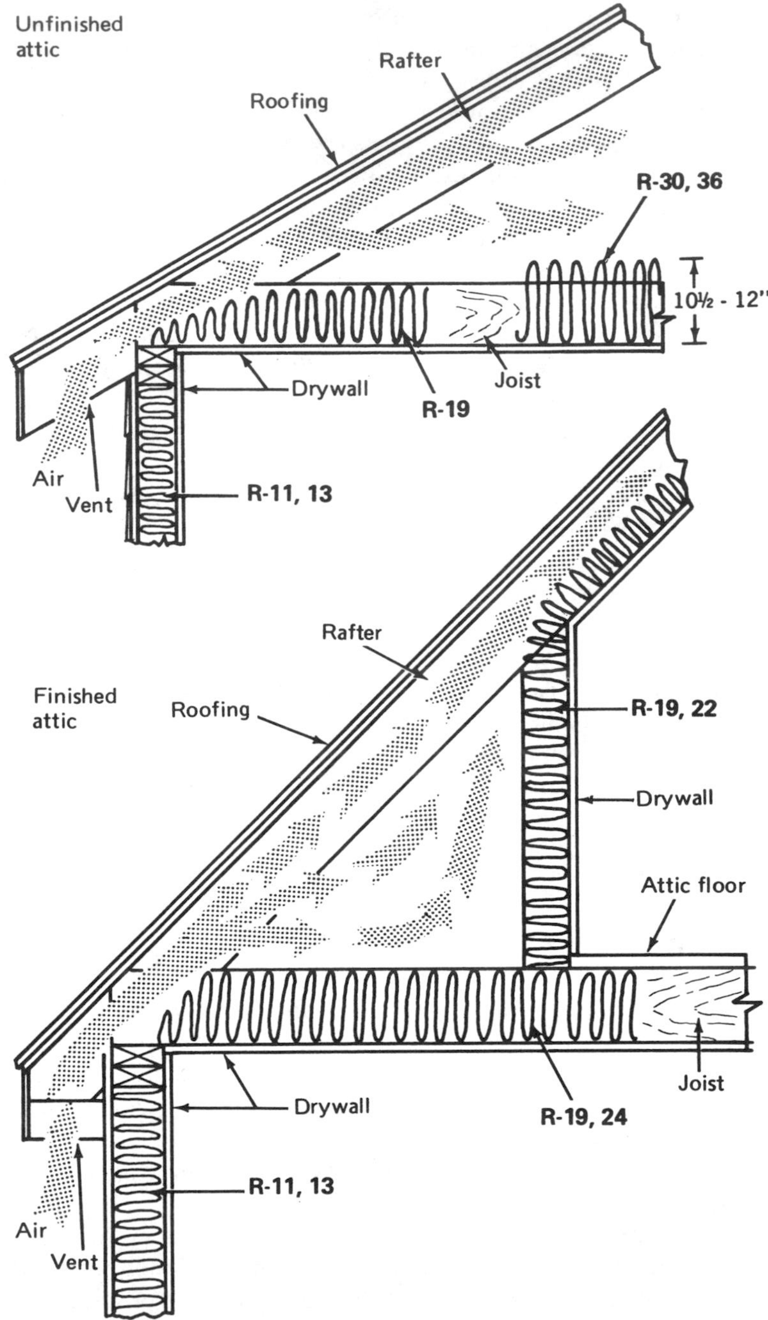

Figure 4–6 Insulating the ceiling and attic room: air passages must remain intact.

can do, and that is to stuff insulation between the window frame and surrounding framing prior to finishing the wall. In this way he or she ensures that no leakage of air occurs around the window.

Doors are the other wall openings that can, if improperly insulated, cause tremendous losses of heat and add problems during summer cooling. Solid doors should be used instead of hollow core, and glass panels should be of the thermal variety. The most important and the most difficult part of a door assembly to seal is the space between the door and the jamb or sill. Again, most new door units are weatherstripped with a compression strip around the top and sides. This is generally quite effective in sealing drafts.

The sill presents a different problem. Quite often the threshold must be removed for a variety of reasons: if the door is used over a concrete floor, for example. The installed metal sill is also removed; in other cases no metal sill may have been installed at the factory. A third reason is that the threshold must be built-up so that the door passes freely over thick carpets on the floor. Regardless of the reason, the carpenter is responsible for installing the metal sill so that a completely airtight surface is obtained. He or she can install either of two types of metal sill. The compression type (Figure 4–7a) uses a plastic compressible strip to contact the door. The ends of the sill should be puttied and the sill should be caulked prior to fastening it to the floor. The second type of sill (Figure 4–7b) requires that a metal hook strip be nailed to the bottom of the door and this hook slides into a channel on the sill. Putty at the ends and caulking under the sill are also required for this installation.

An extremely effective method of improving the R value at a door area is to properly install a storm door on the outside. This door effectively creates a dead air *space* that has insulating qualities. The door itself may contribute some thermal resistance, especially if it is glazed with thermal glass.

STRUCTURAL PROBLEMS IN A SOLAR ROOF ASSEMBLY

Solar energy systems are being used more and more by residential and commercial owners. These units are almost always installed on the south-facing side of a roof. In northern climates the roof pitch must be *12-in.-12* for best results, whereas in the south, lower slopes of *3-in.-12* to *7-in.-12* may be used.

A solar system consists of panels called collectors, piping made from plastic or copper, and water. These parts of the system unite in absorbing heat energy from the sun and carry the energy to a storage section in the basement or underground.

The carpenter must construct a more substantial roof than would otherwise be needed. Where, for example, an *S*/180 deflection factor may have been acceptable under normal conditions, the added weight and force created by the solar device prohibits its use. Either of two solutions must be accepted and employed: (1) a different species and grade of rafter must be selected, having an *S*/240 or *S*/360

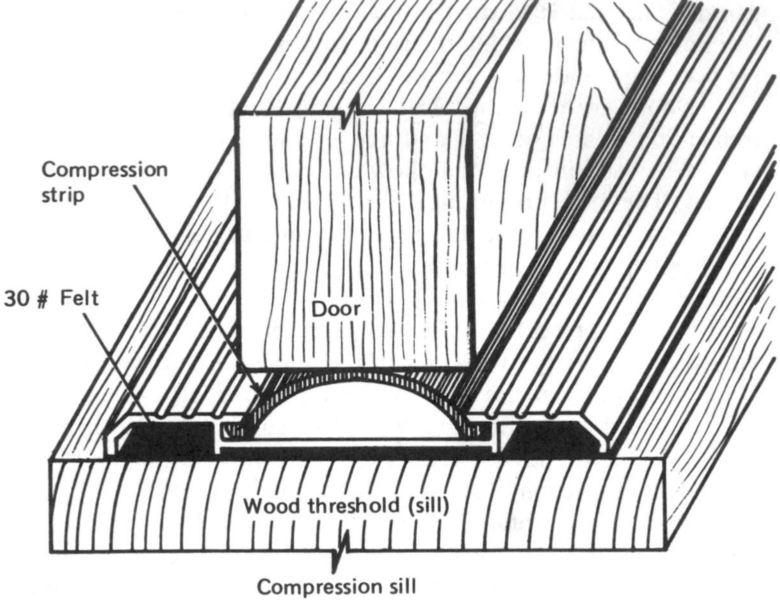

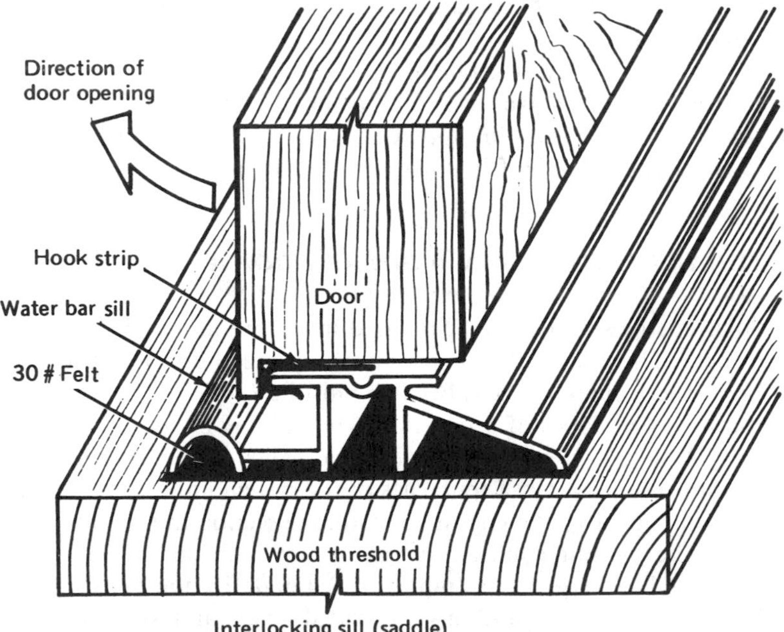

Figure 4-7 Insulating between door bottom and threshold (sill) with metal sills.

deflection factor, or (2) trussing between rafter and ceiling must be included where the rafter is reinforced so that the $S/180$ deflection is improved to $S/360$.

Since the solar panels are similar to shallow boxes, sealing their perimeters against rain, snow, and ice is a very real problem. Nondrying mastics, hot bituminous, or flashing must be used. The panels must also be installed in such a manner that rain, for example, has a free path to flow from the roof. Clearly, the carpenter has a vital role in proper construction of the roof and the solar system being used.

Conclusion

The facts and concepts presented in this chapter have defined the carpenter's role in creating an energy-efficient building. He or she is the one who has the direct responsibility and the opportunity to install insulation, while fabricating the building. You should now understand that certain spaces, like the roof, need air passage, and that certain other spaces, under floors and in walls, must be entirely blanketed with insulation. You should also have learned that structural provisions for bracing must be applied when insulation boards having no strength are used as sheathing. During the insulation of doors and windows you should now know that stuffing insulation between jamb and frame is a must and that door sills must be properly installed.

If the carpenter is required to perform these tasks, it is because the future owners of the building demand it, or government, state, or local regulations require it. It is therefore necessary that the carpenter apply his or her understanding of thermal protection to obtain various thermal resistances (R values).

Broadly stated, Table 4–1 lists R values for various thicknesses of insulation. These are general values; specific products may vary according to manufacture. Of course, the other products used in construction add to the overall R value.

TABLE 4–1 THERMAL PERFORMANCE OF INSULATION BY R VALUE VERSUS THICKNESS RANGE

R value	Thickness range (in.)	R value	Thickness range (in.)
36 +	12–14	13	3½–4
30	10¼–12	11	3–3¾
24-26	6–7½	8,9	2½–3
21,22	6–7½	7	2–2¼
19	5½–6½	3.5	1–1½
14	5–5½		

Note: Type of insulation is disregarded and values presented are representative of those effective within the thickness span.

PROBLEMS AND QUESTIONS

1. What is the unit used to express thermal energy?
   a. E.E.U. **b.** Btu
2. What are the normal areas in a house that are insulated?
3. List two typical places in a house where there is a high loss of heat or air conditioning?
4. If a 4-in. cellular-glass insulating board were used to sheath a house, what would its R value be?
   a. 3.6 **b.** 8.8 **c.** 11.0
5. Is glass fiber with plastic binder rated higher in R value than a cellular-glass board?
6. List the total R value for the following wall.
   a. ½-in. drywall
   b. 3½-in. batts or blankets
   c. 2-in. glass fiber with plastic binder
   d. outside air
   e. inside air
7. What alterations would you make to construct a wall that had an R value of 18 instead of R-12?
8. Why is it essential that air flow through attic spaces at all times?
9. What makes installing an aluminum threshold or saddle so difficult?
   a. the height of the door over the rugs
   b. the need to weather proof against air and moisture
   c. the need to screw it to a concrete floor
10. If a solar panel with water circulating through it is to be installed on the roof, will additional bracing or heavier rafters be needed to support the weight?
11. If a roof were originally designed for rafters using $S/180$ rated and solar energy were to be installed, would you still use $S/180$ or ______ instead? Explain.
    a. $S/240$
    b. $S/360$

Chapter 5

Floor Systems

Every floor system is an integration of different materials that results in a soundly designed, reliable, and secure assembly. A floor system may require many different types of materials for its construction. Most of these are examined in this chapter. The many grades and sizes of these materials are discussed, along with how to use them effectively. The relationship of a floor system to the blueprint and specifications data is so important that we deal with it at length. It is also important that a floor system be designed properly, whether it be the conventional type or other variety.

DEFINITION OF A FLOOR SYSTEM

The opening sentence of this chapter gives a functional definition of a floor system. Any floor system must be soundly designed. It must be strong and solid, and it must carry all predicted loads. Those loads may be continuous, temporary, or impact types. In each situation the system will normally be engineered to withstand all of these load ratings. It must, of course, be reliable for a very long time, since such an assembly is not likely to be replaced. Also, it must be properly constructed if it is to have the expected reliability.

However, this definition of a floor system is conceptual. What is needed is a more practical definition, one that relates to its structure or variety of structure. First and probably most important, is the need to define the live and dead load limits for residential floor systems. Construction industry standards have been established, setting live loads at 20, 30, and 40 psf (pounds per square foot) for various rooms in a house based on the purpose of the room. Figure 5–1, a pictorial floor plan of a two-story structure, shows that the living room, family room, and dining room require a live load of 40 psf. These three rooms are typically the *activity* rooms of a house. They are the most lived in, have the greatest concentration of pedestrian traffic, and sustain the greatest continuous, temporary, and impact loads. Contrast the upstairs sleeping rooms and bath, where a 30-psf live load is required. This 25% reduction is quite easily understood if the reasoning above is applied. Seldom is there heavy traffic. Seldom is there an excessive number of people. Quite likely, impact forces are almost nonexistent. Also included in defining the floor system is a common dead load value of 10 psf.

Figure 5–1 also shows floor plans that indicate a need for two systems, the first-floor assembly and the second-floor assembly. Both are independent, and little difficulty should be encountered during their design and construction. However, consider the single-level floor plan shown in Figure 5–2. The same values of 40 psf for activity centers and 30 psf for quiet zones apply. Yet the floor system is unified and on one level. What would be the appropriate approach to take in defining this system?

The most commonly accepted practice is to use 40 psf throughout. Adopting this practice tends to increase the cost of materials slightly. However, the labor costs savings usually more than offset the materials costs. To illustrate, part of the floor

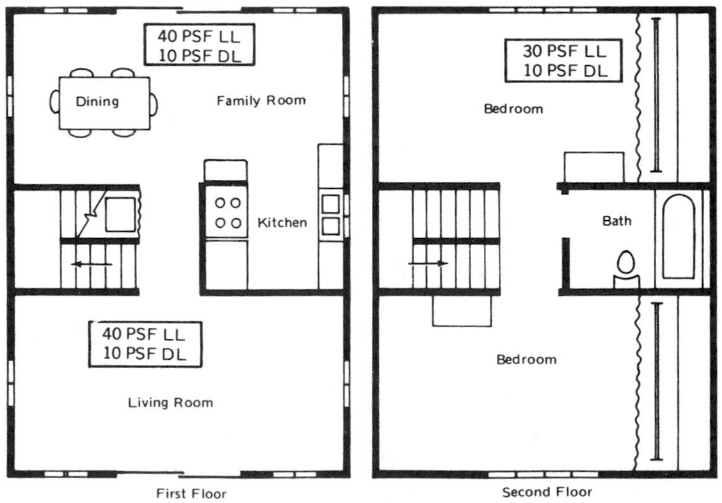

Figure 5–1 Activity areas require 40-psf live load and 10-psf dead load. Quiet areas need 30-psf live load and 10-psf dead load.

Figure 5–2 One-story house has different live load requirements for activity centers and quiet zones.

assembly could be made from 2 × 8s. Part (the quiet areas) could be built with 2 × 6s to meet the 30-psf requirements. Adjustments to girder size and height and sill heights would need to be made so that all floor joists would have the same top edge horizontal alignment. Bridging of different sizes would also be needed. The carpenter's tasks related to these adjustments results in considerable labor expenses. Therefore, adoption of a single value of pounds per square foot usually results in the greatest efficiency of resources. As an added thought, the fact that a slight overbuilding in quiet zones is included could be used as a quality selling feature by the builder or salesperson.

The definition of a floor system includes the parts of the system that make up the assembly. Each is defined later more completely; however, included are the sills, joists, girders, sheathing (plywood), and underlayment, if used. Besides materials there are the construction techniques, including the spacing of joists, the type of sill and girder construction selected and incorporated, and the reinforcements needed to integrate the materials. The definition of a floor system creates immediately a need to examine the parts used to form it.

PARTS OF THE SYSTEM

Girders

It is customary that a girder be installed either within the joists spanning the width or length of a floor or under the joists. Methods A and B shown in Figure 5–3 are used where the top surface of the joist and girder must align. The illustration of

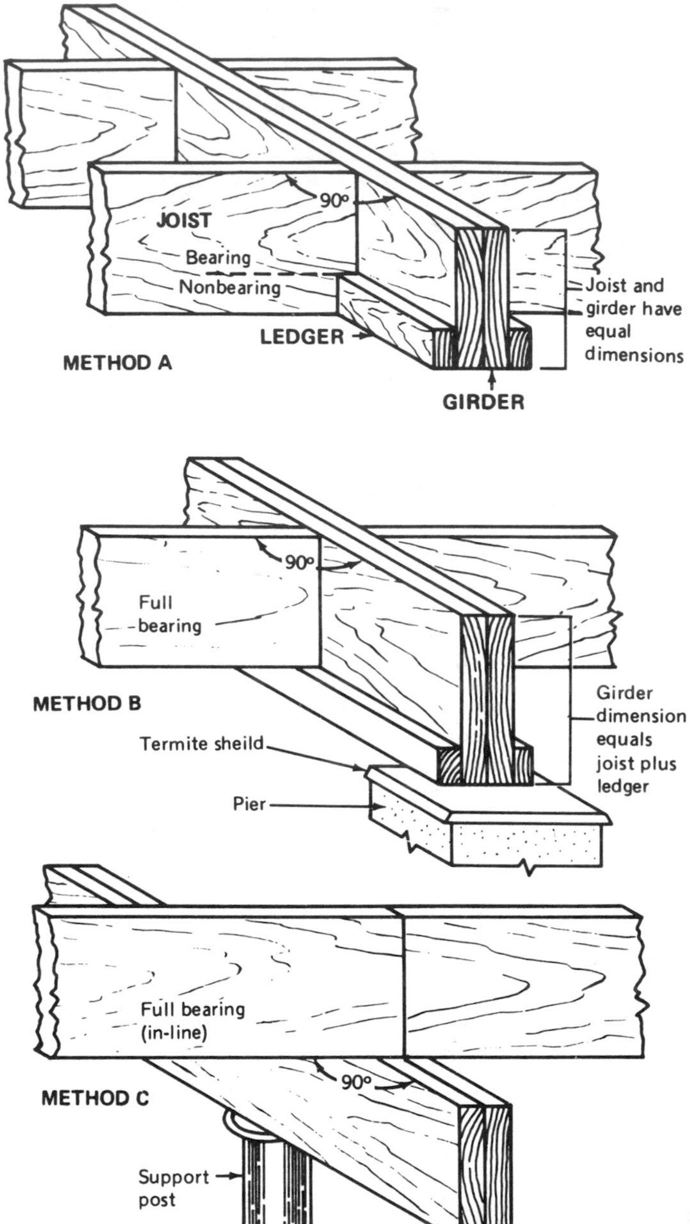

Figure 5–3 Three methods of using a girder; in-line, in-line full joist support, and under joists.

method A shows that each joist must be matched to fit over the support ledger. Employing this technique does not have any adverse effects on the girder, but does reduce the bearing width of the joist.

Contrast method B with A and note that the full joist rests on the ledger. However, this method may cause a girder to be built that exceeds the structural strength and stiffness requirements necessary. Its use also creates a 2-in. or larger offset below the joist.

Method C shows a simple way of including a girder where it may be cut or built-up to the required strength and stiffness without concern for the size of the joists used for the floor. It may be constructed from a solid member or built-up as shown. Although the girder is quite heavy, it is relatively easy to install. The joists will rest on top of this girder and are toenailed to it.

Girders may be made of any size lumber, but the usual sizes are 4 × 6, 4 × 8, 4 × 10, 4 × 12, 6 × 6, 6 × 8, 6 × 10, 6 × 12, and 8 × 8 or 8 × 10. The most common sizes are the 4 × 6, 4 × 8, 4 × 10, 6 × 8, and 6 × 10. These are large members, but they must carry a larger load for many years. One way to reduce the needed size is to limit a girder to the shortest length possible.

In addition to the varieties of lumber sizes, girders may be purchased in the form of steel I-beams and WF (wide flange) beams; both are shown in Figure 5–4.

Floor plans and sometimes foundation plans state the size and placement of girders. It is also possible that the girder size is listed in the specifications or footnotes, or that the carpenter must make the decision about the girder's size and placement. Thus the selection of a girder must be made on the task it must perform. This single member has more weight and stress placed on it than almost any other part of a building.

A girder spans an open area, with each end resting on a solid foundation. It may be that the girder spans the entire building, but the overall span is broken down into several smaller spans by using columns or piers. This procedure is extremely important, as will be shown later. But briefly, the total weight on the girder is reduced greatly if the spans are held to 8, 10, or 12 ft.

A girder must also carry the weight of part of the floor resting on it or sup-

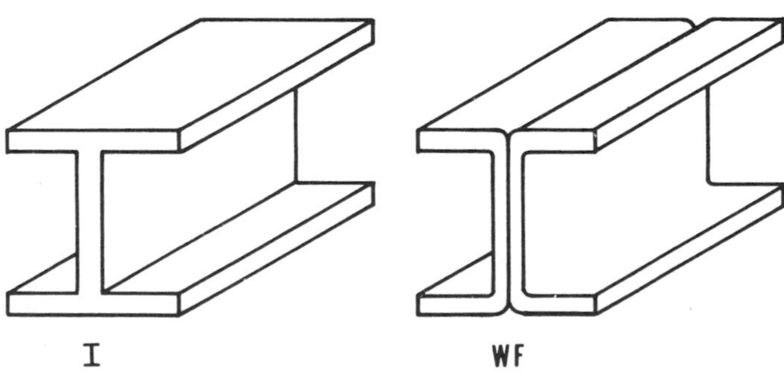

Figure 5–4 Steel girders (beams): I and wide-flange (WF) varieties.

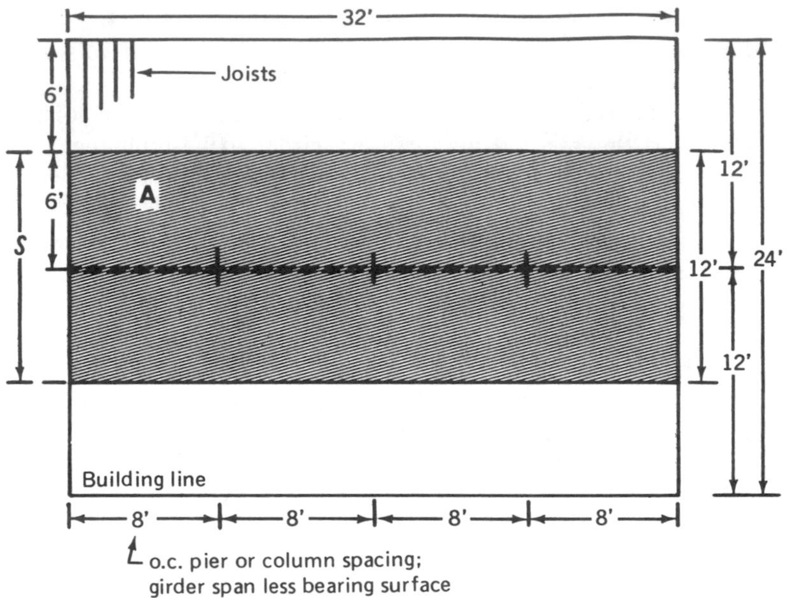

Figure 5–5 Floor area supported by a girder(s).

ported by it. Figure 5–5 illustrates this. The building shown is 24 ft wide by 32 ft long. A girder is built to run 32 ft midway through the building. It is supported on the outside walls and by three piers or columns. The result is a clear span of girder just under 8 ft. Since the joists are 12 ft long on either side of the girder, one-half of their length is supported by the outside wall. The other half (6 ft) is supported by the girder. In this example the girder supports 6 ft of joists each side of itself for a total of 12 ft. Notice that the amount of support required of the girder is equal to one-half of the building's width. This condition exists everywhere a single girder is installed between outer walls. The shaded area in Figure 5–5 represents a potential 50-psf load.

Figure 5–6 represents just section A of the shaded area of Figure 5–5. The first pier is set 8 ft from the end wall and sets the span of the girder at this distance. Since the *S* distance is 12 ft, a grid is made of 1-ft squares, each having a combined load of 50 psf. This means that each 1-ft run of the girder over its clear span must support 600 lb. In the 8-ft span the total weight on the girder is 4800 lb. This is considerable and would indicate the use of a 4 × 8 or 6 × 6 girder made of No. 2 grade lumber and possibly some No. 3 grade species of sufficient strength.

This is not all there is to selecting the proper-size girder. Suppose that there is a load-bearing wall placed on top of the girder. This means that there will also be ceiling joists or rafters that bear on the partition. Figure 5–7 shows a set of conditions where there is a bearing partition, ceiling joists, and an unfinished attic space. In this situation the girder must support the weight across the *S* distance of the

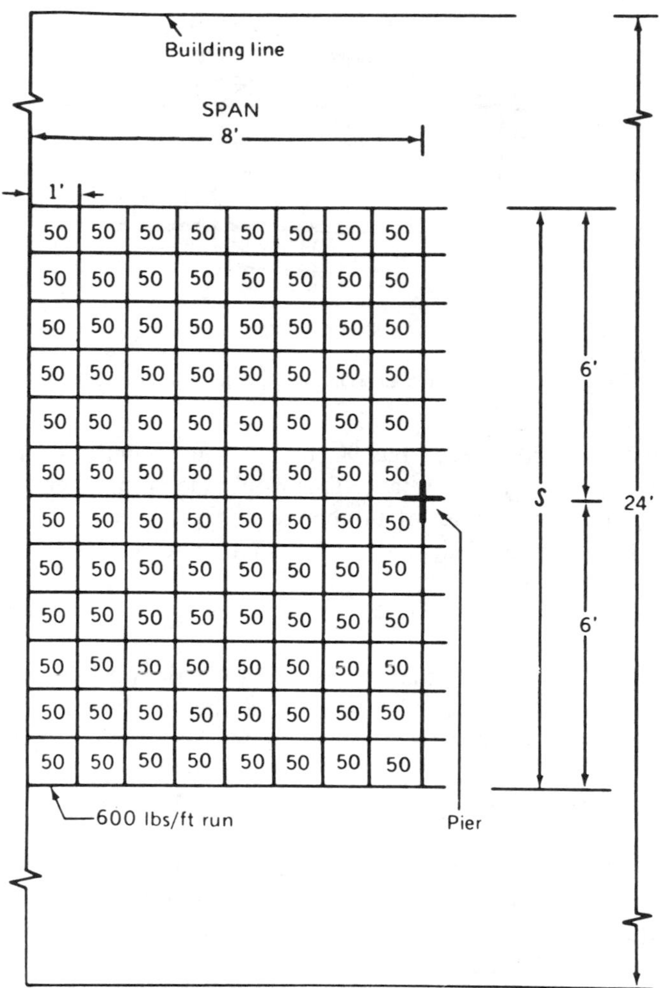

Figure 5–6 Defining the load on a girder using a 1-ft² grid layout.

ceiling joists and drywall or other materials, plus the partition and all live loads *in addition* to the first floor. Notice how quickly the different loads add to the total load the girder must carry.

Figure 5–8 shows in more detail how the loads of each part contribute to the whole load. Beginning with the attic, a live load of 20 psf is anticipated primarily from storage of personal belongings. Next, a 10-psf dead load for ceiling joists and ceiling materials is added. The load-bearing partition and its wall coverings add another 10-psf to the load on the girder. The first floor is designed for a 40-psf live load and a 10-psf dead load. This brings the total load on the girder to 90 psf. To imagine the effect this has, consider Figure 5–6 again; but instead of each square

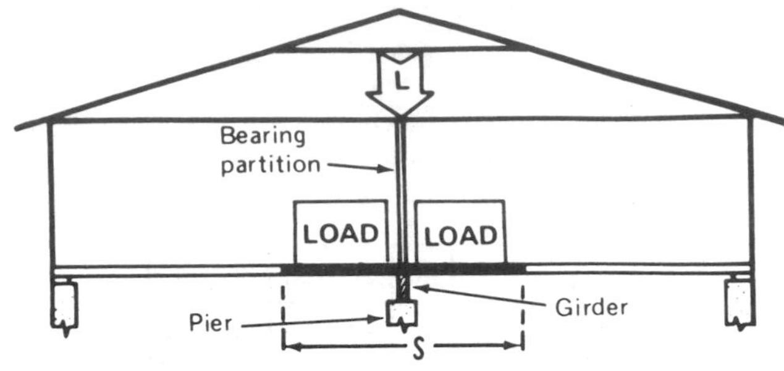

Figure 5–7 Defining the loads that a girder must bear.

foot having 50 lb, each has 90 lb. The same length of girder would need to carry almost twice as much weight. This would mean that a girder made of a higher quality (grade) and species of lumber or a larger size would be needed. Several tables for the selection of girders are provided in Appendix D.

Beams

A beam is a member that is usually larger than a joist. Although a beam may be used as a joist on, say, 24-in. centers, they are usually spaced farther apart. The beam and plank floor system is a common example of the use of a beam. As with joists, the beam is measured in nominal widths and thicknesses and is rated for stiffness (E) and strength (F_b) much like a joist. This means that live and dead loads similar to those on living areas, quiet areas, and even roofs may be applied. Several examples of beams of different sizes are shown and explained in the following figures so that some understanding of how the size and length of a beam play a significant role in the grade selected. These examples represent floor beams where a stiffness of $S/360$ is needed.

The stiffness or *modulus of elasticity, E,* and strength or *fiber in bending, F_b,* are the two characteristics necessary to determine the proper grade of lumber that should be used as a beam. Table 5–1 is developed for a nominal 4-in. by 6-in. beam

TABLE 5–1 4 × 6-IN. BEAM; STRENGTH AND STIFFNESS REQUIREMENTS (40-psf LL; 10-psf DL)

	Length of span (ft)								
	8	9	10	11	12	13	14	15	16
$E^{a,b}$	0.28	0.41	0.56	0.74	0.96	1.25	1.51	1.9	—
F_b^{b}	275	345	430	510	610	720	840	960	1090

[a]Multiply value in table by 1 million to obtain stiffness (modulus of elasticity).
[b]E and F_b are both measured in pounds per square inch (psi).

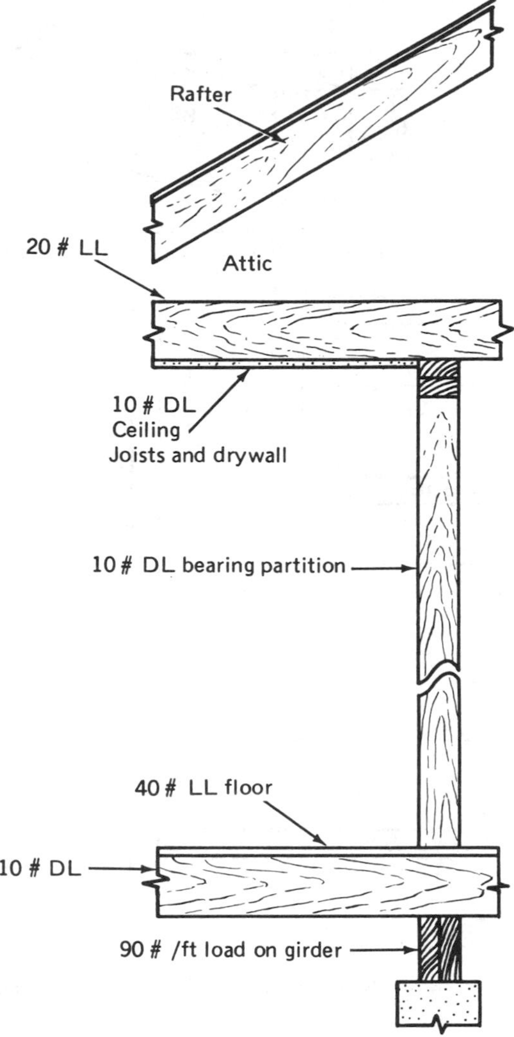

Figure 5-8 Adding up the loads that a girder must carry; attic, framing, walls, etc.

so that the reader may begin to make a comparison of the relative usefulness of the various sizes.

First note in the table's title that the conditions of "40-psf LL and 10-psf DL" are the initial parameters selected. The reason for selecting these values reverts back to the fact that all activity areas in a house need these values. A live load of 40 psf and a dead load of 10 psf are established by most of the local and national building codes. The modulus of elasticity is developed from the live load value only, in this case 40 psf. The fiber in bending is a combined value of the live load (40 psf) plus dead load (10 psf) resulting in a combined value of 50 psf.

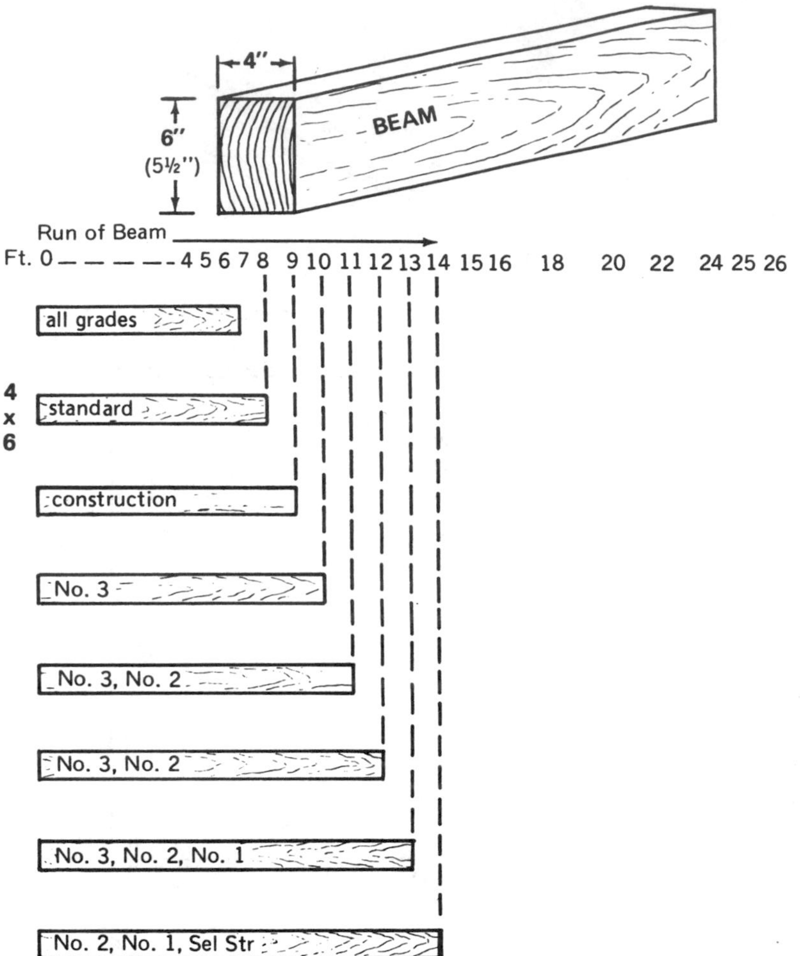

Figure 5–9 4 × 6 beam: 40-psf LL, 10-psf DL, span and grade to use for various spans.

The first observation made from the tabulated data is that this size of beam will adequately support any span up to 8 ft made from any grade of construction lumber provided that its E equals or exceeds 280,000 psi and F_b equals or exceeds 275 psi. Figure 5–9 shows the girder and lists all the minimum grades acceptable for each length. A beam of *standard* may be used for a 4 × 6 beam spanning 8 ft. It must have an E and F_b as stated before.

If the span of the beam is 13 ft, for example, Figure 5–9 lists a grade No. 3, No. 2, or No. 1 as appropriate. Table 5–1 lists and E of 1.25×10^6 or 1,250,000 psi, and F_b of 720 psi. A search through Table W-1 in Appendix A, which is designed primarily for joists and rafters but is accurate for beams as well, provides the following:

Grade no. 3	Grade no. 2	Grade no. 1
Douglas fir–larch	Douglas fir–larch	Eastern hemlock
Douglas fir–(north)	Douglas fir (north)	Eastern hemlock (north)
Southern pine	Douglas fir (south or hem–fir)	Eastern spruce
	Hem–fir (north)	Englemann spruce
	Southern pine	Alpine fir

The listing shows that 10 species of lumber qualify, and of those only three may be used with a grade No. 3. All other species must be of a better than No. 3 grade. Also note that the grade for the eastern woods must be No. 1.

Finally, the 4 × 6 beam is effectively used with spans of 5, 6, 7, up to 14 ft, and in some cases 15 ft. When one of 15 ft is used, Douglas fir–larch or southern pine species must be used. It might be better to select the next larger girder size if a span of 14 ft is called for.

Table 5–2 and Figure 5–10 combine to provide a picture of the requirements for a 4- by 8-in. beam of various lengths. The dimensions of this beam provide considerable strength to a floor assembly.

Table 5–2 shows that this beam may be used for any short spans; but only when spans of 10 ft or more are expected do stiffness and strength values become important in the selection of species and grade. The table indicates that the stiffness values are within various grades up to 18 ft, where a modulus of elasticity of 1.4 million psi is needed. However, if an exceptionally fine grade of Dense Sel Str Douglas fir–larch or southern pine is available, a 20-ft beam may be used. Thus far the tabular data have illustrated that strength requirements are most easily met.

In every case the member's strength exceeds the minimum required pounds per square inch. Not so with the stiffness value E. The upper limit is reached long before the strength value is maximum. Figure 5–10 illustrates that in almost all spans the beam may be made from No. 3 graded members. Careful selection of species within these assigned grades is necessary beginning with spans of 15 ft and longer. Al-

TABLE 5–2 4- × 8-IN. BEAM; STRENGTH AND STIFFNESS REQUIREMENTS (40-psf LL; 10-psf DL)

	Length of span (ft)									
	10	11	12	13	14	15	16	18	20	22
E^a	0.25	0.32	0.42	0.54	0.68	0.8	1.0	1.4	1.95	—
F_b	245	295	350	410	460	550	625	800	998	1200

[a]Multiply value by 1 million.

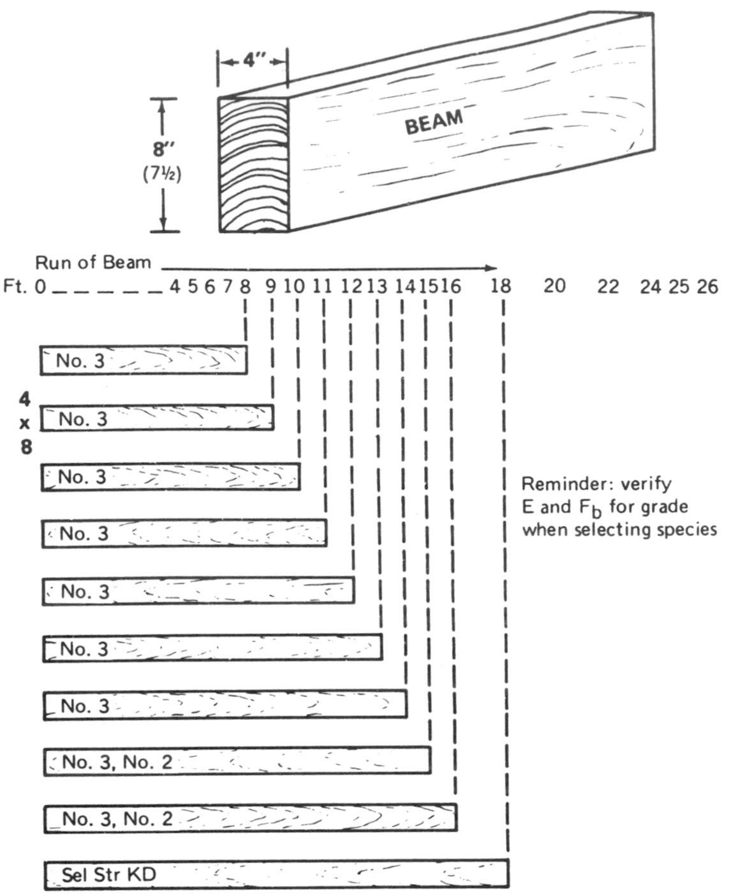

Figure 5–10 4 × 8 beam: 40-psf LL, 10-psf DL, span and grade to use for various spans.

though the fiber in bending values are relatively small, the modulus of elasticity values rise rapidly.

The 4 × 10 beam shown in Figure 5–11 and detailed in Table 5–3 is one generally used where extremely long spans are a design feature. For spans below 16 ft this size of beam would be very expensive, especially when several other types such as the 4 × 8 or 3 × 10 would provide adequate stiffness and strength. As Table 5–3 indicates, beams up to 26 ft are well within the stiffness and strength parameters of the various grades of lumber available today.

A review of Appendix A shows that the selection by species narrows quickly when beams of 25 ft and 26 ft are used. For example, only the southern pine (KD) in Dense Sel Str meets the modulus of elasticity requirements needed for the 26-ft beam. However, the Douglas fir–larch, the Douglas fir (north), and the southern pine are adequate for the 25-ft beam and the shorter lengths.

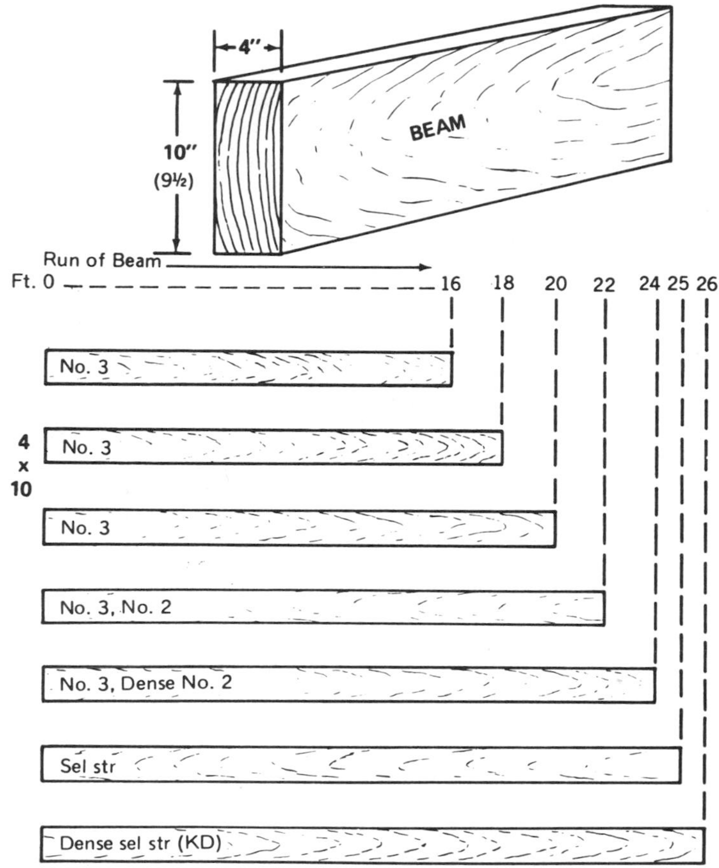

Figure 5–11 4 × 10 beam: 40-psf LL, 10-psf DL, span and grade to use for various spans.

The data on beams that have been presented thus far are restricted to a 40-psf live load and a 10-psf dead load. Therefore, the beams would not be reliable if loads above these were required.* It also follows that beams cut on these specifications would exceed the minimum requirements if loads less than those listed were designed into the structure.

Table 5–4 and Figure 5–12 combine to provide data on a 4 × 8 beam that has stiffness and strength requirements of 30-psf live load and 10-psf dead load. Reducing the load on a 4 × 8 beam means that less demand is placed on it. Therefore, the beam can span lengths more easily, and lower-quality lumber plus greater varieties of acceptable species make it easier to obtain. Table 5–4 shows that spans less

*Complete tables are published in many forms for beams by various construction industry and lumber associations. See the reference list on page 367.

TABLE 5-3 4- × 10-IN. BEAM; STRENGTH AND STIFFNESS REQUIREMENTS (40-psf LL; 10-psf DL)

	Length of span (ft)						
	16	18	20	22	24	25	26
E^a	0.48	0.7	0.95	1.25	1.6	1.85	2.0
F_b	380	480	600	720	860	950	1100

[a]Multiply value by 1 million.

than 10 ft may be made from any construction grade since the E value is less than 200,000 psi and the F_b value is only 200 psi. At the other extreme of the table, a beam 20 ft long only requires an E of 1.4 million psi and F_b of 800 psi. This means that a No. 2 or No. 1 grade lumber may be used. Again, this does not apply to certain species.

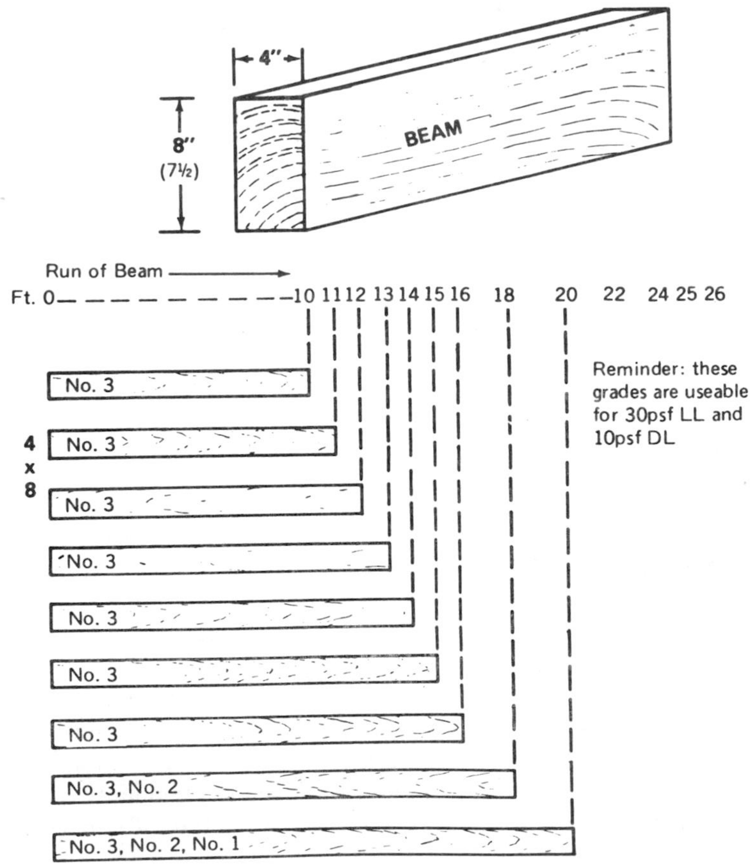

Figure 5-12 4 × 8 beam: 30-psf LL, 10-psf DL, span and grade to use for various spans.

TABLE 5–4 4- × 8-IN. BEAM; STRENGTH AND STIFFNESS REQUIREMENTS (30-psf LL; 10-psf DL)

	Length of span (ft)								
	10	11	12	13	14	15	16	18	20
E^{a}	0.2	0.24	0.31	0.4	0.5	0.62	0.75	1.05	1.4
F_b	200	240	280	330	380	440	500	640	800

[a]Multiply value by 1 million.

The fifth and final example of beam characteristics is shown in Figure 5–13 and detailed in Table 5–5. This is a 4 × 10 beam that will support a 30-psf live load and a 10-psf dead load. This beam would be included in floor design and construction in quiet areas of a house, most probably in the split-level and two-story house. Lumber graded No. 3 should be used for all lengths up to 22 ft. Lengths of 24, 25,

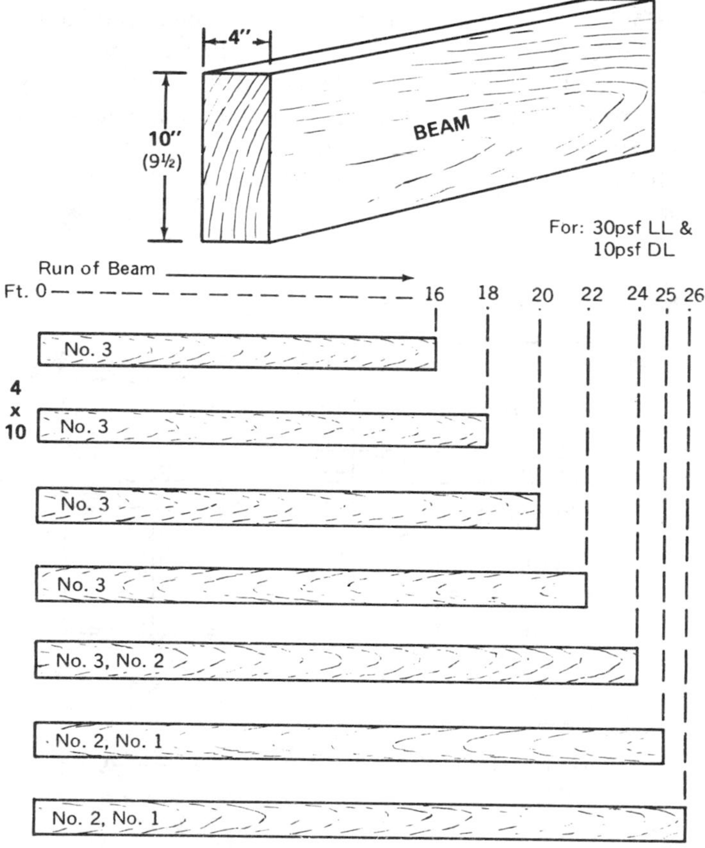

Figure 5–13 4 × 10 beam: 30-psf LL, 10-psf DL, span and grade to use for various spans.

TABLE 5–5 4- × 10-IN. BEAM; STRENGTH AND STIFFNESS REQUIREMENTS (30-psf LL; 10-psf DL)

	Length of span (ft)									
	13	14	15	16	18	20	22	24	25	26
E^a	0.2	0.24	0.3	0.36	0.52	0.7	0.94	1.2	1.4	1.55
F_b	205	240	270	310	390	480	580	700	750	810

[a]Multiply value by 1 million.

and 26 ft require a closer examination of E and F_b within grades and species to ensure adequate stiffness and strength.

Anchoring Girders or Beams

A girder or beam must span the width or length of a building or at least a large segment of the floor area. It must therefore be anchored at both ends. Figure 5–14 shows several typical methods. If installed between masonry units, termite shielding must be used. If inserted or made part of the joist assembly, anchoring is made to header joist and L-type sill. A sound installation means that (1) the girder rests on a minimum surface area of 1½ in. wide, (2) it is properly protected against decay and infestation from termites, and (3) it is soundly anchored to prevent shifting and twisting.

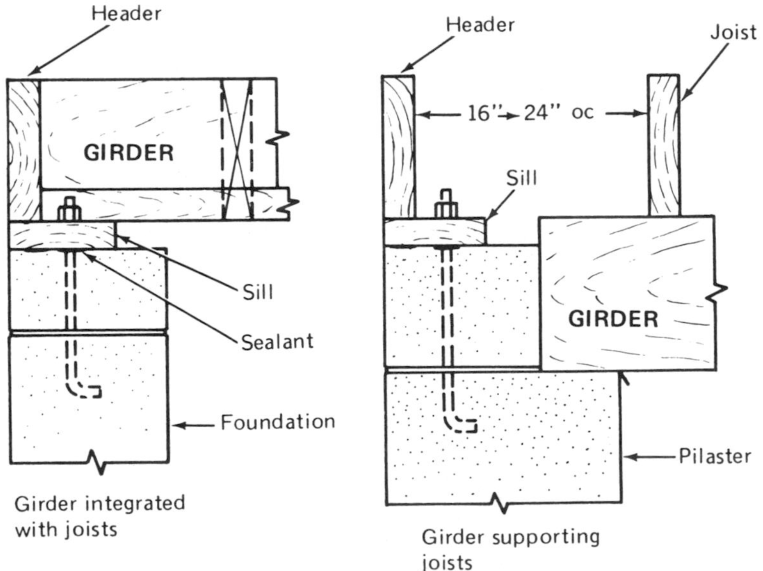

Figure 5–14 Installing girder (or beam) on a bearing surface; anchoring and adequate seating.

Sills and Types of Sills

In earlier days of construction the member that was attached to the foundation was commonly called a *bed plate*. This term has given way to the now common *sill*. However, the term "sill" is just as frequently referred to as an assembly consisting of a flat member resting on a foundation plus a vertical member usually the size of a joist. Where such combinations of members are used, the assembly is called an L-type sill. There is also a box-sill type that is sometimes used. It is shown and explained later.

L-type sill. Figure 5–15a shows a built-up L sill. This arrangement of horizontal sill (plate) and beam or header is the one most commonly used by carpenters. It is easily assembled and reliable. It provides ample surface for joist end support and makes vertical alignment of joists a relatively simple task. Notice the symbol L drawn over the joist. The combination of sole plate and header form the letter L and hence the name L sill. Subflooring, such as plywood, is usually laid flush with the outer edge over the joist and sill assembly. Walls are erected on top of the plywood or other sheathing. This method of framing a sill does not create a fire-stop,

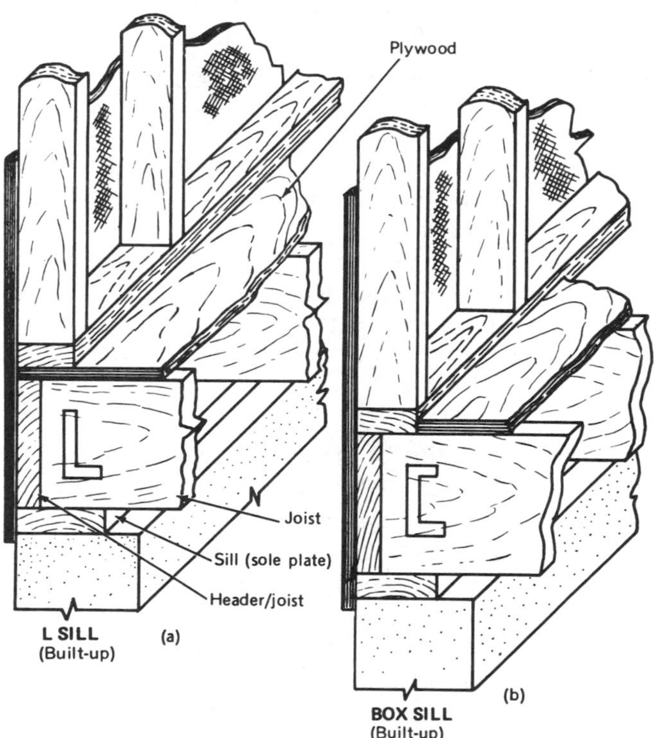

Figure 5–15 L- and box-type sills.

but when the sheathing and wall sill are installed a fire trap is constructed. Flames in the basement area cannot pass into the wall until they first burn through the subflooring and wall sole plate.

Box-type sill. The box sill assembly is a variation of the L sill and derives its name from the concept of a box with a bottom side and top. Figure 5–15b shows a box sill assembly. Note that it is built up from the sill header-joist and sole plate of the wall. The sole plate of the wall is nailed directly to the top of the header and each joist. Subflooring is then cut to fit against the wall's sole plate.

Installation of sills. Most experienced masons can build a wall that meets most specifications for plumb and level. The specifications do, however, allow for varying degrees of error. The one of most concern to the carpenter is the level accuracy of the wall's top course. Masons are allowed ½ in. from level in each 20-ft run. It would be nice if their accuracy were ½ in. from corner to corner, but there are no guarantees for the carpenter that the mason will achieve this. Therefore, the carpenter must be prepared to install a sill that is level, but must fully expect the masonry wall to be unlevel.

Two tasks are accomplished simultaneously when a level sill is installed. First, a new standard of level is established for the house to be built upon. Second, the grout or mortar bed used to fill in irregularities between sill and foundation seals the entire length and adds both an insulation factor as well as reduces the opportunity for insect infestation.

The mason or his or her laborer should prepare a batch of mortar so that after sills have been precut and drilled and are ready for installation, they can apply sufficient mortar along the wall. Bolting the sills should spread the mortar and excesses should be troweled away.

If the wall's level is exceptionally good, caulking can be used in place of mortar. It can be placed in ribbons along the wall and it will seal the spaces between foundation and sill.

Joists

Following the construction of the sill, the joists are cut and installed. The selection of joists is, of course, dependent on the span they are to cover and their live and dead load requirements. Appendixes B and C provide data for various-size joists with on-center spacing as well as grades to use for numerous spans.

Recall that living or activity areas within the house need 40-psf live load and 10-psf dead load characteristics. Therefore, selection of the joist size and grade must be taken from the tables having these values. Also recall that sleeping quarters may have their floors framed from members that properly carry a 30-psf live load and a 10-psf dead load.

Selection of a joist must include *stiffness.* Recall that stiffness is a measure of the modulus of elasticity, and *strength* is a measure of fiber in bending. If a floor is to remain sound for many years, both qualities must be possessed by each joist and must be properly defined. The stiffness value, *E,* must be sufficient so that a floor can deflect and return to its original condition time after time. Any member included in a floor assembly that cannot meet these requirements degrades the entire assembly.

Having the qualities included in each joist is important, but the carpenter must take the opportunity to make use of the stiffness and strength. For example, each member that is milled as a joist has a *crown side.* This crown side indicates *the area to be compressed.* The concave underside should be the one to be put into tension. The crown side is generally placed under pressure by live and dead loads that tends to nullify some of its tension by compression. The underside is already compressed and the loads stretch the compressed fibers releasing some of the compression.

Consider now what problems can occur if one or more joists are placed crown side down and the remaining ones are placed crown side up. Figure 5–16 shows that the loads cause the crown-up joists to deflect downward and recover when the load is removed. However, the joist with crown down is already deflected and cannot easily deflect more. So what happens? Uneven movement causes unequal strain and stress on subflooring and finished flooring; squeaks result and sagging often accompanies the squeaking. Worst of all, people walking across the floor get the sensation

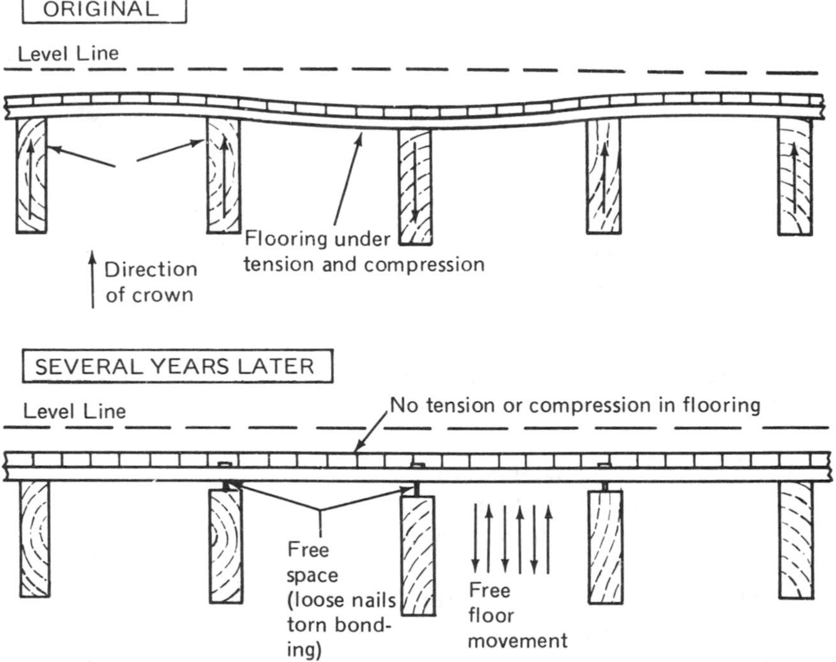

Figure 5–16 Improper installation of joist (crown), resulting in defective floor assembly.

of the floor giving way. Details in Figure 5–16 are, of course, somewhat exaggerated so that the concept can readily be seen. Separation of flooring from joist rarely exceeds ¼ in. at the most negative (lowest) point in the span. Nevertheless, careful alignment of joist is absolutely essential to sound construction.

Headers, Tail Joists, and Trimmer Joists

Headers are frequently needed to complete the framing of an opening in the joist assembly. The primary functions of headers are to support the *tail joists* and to define the width or length of the opening. Frequently, a bearing wall is placed across headers near stair wells, for example. This condition would make it extremely important that the headers be doubled and supported if necessary by hangers or support beneath the header or well nailed to other floor members as Figure 5–17 illustrates.

Both drawings in Figure 5–17 include headers; notice that in Figure 5–17a the headers are short members that span the width of the openings. However, in Figure 5–17b they are relatively long and span the length of the openings. These two illustrations help define the word *header.* Another way to identify a header is to notice that tail joists are always nailed to them.

Tail joists are pieces of regular joist stock that are cut to fit from header to outer wall or girder or some form of bearing surface. Tail joists must be well nailed to the header. Customarily, three 16d common nails are used to secure the tail of the joist to the header. Although not shown, the tail joists can be made with a cutout so that they will rest on a ledger board attached to the lower vertical edge of the header. (It would be similar to the ledger shown in Figure 5–3.)

Trimmer joists are the third and final member of framed opening in a floor assembly. These members are regular joist stock cut to fit between supporting surfaces. They are nailed soundly to the joists that support the headers. In Figure 5–17a the trimmer is nailed either side of the full joist that the headers are nailed to, and run perpendicular to the header. In Figure 5-17b the trimmers also run perpendicular to the headers and parallel to the floor joist run.

Support Joist for Load-Bearing Wall (Parallel to Joist Run)

There is often a need to have bearing walls run parallel to the run of floor joists. In customary framing and design methods these walls may be placed over a standard on-center spacing joist, as Figure 5–18a shows. However, many times the wall is located between joists. Since the wall is load-bearing, it means that ceiling joists and maybe some portions of the roof's force may be concentrated on the wall. If no additional support is added to the floor assembly, the sheathing or subflooring must carry this load, and it is not strong enough to do so. Therefore, another joist must be placed between two normally spaced joists so that it is centered under the bearing

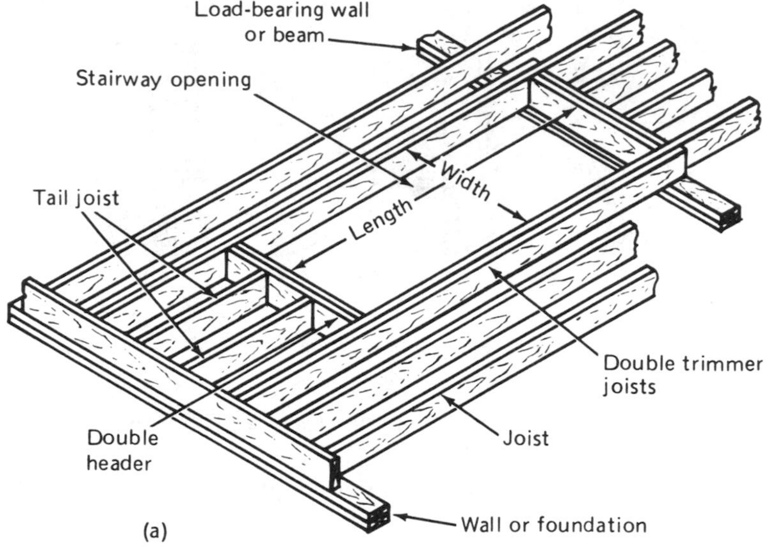

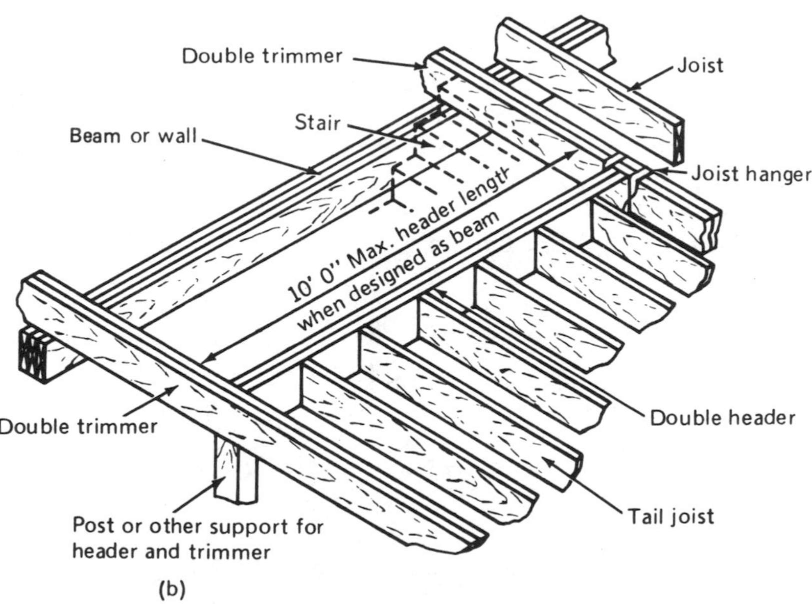

Figure 5–17 Framing for floor openings: headers, tail joists, and trimmer joists.

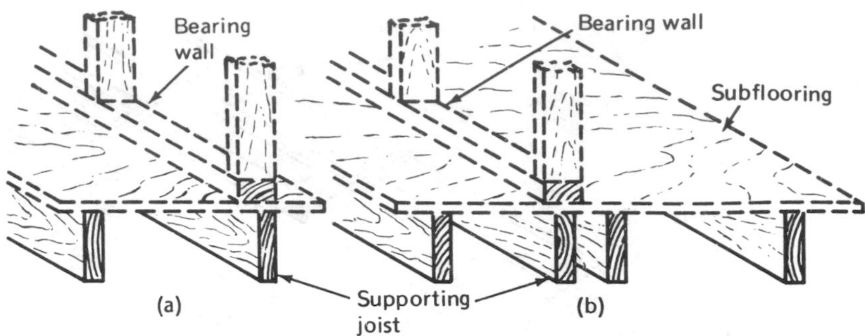

Figure 5–18 Load-bearing joists perpendicular to run of joists.

wall (Figure 5–18b). In this manner there is full support for the full length of the wall.

If a modular or 24-in. system of design were used instead of the conventional construction method, all bearing walls would fall over normal on-center spacings of joists and the need for the additional joist is eliminated. This means that each bearing wall would be placed as shown in Figure 5–18a.

Support Framing for Non-Load-Bearing Wall (Parallel to Joist Run)

Since the wall is not carrying a load, such as ceiling joists or roof, it is classified as non-load-bearing. This means that support within the floor assembly need only be designed to carry the dead load of the wall. The customary method of providing support where the wall is located between on-center spacings of floor joists is shown in Figure 5–19. Nominal 2 × 4 or 2 × 6 members, frequently referred to as "cats," "solid bridging," or "blocking," are securely nailed between joists on 3- to 4-ft centers.

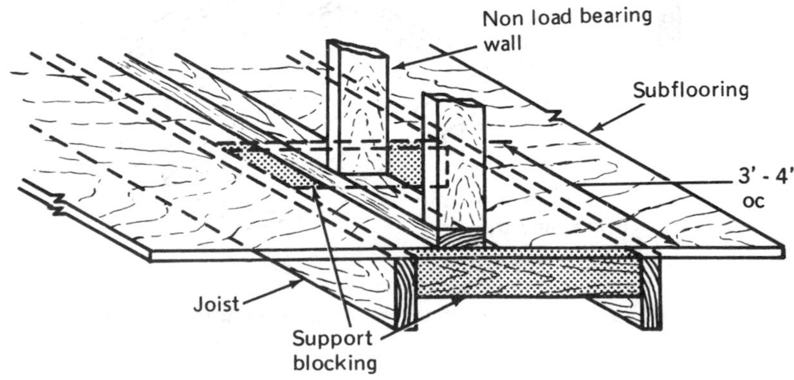

Figure 5–19 Non-load-bearing wall supported by blocking in joist assembly.

Bridging

Industry standards prescribe the use of bridging where the span of a joist exceeds 8 ft. A row of bridging made from nominal 5/4 × 3-in. stock, joist stock or metal straps should be nailed midway between the end supports of the joist where spans are 16 ft or less. Where spans exceed 16 ft, two rows of bridging must be used to comply with the standard.

There may be, on the part of some builders, a desire to omit bridging because of the labor and material costs. Should such a plan be followed, the result would be a floor assembly that gives with concentrated loads. The function of bridging is to cause a concentrated load on the floor assembly to be distributed throughout the joists nearby.

The downward pressure caused by the weight is not absorbed by a single joist or two adjacent joists. Rather, the energy of the pressure is transferred laterally into bridging and thereby to adjacent and remotely adjacent joists. True, the subflooring also aids in distributing the force, but bridging is vitally important. By distribution of the pressure as described, the joists do not deflect to their maximum, so avoidance of damaging stress is achieved.

Subflooring

The most commonly used material for subflooring is plywood. It is a unique material possessing many properties that allow a variety of construction techniques to be applied. It is available in many sizes, from the standard 4- × 8-ft sheet to those ranging up to 16 ft long. It is also available in many different thicknesses, grades, and types. All of these properties make the carpenter's job somewhat more difficult rather than simplifying it.

As each floor system is designed, the selection of plywood for a subfloor must meet all necessary requirements. Or the opposite view may be taken, which is that the selection of plywood may in large part dictate the construction method of the joist assembly. Several adaptations are shown in the next section of this chapter. For the moment, though, the characteristics of plywood available for subflooring are illustrated in Table 5-6 and examined.

The most widely recommended types of plywood used for subflooring and underlayment are listed in Table 5-6. The correct type must be selected to meet the building requirements conditions and also include environmental considerations. For example, interior types should be used only where moisture is not expected to be a problem. Such places as subfloor over a heated basement or between first and second floors are appropriate for this type of plywood. Subflooring plywood over open crawl spaces subject to collecting moisture and other applications where moisture is definitely a problem must be made from exterior types. Finally, extra thick panels with tongue and groove edges must be used where joists are spaced wider then 24 in. o.c.

The letters CC and CD denote the surface condition of the plywood. The first

TABLE 5-6 PLYWOOD FOR FLOOR SYSTEMS

Subflooring

APA Rated sheathing Exp 1 or 2
APA Rated sheathing Ext.
APA Structural I rated sheathing Exp 1, or
APA Structural I rated sheathing Ext.

Joist Span (in.)	Panel Thickness (in.)
16	7/16,, 15/32, 1/2, 5/8
20	9/16, 19/32, 5/8, 3/4, 7/8
24	23/32, 3/4, 7/8
48	1-1/8

Combined Subfloor-Underlayment (Under carpet and pad)

APA Rated Sturd-I-floor Exp 1, or 2, or
APA Rated Sturd-I-floor Ext.

Joist Span (in.)	Panel Thickness (in.)
16	19/32, 5/8, 21/32
20	19/32, 5/8, 23/32, 3/4
24	11/16, 23/32, 3/4, 7/8, 1
32	7/8, 1
48	1-1b

Underlayment (Over Subflooring)

APA Underlayment Int.
APA Underlayment Exp 1, or
APA Underlayment C-C plugged Ext.
APA C-C Plugged Ext.

Application	Thickness (in.)
Over smooth subfloor	1/4
Over lumber subfloor or uneven surfaces	11/32
Over lumber floor up to 4 in. face grain perpendicular to boards	1/4

Source: American Plywood Association's *Design/Construction Guide Residential and Commercial* (with permission)

Notes: [a]APA stamps on each panel specify the name, use, and other factors necessary to make sound decisions.

[b]Underlayment panels may also be obtained sanded, specified by group number for species and have rating A-C surfaces.

[c]Exp. = Exposure 1 (one surface) 2 (two surfaces); Ext. = Exterior glues used; Int. = Interior non water proof glues used, not suited for damp, wet places.

letter denotes the best surface and in C grade small knots are allowed as well as discolorations, stains, and some splits. The surface may be unsanded or touch sanded. The back surface in CC is the same as the front. In CD more defects are permitted; larger knots are allowed and more severe splits may be included.

Groups 1, 2, 3, 4 and 5 shown in Table 5–7 refer to group species of woods that are used to construct the plywood. Group 1 species are the best types and provide the greatest strength and stiffness. Group 2 species also contain many fine types that make strong and stiff plywood but are slightly less strong and stiff. Groups 3 and 4 contain the remainder of the species, many of which are found to have a low modulus of elasticity and fiber in bending.

Finally, a selection of thickness must be made by using the data in Table 5–6. Close examination of the table shows that the minimum thickness of plywood usable for subflooring is 7⁄16 in. The data also show that if joists are spaced 24 in. o.c., 23⁄32-, 3⁄4- and 7⁄8-in. plywood may be used.

In addition to the selection process needed, a special nailing sequence is important. It should conform to 8-in. spacing on intermediate joists and 6-in. spacing around the perimeter of the sheet. Expansion provisions of 1⁄16 in. must be included in high-humidity and other damp areas; spacing of the plywood sheets is easily accomplished by using 4d finish nails as separators.

Several types of plywood are made tongue and groove. Where this type is used, no supports need be included between joists where plywood edges join (Figure 5–20). However, where square-edge types of plywood are used and no underlayment plywood is scheduled for use, blocking, as shown in Figure 5–20, provides nailing as well as adhesive bonding surface. This blocking is made from nominal 2 × 4 or

TABLE 5–7 SPECIES USED TO CONSTRUCT PLYWOOD

Group 1	Group 2	Group 2 (cont.) / Group 3	Group 3 (cont.) / Groups 4, 5
Group 1	*Group 2*	Meranti	Spruce, Englemann
Birch, sweet	Cedar, Port Orford	Pine, pond	Spruce, white
Birch, yellow	Fir, California red	Pine, red	Fir, subalpine
Douglas fir 1	Fir, grand	Pine, western pitch	Hemlock, eastern
Larch, western	Fir, noble	Pine, white	*Group 4*
Maple, sugar	Fir, Pacific silver	Spruce, sitka	Aspen, bigtooth
Oak, tanbark (tanoak)	Fir, white	Sweetgum	Aspen, quaking
Pine, loblolly (swamp)	Hemlock, western	Tamarack	Birch, paper (canoe)
Pine, longleaf (Georgia)	Lauan, almon	*Group 3*	Cedar, incense
Pine, shortleaf	Lauan, Bagtikan	Alder, red	Cedar, western red
Pine, slash (Caribbean)	Lauan, red	Cedar, Alaska	Pine, eastern white
Pine, southern[a]	Lauan, tangile	Jackpine (gray)	Pine, sugar
	Lauan, white	Pine, lodgepole	*Group 5*
	Maple, black	Pine, ponderosa	Poplar
	Menkulang	Pine, spruce	Basewood
		Redwood	

[a]"Southern pine" is sometimes used to designate all the pine trees listed here, including the Caribbean pine since it grows in Florida, as well as in Cuba and the Bahamas.

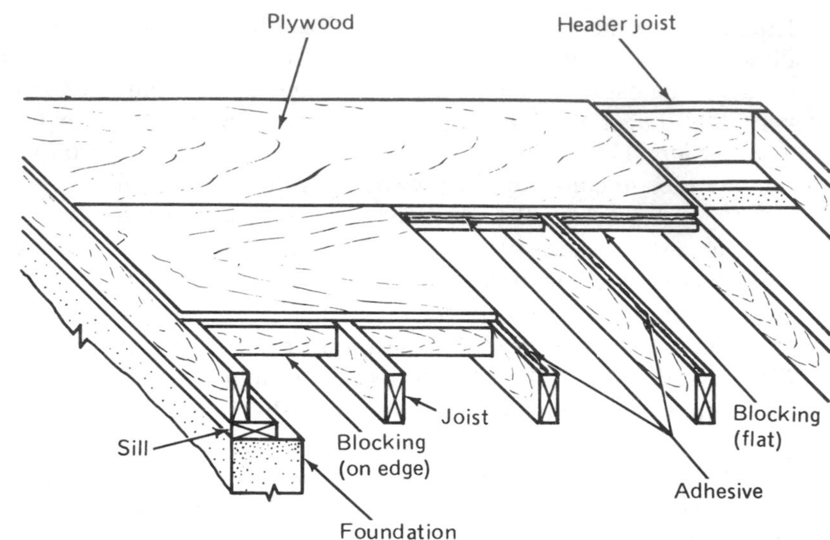

Figure 5–20 Blocking to provide edge support for plywood sheathing.

2 × 6 stock. It may be installed flat surface up, providing ample bonding and nailing surface for securing the plywood. It may also be installed on edge where a ¾-in. surface is available for each plywood sheet to be secured to. The selection of method is arbitrary. Consider the possibility that bridging is necessary with solid blocking. The span of joist is close to 16 ft. A row placed 8 ft from the outside wall or on center could perform a double function. It could be the blocking as well as the edge nailing surface for the plywood sheathing.

PLANNING THE FLOOR ASSEMBLIES

The preceding pages have provided the foundation needed for planning the construction of a floor assembly. A simple one-and-a-half-story house has been selected for the planning. The upstairs is a finished attic with four bedrooms where the ceiling follows the rafter line partway and the walls are built inward several feet from the outer walls, front and rear. The downstairs floor must be designed for activity and rest areas, and is to be built over a block foundation. Figure 5–21 shows the exterior view. Notice that the gable roof has gable ends on the sides of the building. This means that the rafters run front to back. If the floor and ceiling joists run front to back, construction will be easy. This alignment of joists and rafters is not absolute, but it is the rule rather than the exception.

Criteria

1. Design the first-floor assembly so that when constructed it will support a 40-psf LL and a 10-psf DL.

Figure 5–21 Framed house with expanded finished attic.

2. Design the second-floor assembly so that when constructed it will support a 30-psf LL and a 10-psf DL.
3. Include provisions for supporting non-load-bearing walls that run parallel to the joists.
4. Construct a girder that will carry the load and span 11 ft 6 in. for the first floor.
5. Construct a girder to be flush with joist edges for the second-floor assembly to span the dining room, and the kitchen or living room.
6. Ensure that load-bearing partitions are aligned over first-floor girder(s) as closely as practical.
7. Install bridging as required.
8. Frame the stairwell opening in the second-floor assembly.

First-Floor Assembly Plans

Figure 5–22 shows the floor plan for both the first and second floors. Plans (a), (b), and (c) of the figure show possible designs of the floor assemblies. Each design differs, as is explained.

In plan (a) the girder is placed off center of the 24-ft span such that it is centered 14 ft back from the front of the house. This places it directly under the short kitchen wall, and the hall wall leading into the bathroom and back bedroom. The partitions that will rest on top of the girder are load-bearing, so this indicates a proper position for the girder. Notice, however, that the partition between bedroom closets is centered in the building and is off the girder. This does not present a serious problem because the length of the partition is roughly 6 ft and only three or four ceiling joists will bear on this wall.

The piers that will support the girder should be placed 8 ft o.c. so that the girder spans equal distances. Thus the load will be uniform along its length. This allows for a selection of a 4 × 8 or 4 × 10 girder.

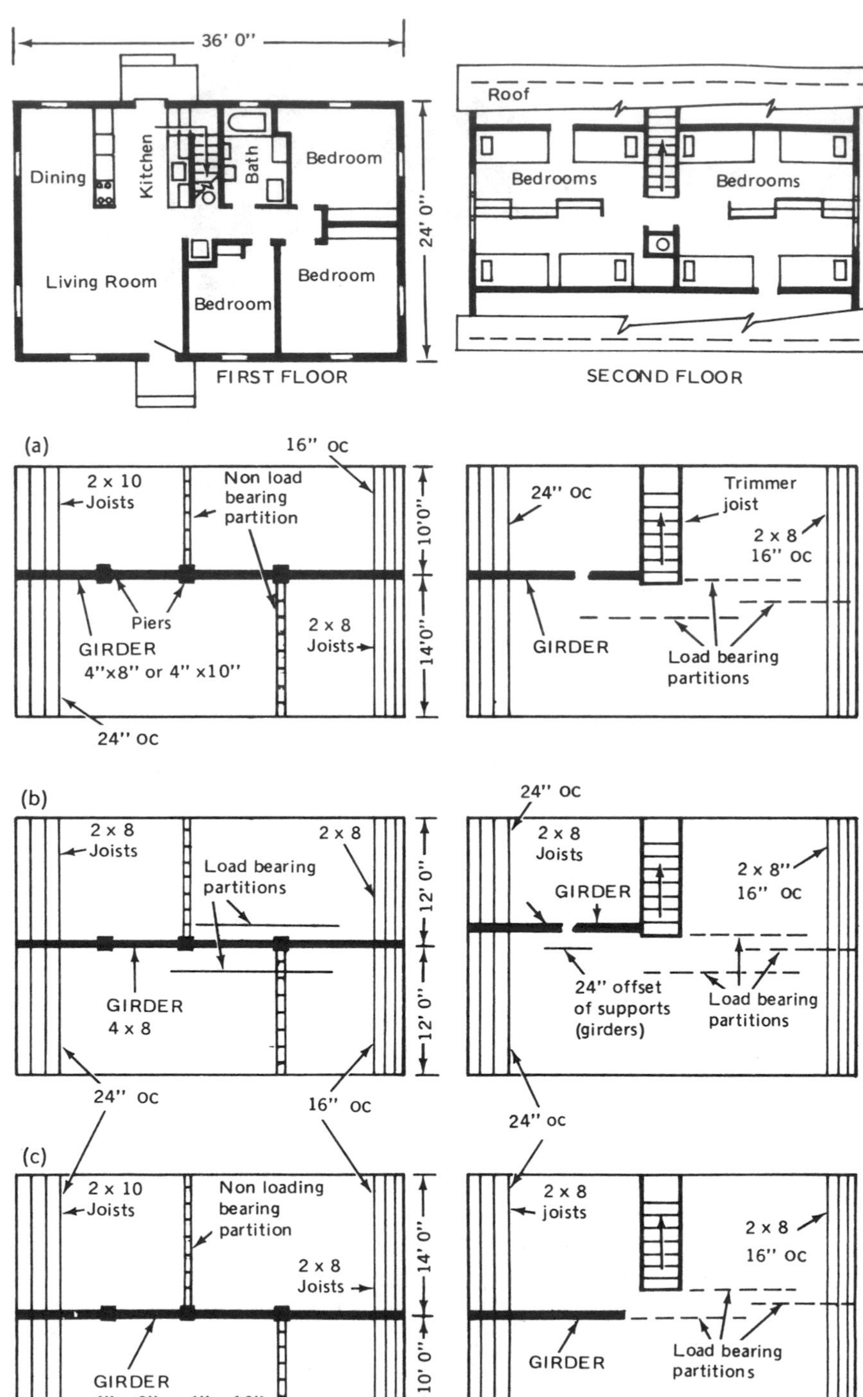

Figure 5–22 Designing the floor assembly for first and second floors.

If the girder is placed under the floor joists, as *method C* of Figure 5–3 shows, a 4 × 8 girder is adequate. Since the floor assembly is built on a foundation (over a crawl space or basement), such an arrangement is possible and practical. If used, the floor joists will rest on top of the girder and will be nailed to the girder.

If for some reason the joist and girder must have flush surfaces, a restriction is imposed on the size of girder. The longest span of joist is just under 14 ft (from outer wall to girder center). This means that the joist used for the floor must be either a 2 × 10 if 24-in.-o.c. spacings are used or a 2 × 8 if 16-in.-o.c. spacings are used. A careful look at Appendix C shows that the best (Dense Sel Str)-quality 2 × 8s must be used. But a lower-quality (No. 2) 2 × 10 can be used even though on-center spacings are farther apart. This means that the girder must be made from either 2 × 8s for 2 × 8 joists or 2 × 10s for 2 × 10 joists.

The carpenter must make a trade-off since a variety of possibilities exist. Clearly, though, the girder placed as shown in plan (a) is ideally located. It provides proper support for floor joists as well as load-bearing walls.

The balance of construction techniques, such as L-type sill, nailing schedule for joists, and a row of bridging midway between the outer wall and girder for both the 10-ft and 14-ft spans is required. In addition, blocking for two non-load-bearing partitions must be included as shown. All other non-load-bearing partitions fall over joist placement.

In plan (b) thc girder is centered 12 ft from front and back. This arrangement eliminates the need for using 2 × 10 floor joists whether on-center spacing of 16 or 24 in. is used. However, a problem does exist; the load-bearing partitions rest on the joists approximately 2 ft on either side of the girder. This places horizontal shear forces on the joists. The entire weight of the second floor is distributed on these load-bearing walls and first-floor joists and is not borne directly onto the girder.

There is a material savings in dollars if this system is used, but because of the offset of load-bearing walls the placement of the girder is poor. Further, the girder used on the second floor bears on the joists of the first floor rather than on the lower girder.

In plan (c) the girder is off-center toward the front of the building. This arrangement parallels the arrangement in section *A*, where load-bearing walls bear directly on the girder. All parameters defined in plan *A* apply here as well. It thus follows that larger joists (2 × 10s) should be used if 24-in.-o.c. spacings are employed; the girder may be a 4 × 8 or 4 × 10, depending on how it is used and the sizes of the joists. Bridging as well as solid blocking for non-load-bearing partitions must be included. A problem will surface when the second floor is installed.

Second-Floor Assembly Plans

The design criteria for the second floor begins with a 30-psf LL and a 10-psf DL. The alignment of joists is directly over the first-floor joists, front to back of building, and at the same on-center spacings. Since there are spans over the dining area, kitchen, and living rooms that have no partition for supporting ceiling joists, girders must be included in the design. Further, a stairwell must be built. The right side of Figure 5–22 shows the second-floor plan and plans (a), (b), and (c).

In plan (a) two short girders may be built of 2 × 8s and placed on the load-bearing wall. A more desirable arrangement could be a single built-up 4 × 8 girder that spans both rooms. The girder or girders must be made flush with joists as method A in Figure 5–3 shows, so that ceiling drywall and subflooring can be installed evenly on opposite edges of the girder and joists. A review of the Table 1 data in Appendix C shows that No. 2 grade 2 × 8s at 16-in. o.c. spacing should be used, or 2 × 10 grade No. 3 joists should be used if 24-in.-o.c. spacing is desired. Notice that the maximum span for 2 × 8s spaced 24 in. o.c. is 13 ft 4 in. and that a Dense Sel Str or No. 1 dense grade is called for. It is quite possible that the span may fall within the limits considering the width of the outer wall (6 in.) and exact replacement of the girder. In the event that the span is not exceeded, the second floor may be framed with 2 × 8s. Where the span is a full 13 ft 4 in., a few special joists need to be used. Elsewhere, No. 2 and No. 3 grade joists may be used.

Plans (b) and (c) in Figure 5–22 do not offer any better solution than plan (a). But notice also that the load-bearing walls are not over the first-floor girder except in the closet area at the right. In plan (c) a long, 14- to 15-ft girder is placed over the living room area. This is a long span for a girder and should be avoided. Since 2 × 8 joists are expected to be used, a steel *I-beam* or larger (6 × 8) wood beam needs to be used.

The second floor joists need to be bridged just as the first-floor joists were. Two rows of bridging are installed, one each in the span between outside wall and girder or load-bearing wall. The stairwell needs to be framed in with short headers perpendicular to the joist run and trimmer joists either side of the opening. These are necessary because partitions are installed on each side of the opening.

Conclusion

The plan for the design of the first- and second-floor assemblies of the house in Figure 5–21 points to using plan (a) as the better alternative even though there is a greater cost factor. Preplanning of this kind would have included this study. Estimates for cost should have included proper selection of material by grade and size. Setting aside cost for the moment, the designs in plan (a) and even (c) provide a more sound, reliable floor assembly than (b) because the forces on the load-bearing walls are more uniformly aligned. Recall that the second floor bears on the first through the load-bearing walls. If the plan (b) design were selected, loads would be placed on joists instead of girders. This places horizontal shear stress on all joists, which is not desirable. Finally, the selection of joists includes the *load span* and *on-center spacing*. A proper grade selection can be made only after these three criteria are set. However, if the grades needed for 24-in.-o.c. spacing are not available in the selected size, the on-center spacing may be reduced to 16 in. to allow use of available lower-grade joists.

PROBLEMS AND QUESTIONS

1. List the three types of loads bearing on a floor assembly.
2. What is a typical live load per square foot for a living or activity center?
   a. 20 psf **b.** 30 psf **c.** 40 psf
3. With reference to figure 5–1, what is the total pounds per square foot design for upstairs?
   a. 40-psf LL + 10-psf DL **b.** 30-psf LL + 10-psf DL
4. When would a girder such as one described in method A (Figure 5–3) be used?
5. List some typical sizes of girders.
6. A building is 24 ft wide × 32 ft long, and a girder is placed midway (12 ft) for the full length of the building and three piers are used. How many feet on each side of the girder is supported by the girder? If the live and dead load equals 40 psf, what would the total weight be on the girder in each 8-ft section?
7. What grade of lumber would you select if a 4 × 6 beam were to bear a load of 40-psf live load and 10-psf dead load and span 10 ft?
   a. standard **b.** construction **c.** No. 3
8. What would be the reason for using a grade No. 2 4 × 8 beam if its loads would be 30-psf live load and 10-psf dead load?
9. What is the basic difference between an L-type sill and a box-type sill? Which is used more frequently?
10. Why is mortar used sometimes when a sill is being fastened to a block wall?
11. What causes a squeaky floor assembly?
12. Draw a framed opening in a floor assembly and label the headers, tail joists, and trimmer joists.
13. Why could "cats" be used under floor sheathing and perpendicular to joists?
14. What is the building industry standard for the on-center separated maximum limit for bridging?
    a. 6 ft **b.** 8 ft **c.** 10 ft **d.** 12 ft
15. What grade of plywood would you use as a subfloor if its thickness were ¾ in. and the joists were spaced 24 in. o.c.?
    a. group 4 CC exterior **b.** group 2 CD interior
    c. group 1 structural II
16. Refer to Figure 5–22. Which condition, (a), (b), or (c) allows for the shortest girders upstairs *and* aligns the girders downstairs and upstairs?
17. Explain why 2 × 10s are used for joists in option (a) of Figure 5–22.

PROJECT

You are planning to build onto your house a porch that is 20 ft wide by 30 ft long. You need to make a rough plan of the floor assembly, define the materials to use and generally come up with a construction schedule. Let's see if you can do the planning in the project using information from Chapter 5 and the appendixes. You will need to complete each of the steps in this project to define the answers to the three parts of the plan.

A. Rough Plan of the Floor Assembly

Use the basic plan shown in Figure P5–1 to rough out the plan for building the floor assembly. Specifications need to meet 40-psf live load (LL) and 10-psf dead load (DL). Later, the porch will be roofed and screened in and a set of steps leading to the ground will need to be built.

1. Based on the elevation plan shown in Figure P5–1, it looks like you made several decisions. Let's be sure that we agree on them.
   a. The piers are set for a slope in the finished porch floor. How much is this?
   b. We will need a girder. Looking at Figure 5–3, which type will be selected?
   c. How large a girder do you need? This depends on several decisions:
      (1) First, which way do you want the deck flooring to run: parallel with the exterior wall (30-ft run), or perpendicular to the exterior wall (20-ft run)?
      (2) Floor joists will need to run perpendicular to the flooring. What direction will these be?
      (3) The girder will, therefore, run parallel with the flooring and perpendicular to the joists. Based on this information, what will be the direction and approximate length?
      (4) Which set of piers ghosted in Figure P5–1 will you use to support the girder, set A for plan A or set B for plan B?
      (5) You can now determine the size of girder needed. Complete either (a) or (b).
         (a) Plan A, girder parallel with the house (30 ft) using pier set A.
            (i) What is the span between piers?
            (ii)What minimum-size girder is suited? (Use Figures 5–9 to 5–13 and Tables 5–1 to 5–5.
            (iii) What grade and species is the lowest acceptable?
         (b) Plan B, girder perpendicular to house (20 ft), using pier set B.
            (i) What is the span between piers?
            (ii)What minimum-size girder is suited? (Use Figures 5–9 to 5–13 and Tables 5–1 to 5–5)
            (iii) What grade and species is the lowest acceptable?
   d. We may be able to use the girder you have just selected or you may need to increase its size to 4 × 8 because of the floor joist selection. Let's find out. Proceed with either (1) or (2).
      (1) Plan A, floor joists where the girder is parallel with the exterior wall (30 ft).
         (a) What is the approximate length of each floor joist?
         (b) Using Table C-2 in Appendix C, we find that 10-ft 2 × 6 joists for:
            (i) 24.0-o.c spacing are (listed/not listed).
            (ii) 16.0-o.c. spacing are (listed/not listed) and (iii) require that we use grades ______________ or ______________.
         (c) Can we use a 4 × 6 girder for this construction?
         (d) How many joists will we need? At what length?

FORMULA: Number of joists = length of porch in feet × 12 in. ÷ o.c. spacing + 1 × number of rows
= 30 ft × 12 in. ÷ ______________ + 1
× 2 rows = ______________

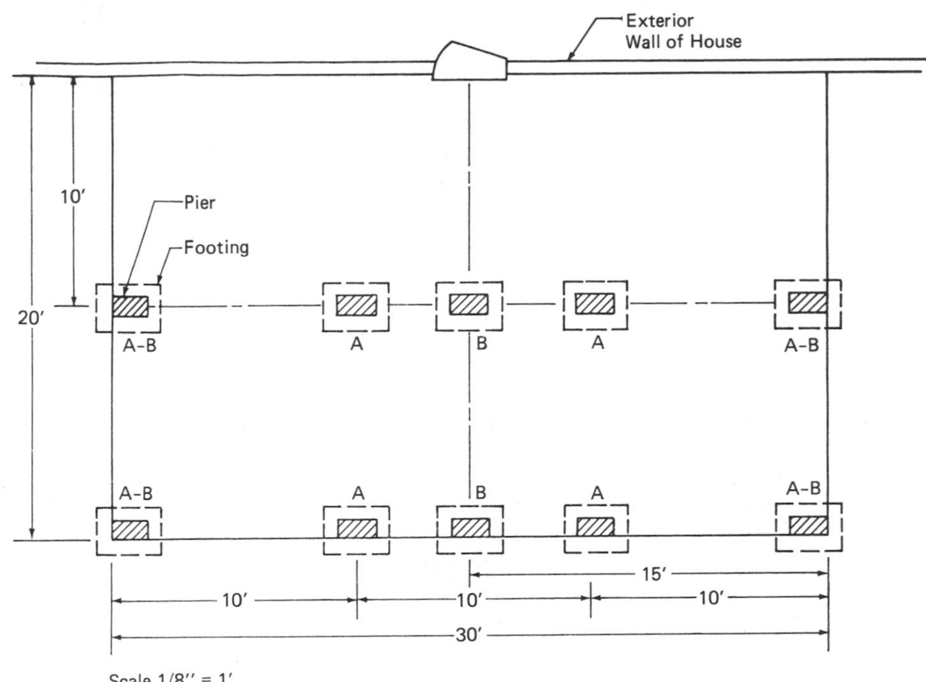

Floor Plan

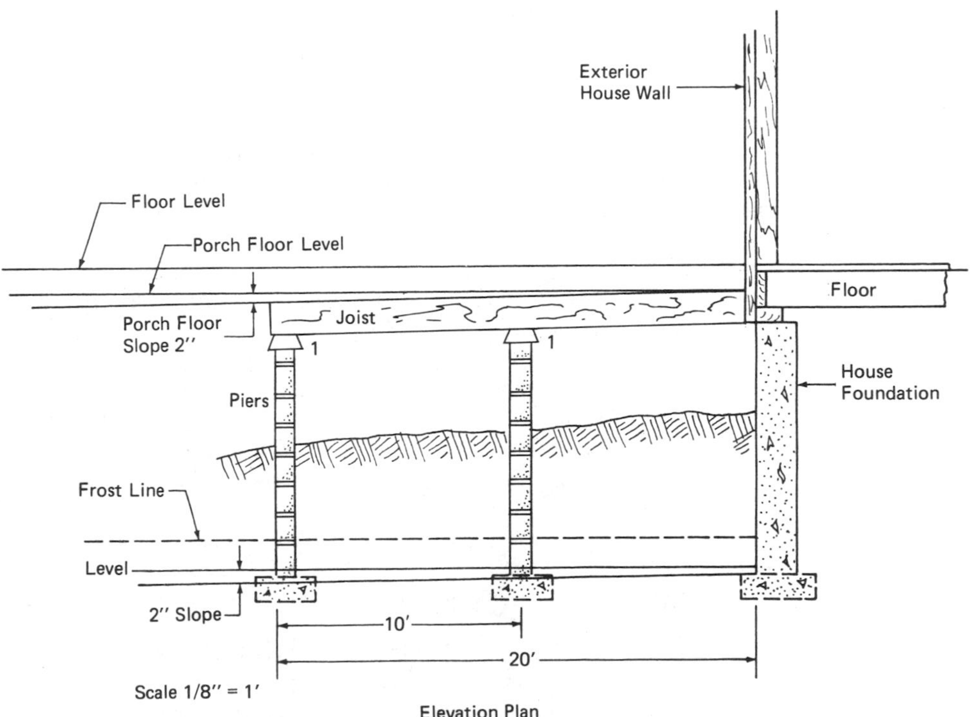

Elevation Plan

Figure P5–1

(2) Plan B, floor joists where the girder is perpendicular to exterior wall (20 ft).
   (a) What is the approximate length of each joist?
   (b) Using Table C-2, we find:
      (i) that 15-ft 2 × 6 joists are (listed/not listed).
      (ii) that 15-ft 2 × 8 joists are (listed/not listed).
      (iii) 24-in.-o.c. spacing (yes/no).
      (iv) 16-in.-o.c. spacing (yes/no).
      (v) Grades acceptable are ______________, ______________, or ______________.
   (c) Can we use a 4 × 6 girder? Should we use a 4 × 8 girder to ease the construction effort?
   (d) How many joists will we need? (Use the formula above).

2. Now let's do some more things. Let's identify some more materials needed for the floor assembly.
   a. *Sills.* The purposes for a sill are to secure the joist assembly to the perimeter piers and to attach to the house. They will carry only a little of the weight of the joists. Therefore, 2 × 6 or 2 × 8 No. 2 common grade fir or yellow pine can be used. Short pieces should also be used on internal piers to fill the space between girder and pier.
      (1) Number of linear feet needed for
         (a) Two ends = ______________
         (b) Long side = ______________
         (c) Pier(s) = ______________
         (d) Total ______________
   b. *Termite shields.* A termite shield needs to be installed on top of each pier to eliminate the threat of termite infestation to the porch floor materials. How many shields are required in each plan?
      (1) Plan A: ______________
      (2) Plan B: ______________
   c. Box-end joist materials. The open ends of the joists must be boxed in with joist materials. How many pieces of what length and size must we buy?
      (1) Plan A: ______________ pieces, ______________ ft long
      (2) Plan B: ______________ pieces, ______________ ft long
   d. *Cross bridging.* Since the span of each joist exceeds 8 ft in length, a set of cross bridging must be installed midway between ends of each row of joists. The simplest way to get these is to buy them from a supply house since they are made for 16-in.-o.c. and 24-in.-o.c. joist spacing. They can, however, be made from 5/4 × 3 spruce lumber. Making them by hand requires calculating their length and with the use of a bevel square or framing square, marking the angle of cut on each end. (Generally, a portable power saw is set for the bevel and all pieces are cut at once. Then several are trimmed for special lengths later.)
      (1) Plan A: ________ rows × 2 × joists/row − 1 ________ = ________ pieces
      (2) Plan B: ________ rows × 2 × joists/row − 1 ________ = ________ pieces
      (3) If wood is used, number of lineal feet of 5/4 × 3 = Plan A ________, Plan B ________.
   e. *Flooring.* On a deck the flooring is usually installed with spaces between each board. But on a porch floor the flooring is installed without spaces. Normally, fir tongue-and-groove stock is used. Also, a 1-in. overhang on outer edges is used. We

need to calculate the amount needed for this job. Formula: Porch length × width + a 10% allowance for waste. Estimate flooring needs: ______________ square feet.

B. Summary of the Material Needed.

For plan A:

Item	Type of materials	Quantity and stock
Girder	No. 3 or better Douglas fir or southern pine	60 ft 2 × 6
	2 × 2 ledgers	30 ft. 2 × 4 (ripped)
Ledger	Nailed against house 2 × 2	15 ft 2 × 4 (ripped)
Joists	No. 1 KD or dense fir or pine	46 2 × 6 10 ft
Sills	No. 2 or better fir or pine (includes short pier pieces)	4 2 × 6 12 ft 2 2 × 6 16 ft
Box ends	No. 1 KD or dense fir or pine	2 2 × 6 16 ft
Flooring	Fir or pine tongue and groove	671 ft^2
Termite shields	Aluminum flashing	8 ft 16 in. of aluminum flashing
Nails	Galvanized	10 lb 16d common 10 lb 8d common

For plan B:

Item	Type of material	Quantity and stock
Girder	No. 3 or better Douglas fir or southern pine	40 ft 2 × 8
	2 × 2 ledgers	20 ft 2 × 4 (ripped)
Joists	No. 2 dense or MG KD fir or pine	32 2 × 8 16 ft
Sills	No. 2 or better fir or pine (includes short pier pieces)	4 2 × 6 12 ft 1 2 × 6 16 ft
Box ends	No. 2 dense or MG KD fir or pine	4 2 × 8 10 ft
Flooring	Fir or pine tongue and groove	671 ft^2
Termite shields	Aluminum flashing	6 ft 16 in. of aluminum flashing
Nails	Galvanized	10 lb 16d common 10 lb 8d common

C. Construction Schedule

Two assumptions are made before this construction schedule is made. First, the design of the floor assembly was made from this project. Second, the masonry work has already been completed. In other words, the floor assembly is ready to be made. There are only eight tasks to do. They are listed below in random order. You should easily rearrange them. Check your answers.

a. Remove the lower pieces of siding from the exterior wall where the floor assembly will be attached.
b. Build and install the girder.
c. Install the termite shields.
d. Install the decking.

e. Cut and install the box sills.
f. Cut and install the floor joists.
g. Connect the box sills to the house and if using Plan A, attach a ledger to the house for the joists to rest on.
h. Cut and install the cross bridging.

Answers:

Part A: **1. a.** 2 in.
b. method A
c. (1) parallel with the exterior wall (30-ft run), perpendicular to wall (20-ft run)
(2) Perpendicular to exterior wall, parallel with the exterior wall
(3) Parallel with the exterior wall (30 ft), perpendicular to wall (20 ft)
(4) set A, set B
(5) (a) (i) approximately equal to 10 ft
(ii) 4 × 6 or double 2 × 6s
(iii) No 3 Douglas fir or southern pine;
(5) (b) (i) approximately 10 ft;
(ii) 4 × 6 or double 2 × 6s;
(iii) No. 3 Douglas fir or southern pine;
d. (1) (a) 10 ft;
(b) (i) not listed for 10 ft;
(ii) listed, No. 2 dense or No. 2 MG FD;
(c) yes;
(d) 46, 10 ft long;
(2) (a) 15 ft;
(b) (i) not listed;
(ii) listed;
(iii) no;
(iv) yes;
(v) appearance grade KD, No. 1 dense, or No. 1 KD;
(c) no, the need for 2 × 8 joists precludes the use of a 4 × 6 girder; so we must upgrade the girder to two 2 × 8s or a 4 × 8;
(d) 32;
2. a. (1) (a) 40 ft;
(b) 30 ft;
(c) 1 at 16 in. or 2 at 32 in.;
(d) 73 to 74 lin. ft;
b. (1) plan A, use 8 termite shields;
(2) plan B, use 6 termite shields;
c. (1) two 2 × 6 16 ft;
(2) four 2 × 8 10 ft;
d. (1) 2, 22, 88;
(2) 2, 15, 60;
(3) plan A at 16 in./piece = 118 lin. ft, plan B at 18 in./piece = 90 lin. ft;
e. 671 ft^2

Part C: Construction schedule order: a, c, e, g, b, f, h, d.

Chapter 6

Walls and Wall Systems

Each wall of a house is designed and built to perform a useful function. It may be one of the exterior walls that defines the shape of the house and includes all the different structural requirements for door passages and window openings, for carrying loads or not, and for having the strength and rigidity required to resist forces placed on it. It may also be an interior wall whose function is to divide spaces for the purposes of sleeping, living, bathing, and others. The interior wall may need to be designed for load- or non-load-bearing, for the control of sound, or to accommodate plumbing and heating equipment. The various types of walls that may be included are those used in single-story construction or two-story construction or may be specialized systems.

These purposes or functions of walls are the subject of this chapter. So that their importance is understood, each wall will be described in terms of a system since many parts are used in special ways to create it. First a definition of a wall system is developed. Following this, a look at the elements of the system are defined and discussed. Then different wall systems are explained.

DEFINITION OF A WALL SYSTEM

The opening paragraphs defined the purposes and uses of many walls used in residential construction. If the purpose were the only thing necessary to understanding the construction of a wall, the carpenter's task would be quite simple. In fact, however, the design of a sound wall system requires considerable knowledge on the part of the carpenter.

Therefore, a more practical definition of a wall system is necessary. For example, a definition that relates the elements of design and construction that make up a proper wall assembly would have much more meaning. Thus a wall system by definition is one that includes the proper sizes and grades of lumber; members such as studs, plates, headers, braces, and trimmers; and attaching materials. The definition also includes design data so that when properly constructed the wall forms an assembly that (1) has the proper proportions as dictated by the building plans, (2) has sufficient rigidity to withstand the external forces of wind and earthquakes (where prevalent), (3) provide support for wall coverings, (4) be able to support loads placed on it, and (5) incorporate any and all special features that may be required.

WALL MATERIALS AND THEIR USES

Studs

Throughout the history of house construction the stud has been cut in many sizes and shapes. In early times in this country and in Europe, studs approximately 4 in.[2] were used in the following places (see Figure 6–1):

A. At each corner
B. Each side of a window and door
C. Between long expanses
D. For corner braces

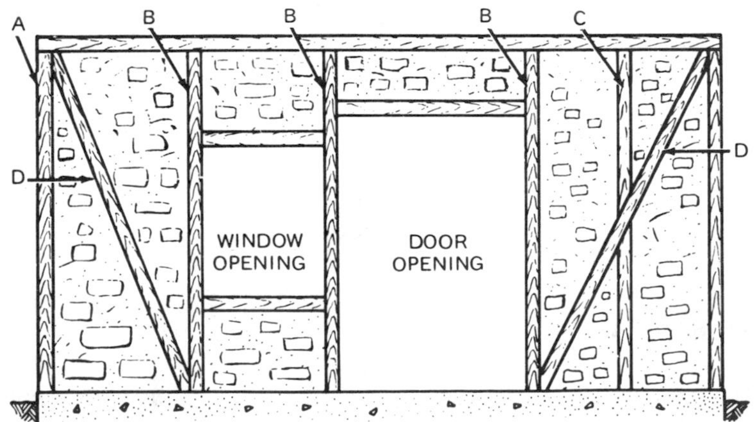

Figure 6–1 Old-fashioned framing with 4 × 4s.

Then mortar, clay, straw, bricks, stone, or any material available was plastered between the studs to complete the wall.

Only recently has the stud become uniformly sized and even more recently its size has been reduced slightly to its present 1½ in. × 3½ in. (approximately 40 mm × 90 mm in metric) as Figure 6–2 shows. Two common lengths are used in wall systems and they are the approximate 8-ft length for single-story, platform, or western framing techniques, and the 16- to 18-ft stud for balloon framing.

Present construction requirement of placing studs at either 16-in.- or 24-in.-o.c. spacings has several desirable features. First, the quality of stud or its grade may be selected from poorer stock since the load it must carry is distributed to many studs. Consider the condition in Figure 6–3, where a stud is in a *load-bearing wall.* This view shows that a single stud would be a poor selection as a support for such a load. This situation is seen frequently where a porch roof is temporarily supported by studs while columns are being readied. However, even the best quality stud would show some form of weakness in this situation.

In the figure the stud is shown under a heavy load that compresses all its fibers downward. Since the grain of the stud may not be uniformly straight and there may be knots and places of tension and compression within the stud, it will *bow.* It is probable that even the best-grade stud made from the strongest species—Douglas fir–larch or southern pine—would bow under such a load.

A common stud wall will not have one stud supporting such a load. As Figure 6–4 shows, where studs are evenly separated and are placed directly under the force of the roof or joists the compression is evenly distributed among the studs. Therefore, the studs are more capable of retaining their original shape and will not, as a rule, bow.

However, there is a possibility that some deflection may take place and the grade of lumber that is used will allow or limit this deflection. Figure 6–5 shows four grades of stud and their relative usefulness in supporting loads. The better the grade of stud, the more stress it can stand; the poorer the stud, the less stress it should have placed on it.

In usual framing practices the *utility* stud suffices for both load-bearing and non-load-bearing walls even though its grade is poor. The reasons for this are the

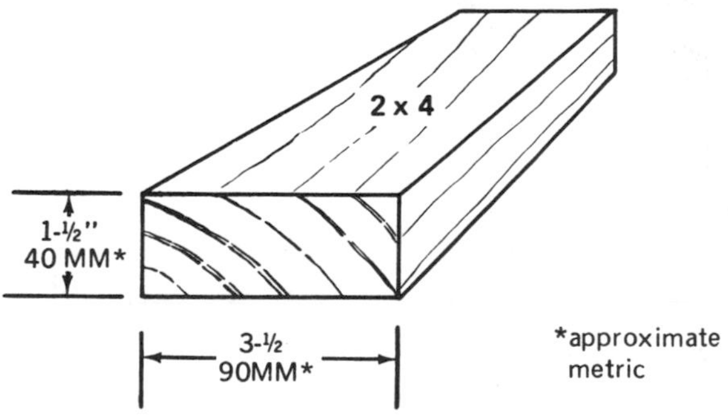

Figure 6–2 Size of a modern 2 × 4 stud.

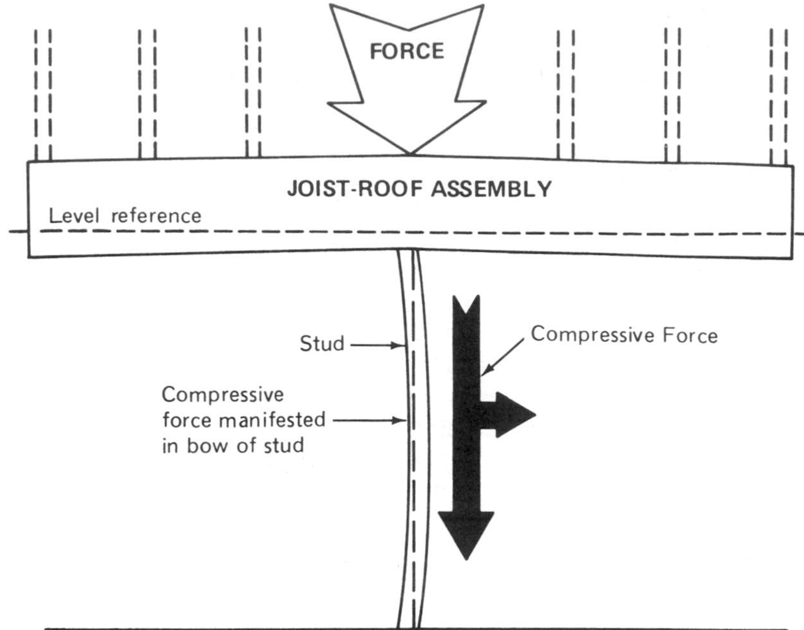

Figure 6–3 Total load bearing on a single stud and the effects on the stud.

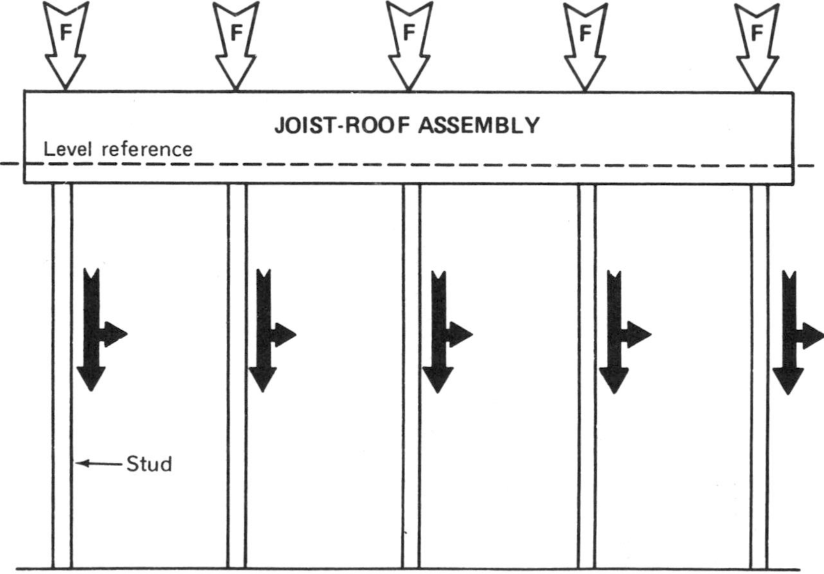

Distributive force in multi-stud wall

Figure 6–4 Total load on a stud wall with equal distribution of force per stud.

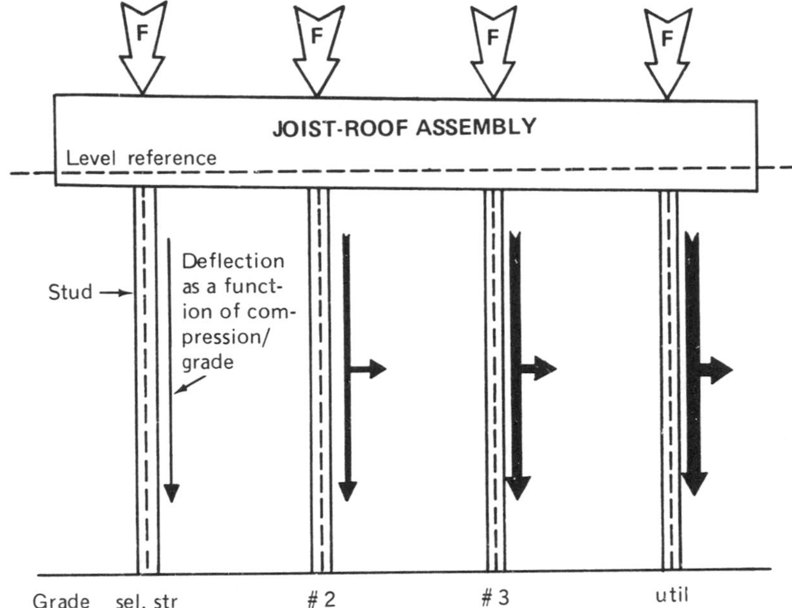

Figure 6–5 Relative ability of various grades of studs to withstand force of loads.

close on-center spacings of the studs, wall bracing, double plates, and relatively small ceiling joists and roof rafters. If larger ceiling joists or roof rafters and heavier live loads are predicted, better grades of studs are used to ensure that the heavier loads will be supported properly.

Seven grades of studs are listed in Table 6–1. Of the seven the first one, No. 1 and appearance, and the second, No. 2 may be used where exposed studs are needed. The other grades usually have objectionable qualities such as knots, stains, pitch pockets, streaks, shakes, splits, and other defects that prohibit their use as finishing materials. The other grades do, however, meet the building requirements of most local, state, and national building codes. There is a restriction on the use of the *utility* stud; its use is limited to 16-in.-o.c. spacing in wall construction. This restriction corresponds to the earlier explanation illustrated in Figure 6–5.

All studs must be dried properly so that their moisture content is either 19% or, if kiln dried, 15%. This is important even though shrinkage in the length of a stud is the smallest shrinkage that takes place. If quantities of moisture exceed these values, as they would in green lumber, *settling* will result as the studs in the wall dry out. This would cause floors and other level surfaces to become uneven. A one-tenth of 1% shrinkage in an 8-ft stud causes less than ¼ in. reduction in length; but in balloon framing, where 18-ft studs are used, the shrinkage of green lumber could exceed ½ in. Therefore, use only cured studs in wall construction and eliminate the problems green lumber could create.

Since studs are nailed to plates as shown in Figure 6–6, they need to be nailed

TABLE 6–1 STUDS BY GRADE AND USE

Grade	Type of wall		Range of:	
	Load-bearing	Non-load-earing	Stiffness, E (10)	Strength, F_b (psi)
No. 1 and appearance	Yes[a]	Yes[b]	0.9–1.8	1050–2050
No. 2	Yes[a]	Yes[a]	0.9–1.7	1000–1650
Construction	Yes	Yes	0.8–1.5	725–1200
No. 3	Yes	Yes	0.8–1.5	475–925
Stud	Yes	Yes	0.8–1.5	550–925
Standard	Yes	Yes	0.8–1.5	350–675
Utility	Yes[c]	Yes	0.8–1.5	175–325

[a]May be used where exposed studs are needed.

[b]May be used when o.c. spacing is not greater than 16 in.

[c]May be used but often exceeds needs for strength and stiffness.

securely. Two methods are used; one uses two 16d common nails driven through the plates into each end of the stud. The other method uses four 8d common nails toenailed into the stud and then into the plate. When toenailing, 60° is the best angle to use as it allows the nail to penetrate the stud adequately, yet permits most of the nail to anchor into the plate.

Plates

The plate or horizontal member in the wall assembly has almost always been used on top of the studs as a way of holding studs in line. The plate has not always been used at the base of the studs. However, the plates are a very important part of the wall assembly in that (1) they do perform the need of keeping the studs in alignment, (2) they provide nailing surface for wall covering materials on both sides, and (3) they perform the function of *fire-stops*.

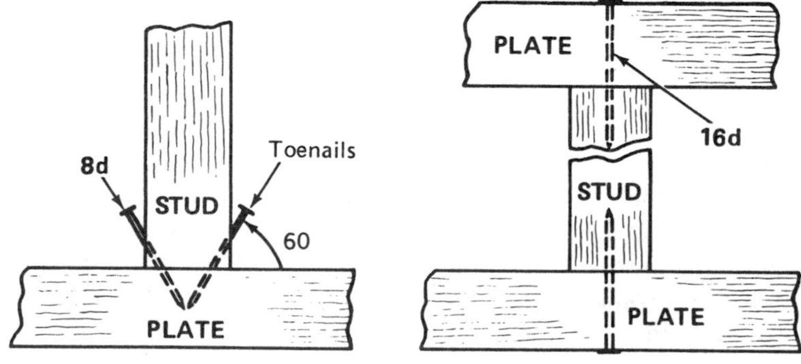

Figure 6–6 Nailing of stud to plate either through plate or by toenailing.

Throughout the history of construction, the plates have been called by various names. Some of these are the *sole, sole plate, shoe, shoe plate, sill plate, bonding plate, connecting plate,* and *top plate.* All but the bonding, connecting, and top plates are names that refer to the plate used at the base of a wall assembly. More currently, the term *plate* refers to either the bottom or top horizontal members of a wall assembly.

Figure 6–7 shows the position of the plates and also shows that the plate performs the three functions cited above. The current trend is to use a double or built-up plate at the top of the wall. This is done not so much for the strength needed to hold the studs in alignment, because a single plate can do this, but because better overall building lines are established and the top plate (bonding plate) is used to unite other wall assemblies through overlap construction (Figure 6–7). In addition, the second top plate also aids in supporting horizontal shear stresses that might occur, and it is usually recommended for load-bearing walls.

These reasons are sound design principles put into practice; yet several studies have shown that the single top plate has been used in non-load-bearing exterior and interior walls with no deficiency in wall quality.

Plates usually are nominal 2 × 4s and should be of a grade equal to or better than the stud grade. They should not be a poorer-quality lumber. Plates must be properly seasoned so that their moisture content is either 15 % (KD) or 19%. The reason for this is to minimize the problem of shrinking. The shrinking shown in

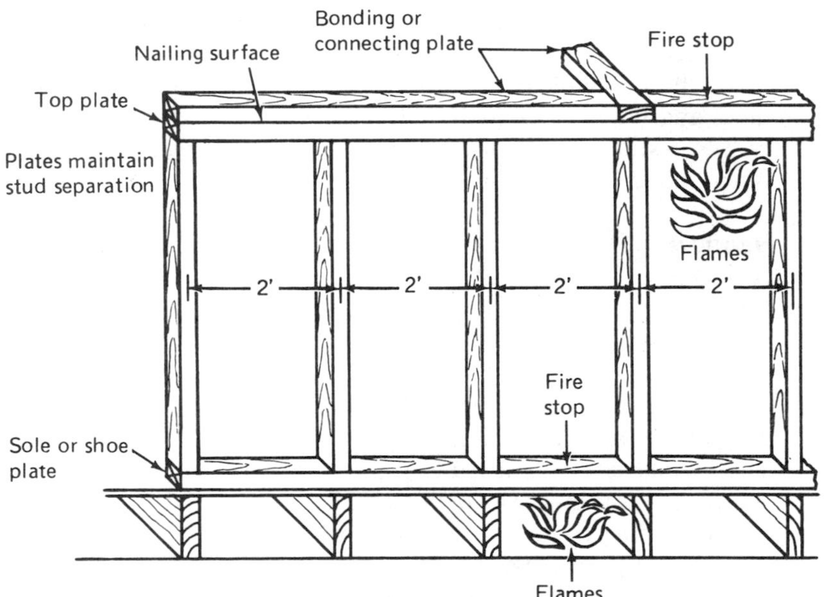

Figure 6–7 Three functions of plates: (1) to maintain stud separation; (2) as nailing surface for paneling; and (3) as fire-stops.

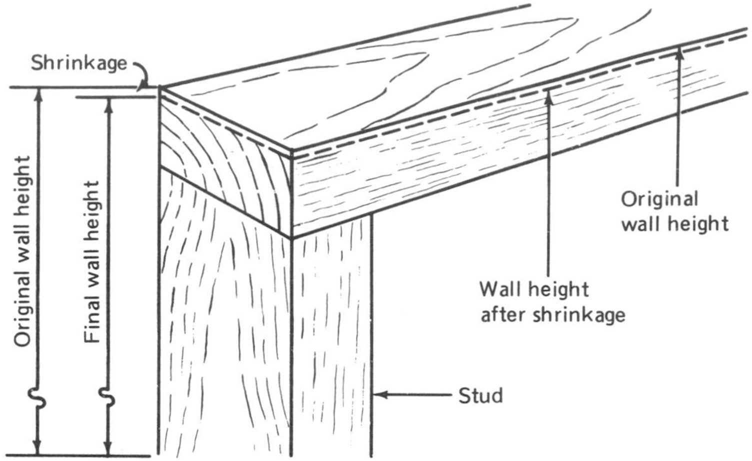

Figure 6–8 Shrinking as a problem from using improperly cured lumber.

Figure 6–8 could create problems with paneling and siding because of the tension placed on these items. Panels cannot shrink; they can only buckle.

Double-wall plates are frequently needed when a soundproof wall is desired. These plates are usually 2 × 6s and, as Figure 6–9 shows, either 2 × 3 or 2 × 4 studs are off-centered so that their surfaces do not contact. Because of their width, the plates allow each stud standing room so that sounds received by one stud are not passed to the other and into the adjoining room.

The 2 × 6 plate may also be useful in wall construction where plumbing vent pipes and sewer pipes are installed above floor level. They will hold 2 × 3, 2 × 4, or 2 × 6 studs in alignment, and may be cut or bored for passage of the plumbing pipes and still retain considerable strength.

Headers

Window and door framing requirements in load-bearing exterior walls and some non-load-bearing exterior walls require the use of a header. Load-bearing interior partitions also need headers where passageways from room to room are framed.

The header, the horizontal member spanning a desired open space, is customarily made from nominal 2-in. stock and installed on edge, as Figure 6–10 shows. The various sizes employed to support the forces and weights resting on them are shown in Table 6–2. The table provides conservative maximum spans for headers. Local building codes may vary somewhat either way.

The importance of a header is illustrated in Figure 6–11. The roofing materials and live roof loads, ceiling joists, ceiling materials, and live ceiling or attic loads bear on the wall. This means that the loads are directed downward to the headers.

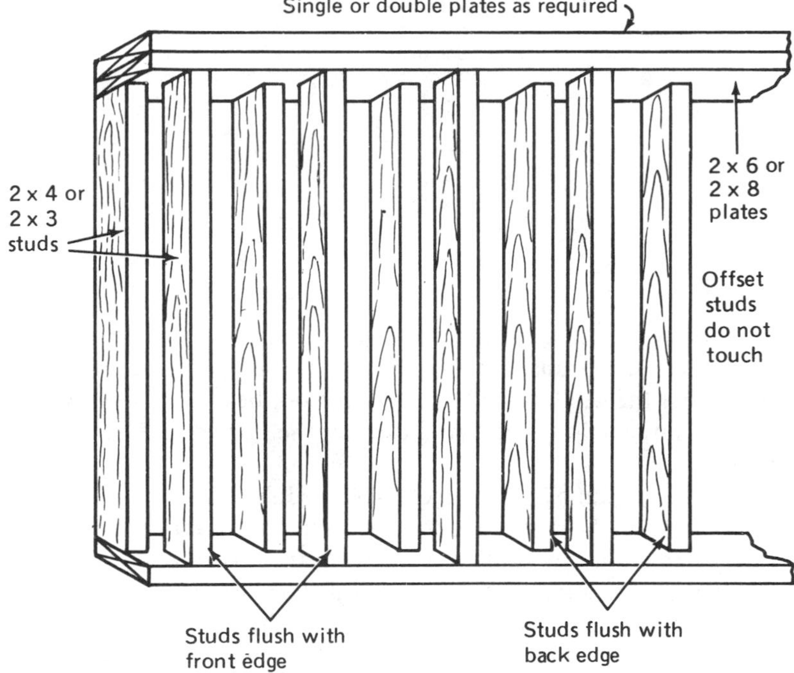

Figure 6–9 Double-wall single plate: 2 × 6 or 2 × 8.

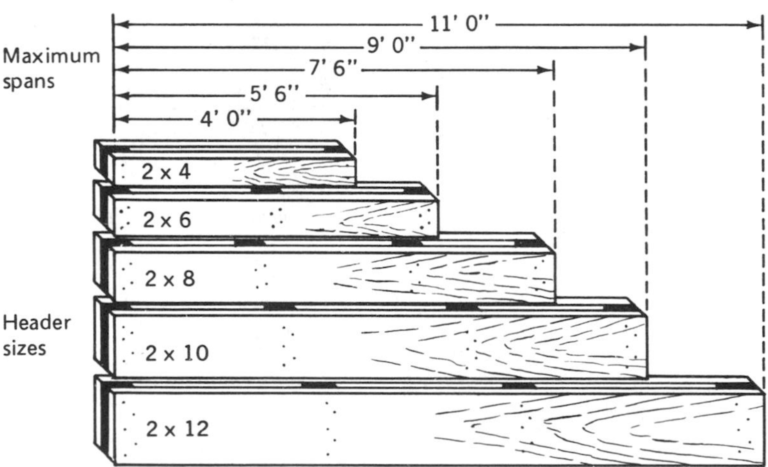

Figure 6–10 Headers: dimension and length.

TABLE 6-2 SPANS FOR HEADERS

Header size	Maximum span
2 2 × 4s	4 ft 0 in.
2 2 × 6s	5 ft 6 in.
2 2 × 8s	7 ft 6 in.
2 2 × 10s	9 ft 0 in.
2 2 × 12s	11 ft 0 in.

Note: All members are placed on edge.

Notice carefully that the concentration of the load is at each end of the header where the force is then transferred to the floor assembly. This conventional approach illustrates that the window or door unit does not bear any of the weight of the wall, and this is as it should be. It also illustrates the importance of using the correct size of header.

As the span of header increases, the load placed on it increases. It has already been shown in the study on girders (Chapter 5) that the total loads multiplied rapidly as the spans increased. The same holds true for the header. Figure 6-12 provides two sample situations that illustrate this fact. In Figure 6-12a the blackened area is only 4 ft wide and represents the header. The shaded area is the load area or total load the header must support. Notice that it is one joist space wider than 4 ft and equal to half the joist length plus the rafter force bearing on the wall and header. In Figure 6-12b the span is increased to 8 ft. Notice how much more total weight

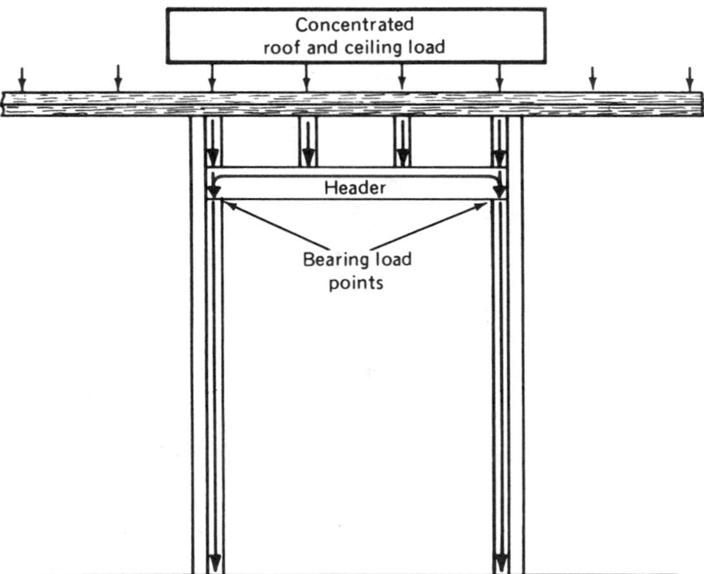

Figure 6-11 Forces on headers and distribution of forces to floor.

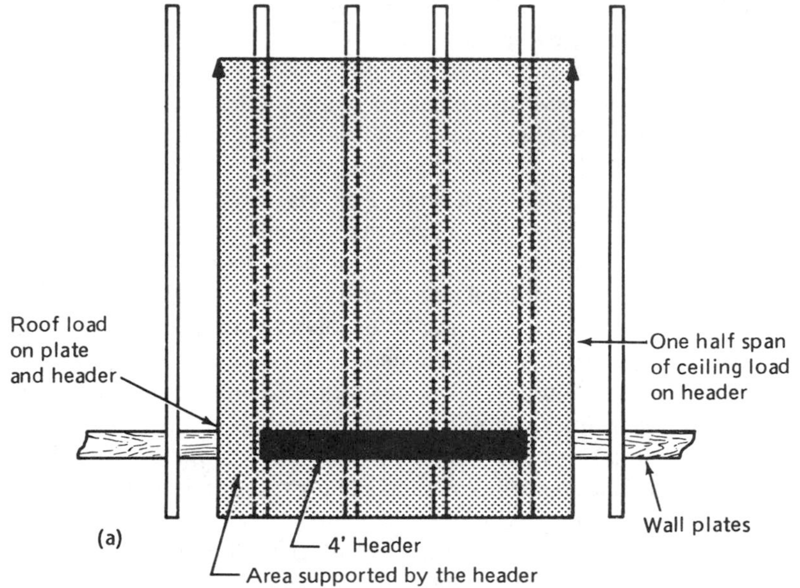

TOP VIEW

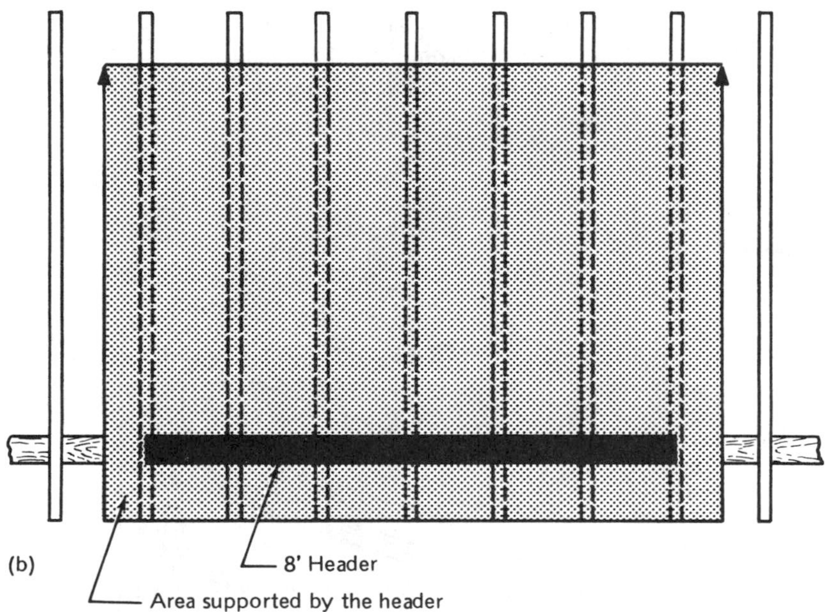

Figure 6–12 Load bearing on headers of different lengths.

the header must carry. Headers made from doubled 2 × 10s must be used to span the 8 ft, as Table 6–2 indicates.

A recent adaptation or modification to the placement of headers is shown in Figure 6–13. Instead of the header being placed at door or window rough opening height, it is placed snugly under the wall plates. The same principles of load bearing are still included in this arrangement. Instead of cutting cripple studs to fill in between headers and plates, a single 2 × 4 is installed across the top flat-wise at rough window or door height to provide nailing surface for wallboard, sheathing, window head nailing surfaces, and window or door trim. This technique may not look as sound as the conventional one, but it is, and it may save labor and material costs.

The preceding discussion on headers has indicated the important function they serve in supporting loads, yet there is one more property of a nominal 2-in. piece of lumber that should be taken advantage of when constructing headers. The *crown* of each header piece should be placed toward the top plates (similar to crown-up on floor joists). As the weight of the load is applied to the header, it will deflect downward. The crown-up method of construction is effective in reducing the deflection caused by the bearing load.

Finally, properly dried lumber of good quality, grade, and species should be used for headers. If this type of lumber is used, shrinkage is minimal. However, if green or partially dried lumber is used for headers, two conditions result: First, the members will dry within the wall and significant shrinkage occurs. This has a severe effect on rigid panels fastened to the wall or brick veneer on the outside of the wall because they do not shrink. Buckling or other structural rupture may result.

The second reason to avoid using unseasoned lumber is that there is a tendency for the fibers in the *bearing zone* to collapse, become crushed, and cause settling. Even before drying out takes place, the damage is done. So to avoid the problem

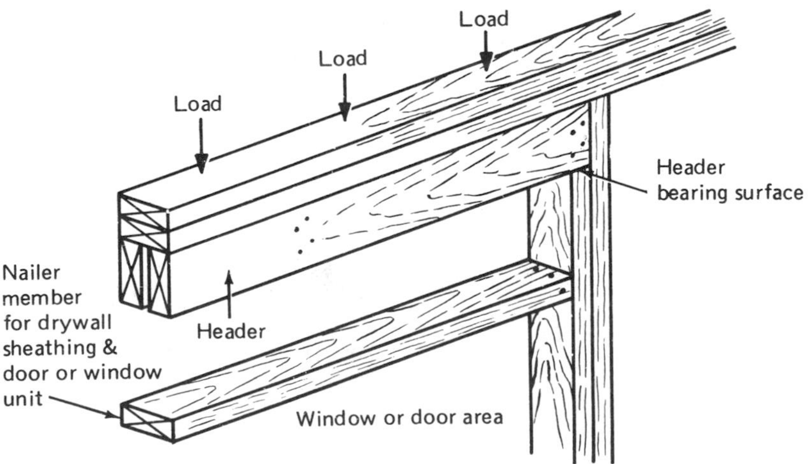

Figure 6–13 Relocation of header snugly under top plate.

of shrinkage for either reason, use only properly dried, good-quality lumber for headers.

Trimmer Studs

The trimmer stud is one that extends from floor (sole plate) to ceiling (top plate) on either side of the openings in the wall. Figure 6–14 shows examples of these studs. A *trimmer stud* performs a function very much like the trimmer joist in a floor assembly around an opening. As a rule, the headers are nailed through the trimmer. The trimmer bears a load placed on it from above in a load-bearing wall, but does not assume a significant part of the load of the header.

Trimmers are used in non-load-bearing walls in similar fashion. There, too, the headers are nailed through them; and, in either type of wall, the trimmer may be a normally spaced stud or an additional stud used exclusively for this purpose.

Cripple Studs

A *cripple stud* is one that is used in several different places in a wall assembly. It is customary to use cripples in vertical positions. They may be spaced at normal on-center separations as in a cripple wall, or above headers as in Figure 6–15. In both

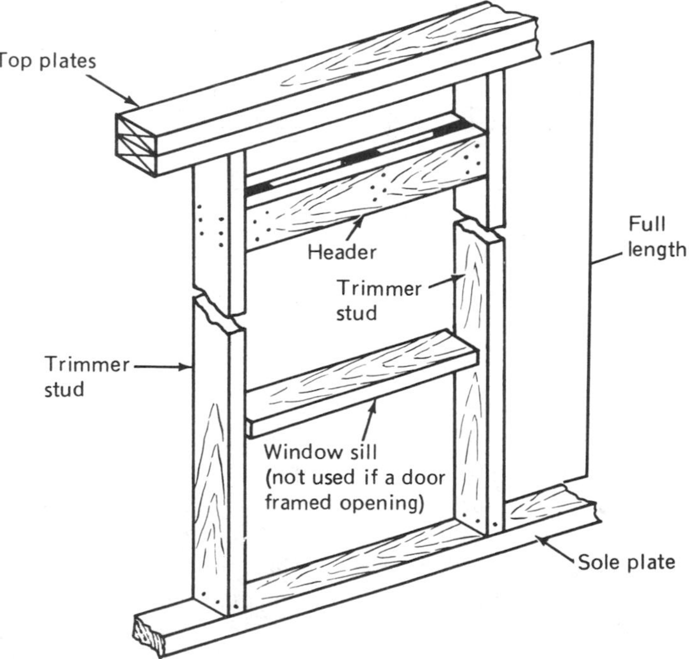

Figure 6–14 Trimmer studs.

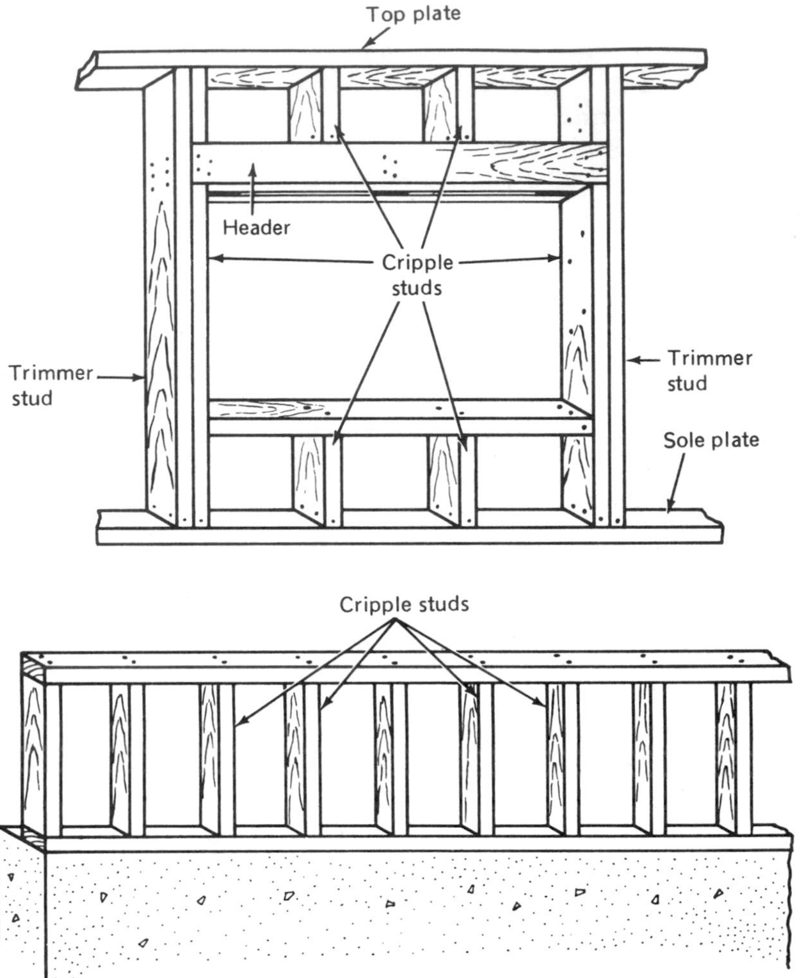

Figure 6–15 Cripple studs.

of these applications they act as short studs. They may be held in place by spiking through plates or by toenailing as any other stud is fastened.

The third purpose for a cripple stud is its use in framing the opening of a door or window as Figure 6–15 shows. The cripple stud (sometimes called a *jack stud*) is fit beneath the header. This means that it is the bearing surface for the header and is also the member that bears one-half the total header load to the lower plate. It is therefore very important that this member be made of good-quality lumber and be properly dried. Even though it is nailed securely to the trimmer, it would shrink if unseasoned, thereby causing stress within the wall and on materials attached to the wall.

In window framing, the 2 × 4 installed as a sill framing member should be toenailed to the cripple. Many times carpenters use short cripples and extend the sill to the trimmer. This arrangement has the effect of making 1½ in. on each end of the sill a part of the load-bearing area from header to floor.

Braces

Almost any material installed within or onto a wall framed of 2 × 4s or larger stock, braces the wall. Bracing is needed to control and prevent a building wall from *racking*. If a wall is out of plumb measured along one end from sill to plate, it is racked. If such a condition is found, racking was allowed to happen because of no bracing or inadequate bracing.

Many times a wall that does not contain diagonal bracing or let-in bracing is mistakenly considered unbraced. Actually, though, ordinary 1 × 6 or 1 × 8 sheathing properly installed provides bracing. If sheathing is installed horizontally, as in Figure 6–16a, it has some ability to resist the racking forces of wind and earthquake. However, its function as a brace is limited and custom has dictated that specific bracing such as the let-in brace of a 1 × 4 be included, as shown in Figure 6–16b.

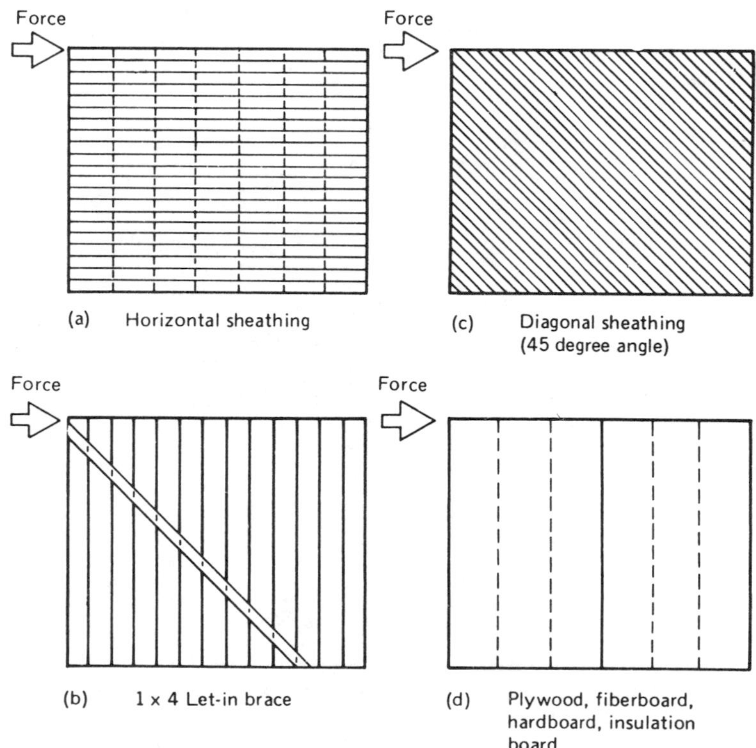

Figure 6–16 Bracing the wall to resist forces.

On the other hand, when sheathing is installed diagonally at about a 45° angle as Figure 6–16c shows, better bracing is obtained. Sheathing properly fit and well nailed with three 8d or 10d common nails per stud crossing does indeed resist the forces that cause racking.

The use of diagonally installed sheathing boards has largely given way to panels made of many materials (Figure 6–16d). Cost of labor is one reason for the switch, since many hours are needed to cut and install diagonal sheathing, but a panel may be installed in minutes. Cost of materials is a second reason, but more important, panels (4 ft × 8 in. or other) properly installed improve bracing up to eightfold.

The first trend in using plywood panels as bracing was to sheath the entire wall with plywood. Sometimes the panels were installed horizontally, other times vertically. Given one way or the other, tests have proven that the vertically installed panels provide more bracing. Next came the use of plywood panels at all interior and exterior corners around the building. Fiberboard panels were installed between

TABLE 6–3 BRACES FOR FRAMED WALL

Bracing material	Relative racking resistance[a]	Relative strength	Notes
Plywood (glued)			Installed vertically
5/8 in.	24	9+	
1/4 in.	14	8+	
Tempered hardboard, 1/4 in.	9.8	6.9	Nailed 3-in. and 6-in. spacing
Diagonal sheathing 1 × 6 × 1 × 8	7.3	8+	45° angle, studs 16 in. o.c.
Plywood (nailed only)			
5/8 in.	4.8	9+	Installed vertically, 16-in.-o.c. or 24-in.-o.c. studs
1/4 in.	4.2	5.2	
1 × 4 let-in brace			45° angle brace
	4.4	3.9	16-in.-o.c. studs
	2.7	2.0	24-in.-o.c. studs
25/32-in. fiberboard	3.0	3.8	3-in. and 6-in. nail spacing
	4.5	6.1	Double nailing
Metal straps	1.9	1.4	2 each, nailed 45° at corners with two nails per stud
Horizontal sheathing, 1 × 6 or 1 × 8	1.0	1.0	Values may be increased up to 50% by increasing nail size to 10d and using more nails

Source: Data extracted from *Guide to Improved Framed Walls for Houses,* U.S. Department of Agriculture, Forest Service, 1965.

[a]Values are stated for solid wall with no window or door cutout in first 4 ft from corner.

plywood panels. But although these had some insulating ability, they had very little racking resistance.

Manufactured panels (other than plywood) have been improved through better combinations of materials and improved manufacturing techniques. Some of these panels have resistance qualities better than horizontal sheathing boards. These are used entirely around a house, not just between corner panels of plywood.

Still, however, the best bracing of a wall frame is obtained by using plywood, glue, and nails. Panels preferably full length and full width are glued and nailed to the studs and plates. Table 6–3 illustrates in descending order the bracing quality of each type of bracing material.

Several obvious factors are indicated in the table, and several are not. Panel type and material are as important as panel thickness. Plywood, for example, may be made from group 1, 2, 3, or 4 species, and each panel made from a different group would have a different relative racking resistance. The group 1 panels have the greatest strength and stiffness, so they would provide the greatest resistance to racking forces. Group 2 panels provide less resistance, as do group 3 panels, and group 4 panels provide the least racking resistance.

Tempered hardboard, which is a manufactured product, has remarkable racking resistance. Fiberboard, which is also manufactured, is only one-third as good as hardboard, but still three times better than horizontal sheathing. Some insulating panels have some rack resistance capability, but nowhere as much as plywood or tempered hardboard.

Nailing and gluing also play an important role, as has been mentioned. Table 6–3 states the nailing schedule for perimeter first and intermediate studs second (note column). Finally, siding may contribute much to the rigidity of a wall if it is a panel type; if it is a lap siding variety, the siding does not.

Sheathing and Wall Coverings

The types of sheathing identified in Table 6–3 were plywood, fiberboard, hardboard, and 1 × 6, or 1 × 8. Insulating panels made ½- to 2¼-in. thick are being used directly over studs. Their function is insulation, which, of course, is vitally important today. As seen in Figure 6–17, let-in corner bracing is used, and the sheathing-insulating board is nailed directly to the studs.

Use of thermal boards as sheathing makes for a wall system that may exceed 6 in. in thickness. Chapter 3 explained the importance of these new products in thermal control. But when siding is installed over thick insulating panels, very long nails are needed to secure the siding to the stud. The insulating panel does not have nail-holding power.

On the inside of the wall a panel of some sort is usually installed. It may be ½-in. gypsum wallboard, ⅜-in. gypsum wallboard and panel boards, lath and plaster, or sound-resistant panels and gypsum wallboard. Each of these products adds:

Figure 6–17 Sheathing a wall with insulating sheathing. (Photo courtesy of Celotex Corporation, Tampa, FL.)

1. To the overall dead load of the wall
2. Strength to the wall
3. Improvement in its racking resistance
4. To its thermal protection and sound control

In every case proper nailing and adhesives improve the total quality of the wall. Every product or part of the wall described previously contributes to the development of a soundly constructed wall system.

EXTERIOR WALL SYSTEMS

All, most, or some of the uses of materials described in the preceding section may be made part of an exterior wall system. If the wall is load bearing, most of the materials will be used and their sizes, grades, nailing schedules, and other factors must be properly incorporated. If the wall is non-load-bearing, not all materials need be used, primarily because bearing loads are not present.

Within the descriptions and illustrations that follow, design characteristics and proper application of materials are presented. Segments of walls are presented first, then a full wall is shown from the floor plan and elevation plan.

Balloon Framing

The studs in a balloon frame extend from the sill to the double plates that the rafters and second floor ceiling joists rest upon. Figure 6–18 shows a corner segment of a balloon frame. Many of the design features identified earlier are incorporated here. They are:

1. *Let-in corner braces.* Let-in braces of 1 × 4 extend outward and downward from the corner in each wall. Before the wall is sheathed, a second pair may be installed in the second-floor segment of stud or may be omitted if plywood or other properly rated panels of sheathing are installed.
2. *Fire-stops.* Two sets of fire-stops are installed; one is above the first-floor joists to trap any basement fires. The other is installed above second-floor joists to trap first-floor fires.
3. *Studs rise two full stories.* Each normally spaced stud and all trimmer studs rise full height. They need to be one of the seven grades listed in Table 6–1. Certain building codes may restrict or prohibit use of the utility stud.
4. *Plates are used at each end of the stud.* A single sill plate is anchored to the foundation. A double plate is constructed around the top of the walls. Overlapping at corners and partitions is accomplished with the bonding plate.
5. *Ribband (ribbon-band).* A 1 × 4 ribbon band is let in horizontally along the wall perpendicular to the second-floor joist run. Second-floor joists (beams) are notched to fit over the ribband to lock wall to joist (see the inset in Figure 6–18). Nailing is important, but the notch prevents shifting and separation. In other words, an interlock union is made between wall and floor assembly.
6. *Second-floor ceiling and roof rafters.* Both the joists and rafters bear on the plates, which are aligned over studs. Good design for load and force is shown.
7. *Common inside–outside corner.* A commonly designed corner is constructed from three 2 × 4s with scrap pieces used as fillers.

The balloon framing method is an excellent one to use for two-story construction. It is also very appropriate for split-level construction. It is also desirable to substitute a balloon frame for a cripple wall supporting a platform wall.

Platform, Single-Story, or Western Framing

The most commonly used framing technique or method is the *platform,* sometimes called single-story or western. Its wall assemblies are one story high; this usually means an 8-ft ceiling height. Its popularity is founded on its simplicity of construction and ease of erection. Many of the design features defined earlier are incorporated in the sample of platform framing shown in Figure 6–19. They are:

1. *Plates.* Sole and double-top plates are used on all perimeter walls. The top plate is held in alignment by the bonding plate.

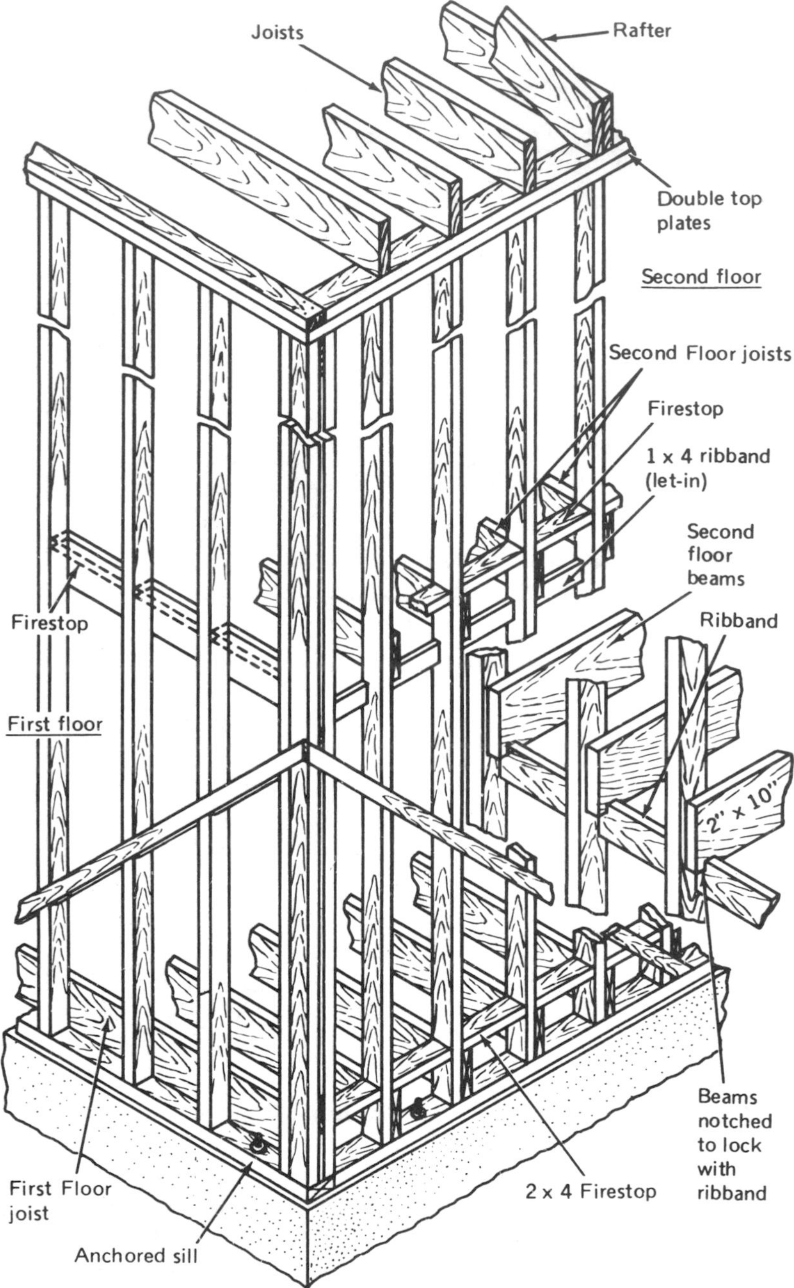

Figure 6–18 Balloon frame wall construction.

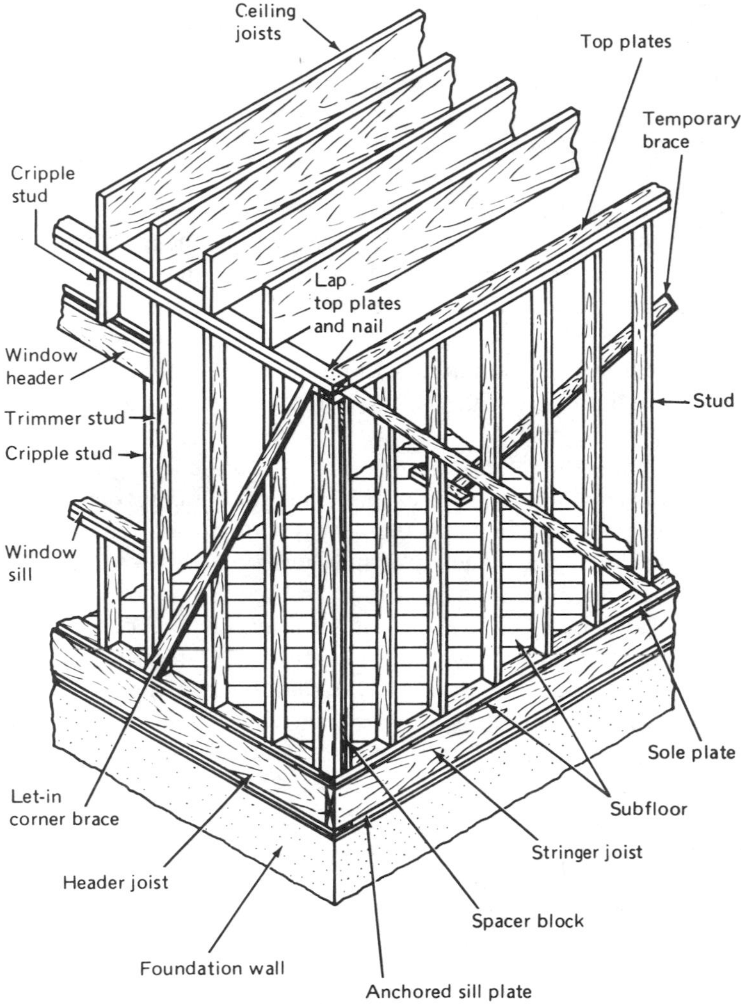

Figure 6–19 Platform, single-story, or western framing.

2. *Corner braces.* Let-in 1 × 4 corner bracing is used on each wall. (*Note:* Panels may be substituted for let-in braces.)
3. *Trimmer stud.* Trimmer studs are used either side of window or door rough openings.
4. *Headers.* Properly sized headers are installed edge up, and supported on each end by cripple studs that extend downward to the sole plate.
5. *Cripple studs.* These are used around the window opening, above the header, below the window sill, and supporting the header.

6. *Common inside–outside corner.* A common three-stud corner is constructed with scrap 2 × 4 spacers between studs.
7. *Ceiling joists.* Align over studs.

Blueprint Data and the Wall

Blueprints are the fundamental link between the architect and carpenter-builder. The set of plans relates the details necessary for the building's shape, size, and other factors. With a set of plans the carpenter constructs the framework of the building. Two of the most useful pages in the set of plans are the floor plan and elevation and detail drawings. However, the plans are not the only link between the architect and the carpenter or builder. Both know that building codes specify certain requirements and both know that the carpenter must fulfill the requirement of constructing a sound wall system.

To show the relationship of the plan to the final wall system, the next several paragraphs and illustrations examine a platform construction technique of an exterior load-bearing wall. First, the wall is shown with conventional 16-in.-o.c. studs. Then the same wall is shown with 24-in.-o.c. studs with modular spacing.

The floor plan segment shown in Figure 6–20 provides necessary dimensions for (1) wall length, (2) position of the windows and door, (3) location of offset in wall, and (4) points where interior partitions intersect with exterior wall. Usually, a scale of inches per foot is provided so that the carpenter can locate the exact position of each window and door. The sizes of windows and doors may either be listed on the floor plan (not usually), written into the elevation plan (sometimes), or provided in a list or schedule attached to one of the plans.

The elevation plan also shown in Figure 6–20 shows (1) the window shape and height, (2) door and adjoining window sidelights, (3) roof shape, (4) cornice line and design, and (5) exterior wall material (brick). Notice that the windows and door heights are aligned under the overhang of the roof. This translates to a lower header edge of 6 ft 10½ in. from the subfloor, since this is the dimension necessary for door unit installation.

The carpenter precuts a quantity of studs based on three per 4 ft of wall for the conventional 16-in.-o.c. stud spacing. In addition, extra studs are cut for trimmers and corners. Recall that trimmers are full-length studs placed either side of door or window openings. Corners are made in various ways, as Figure 6–21a to c show.

When using the conventional 16-in.-o.c. method of wall construction (Fig. 6–20), no special consideration is given to the placement of windows or doors other than to follow the plan. It therefore may be that the trimmer stud is exactly on line with 16 in. o.c. or it may be off-centered any distance. Many additional trimmers, and cripple studs as well, may be needed to complete the wall.

By requirement, the headers for all three windows will be constructed from 2 × 4s set edge up. However, the door opening is 8 ft wide and needs to have a header made from 2 × 10s. The builder may elect to use 2 × 10s for all headers

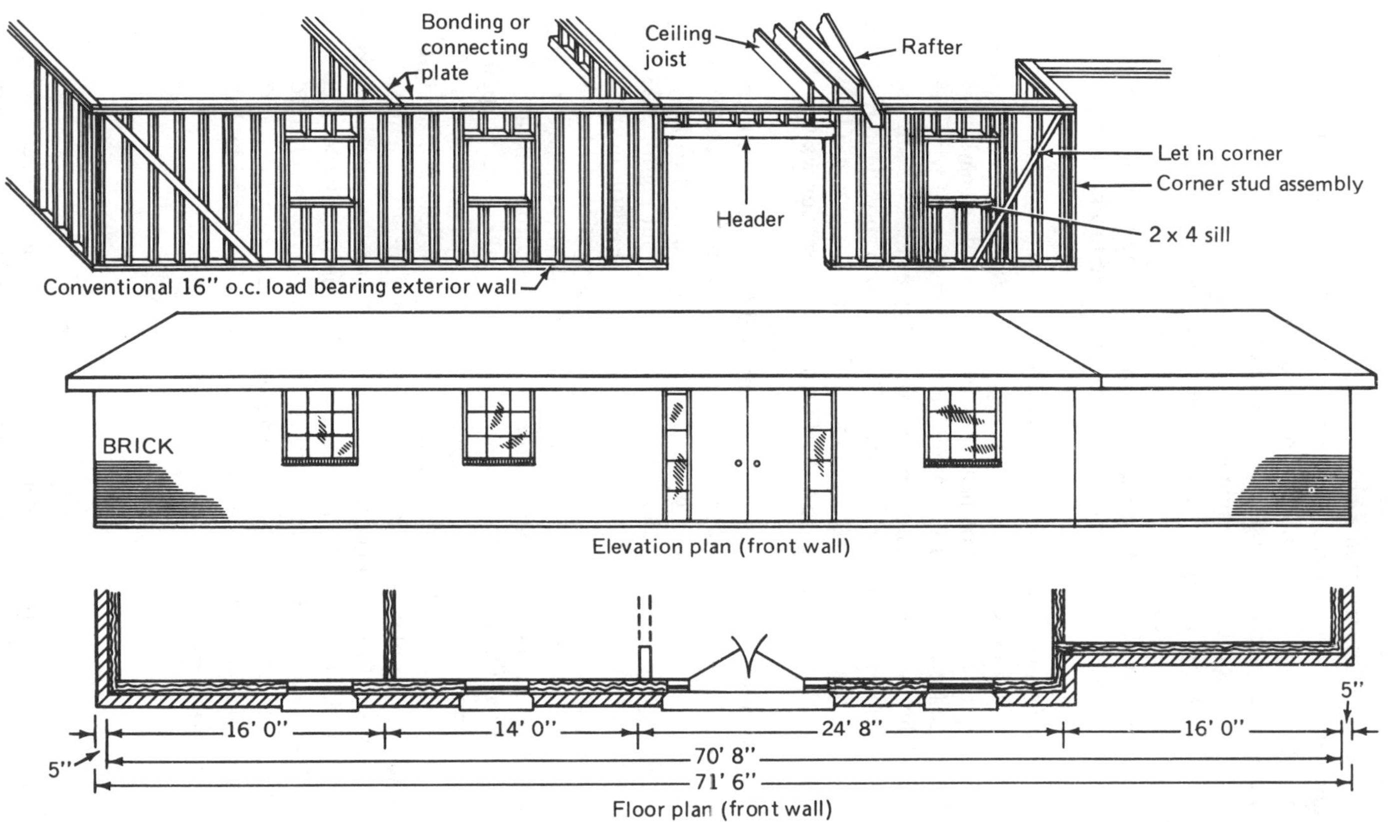

Figure 6–20 Blueprint data and framed wall: conventional 16-in.-o.c. stud separation.

for uniformity of construction even though they may exceed requirements over windows and single door openings. Each header must be end supported by 2 × 4 cripples, and cripples 16 in. o.c. need to be included between header and top plate.

Let-in corner braces are used. The one on the left is at a 45° angle. The one on the right is installed at an angle more than 45° since a window would interfere with it. If the one on the right were installed at a 45° angle, its effectiveness would be lost.

The partitions intersecting with the wall are bonded to the front wall by the *bonding* or *connecting* plate. Interior corners also need to be constructed as Figure 6–21d and e show.

The ceiling joists are aligned directly over studs or not more than a 2-in. thickness from the stud beneath. The rafters are butted to the joists, and they are either resting directly over the stud or just 2 in. from the stud beneath it.

The same floor plan (or one almost like it) can be used with the modular, 24-in. system. This system is slightly more involved than just spacing the studs on 24-in. centers. It involves a design principle of modules. A module may be a minimum of 4 ft wide, or wider, such as 6, 8, 12, or 16 ft wide. Usually, modules are designed as 4-, 6-, and 8-ft widths. Figure 6–22 illustrates the exterior wall of the floor plan in Figure 6–20.

Several changes to the floor plan are needed to make it conform to the 24-in. system. First, the window placement needs to be shifted so that the windows fall within a 4-ft module. The door unit also needs to be shifted to fall within an 8-ft module. Figure 6–22 does not show this idea very well, but Figure 6–23 does. Notice how one trimmer is always at the normal 24-in.-o.c. position in the 24-in. system.

Another change needed to be made is the last segment of the wall (right side). It is a fragment of a module (2 ft 8 in.). This should be changed to either a 6-ft module with the window as part of the module or another full 4-ft module. The plans would need redrawing. Notice that plywood or other suitable panel sheathing is used for corner bracing. If the right end of the wall were expanded to a 4-ft module, a full sheet of sheathing would be effective as corner bracing there also.

A double top plate is used in the 24-in. system for alignment and connecting partitions. A single plate may be used for partitions that are non-load-bearing. Joists and rafters are spaced 24 in. o.c. and they need to be located in line with studs as described earlier in conventional framing. A reminder needs to be made here; with joists and rafters spaced 24 in. o.c., larger members must be used. Be sure to select sizes appropriate for loads. The studs identified in Table 6–1 should carry the loads of these larger members.

INTERIOR PARTITIONS

Interior walls are classified as either load-bearing or non-load-bearing; they are also referred to as *partitions*. A partition therefore may be considered to be an assembly that is designed to serve a particular purpose. If the partition is a load-bearing type,

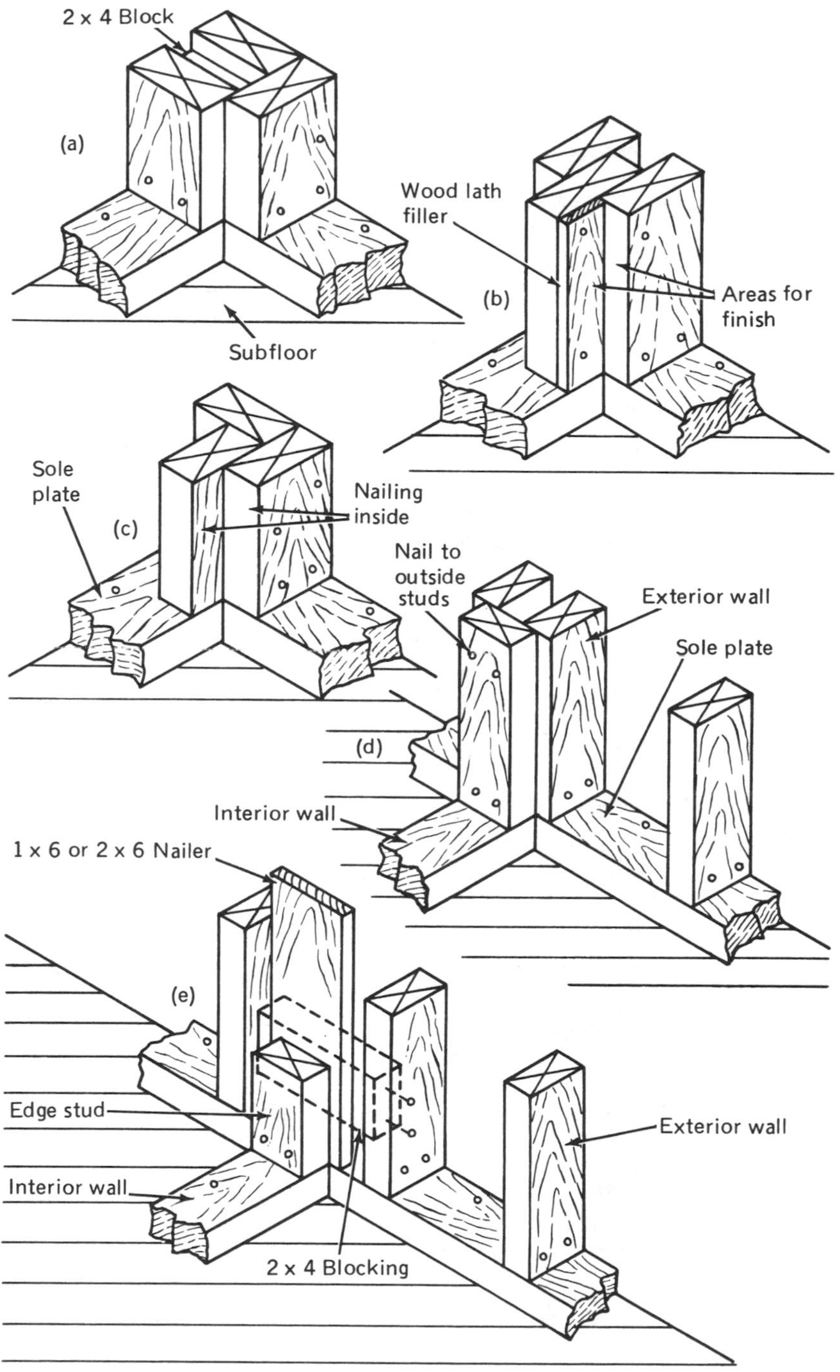

Figure 6–21 Types of corner assemblies for exterior and interior corners.

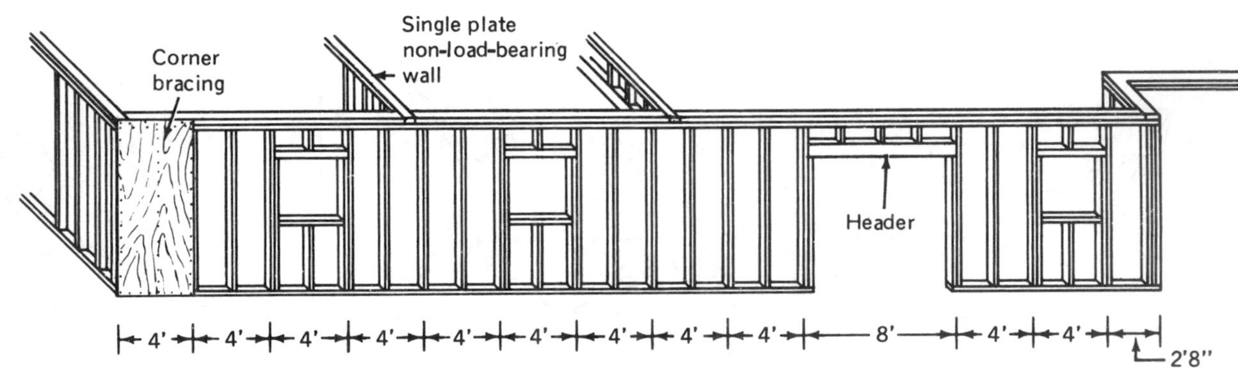

4' Modular units

24" system

Notes:

Window and door openings fall within modular units
2' 8" should be reduced to 24"

Headers for windows - 2 x 4

Header for door - 2 x 10

Figure 6–22 Framing a wall using the 24-in. modular system.

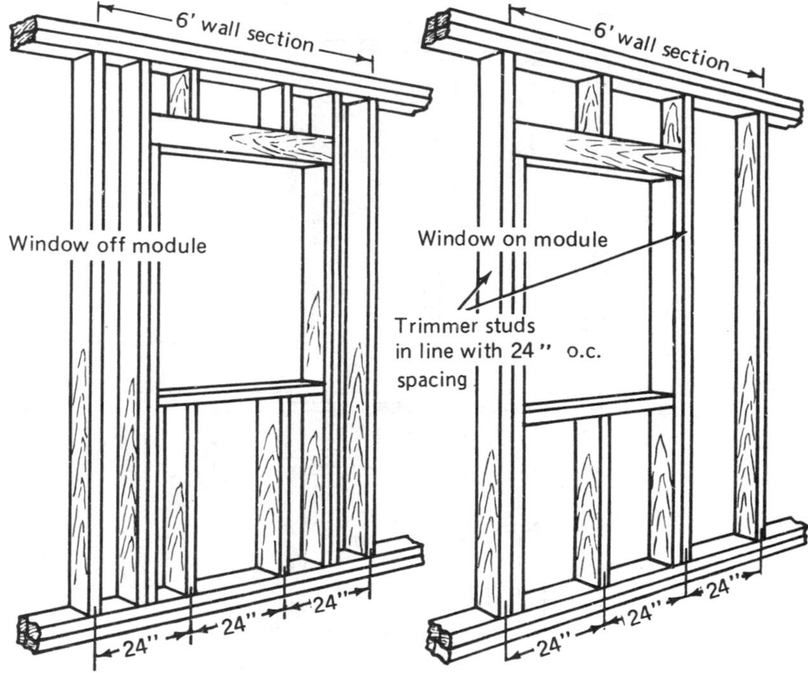

Figure 6–23 Framing window off-module and on-module.

its construction must follow the same design principles as the exterior load-bearing wall design. Headers of proper sizes must be used. Studs must be centered so that ceiling joists bear directly over them when nailed to the top plate. Double plates need to be used. Door or other passage openings must have trimmer studs and cripples installed as parts of the assembly. As a rule, no corner bracing is included in the assembly.

If the partition is non-load-bearing, its construction is simpler. It may be built with single plate at bottom and at top. If this is done, studs must be cut 1½ in. longer to allow for the one top plate versus two top plates. Headers do not support anything, so they may be made from small materials such as 2 × 4s or 2 × 6s. Cripples should be used along side door openings to provide a sound nailing surface for door jambs and door trim (see Figure 6–24). Cripple studs above headers may be omitted if drywall is installed horizontally. On the other hand, if paneling is installed vertically and a joint is made over a header, a cripple stud must be installed (Figure 6–24b).

Interior walls usually are constructed from 2 × 4s; however, they may be built using either 2 × 2s or 2 × 3s and 2 × 6s or 2 × 8s. The 2- or 3-in. wall is found in and around closet and chimney areas. They are not as sound as the normal 4-in. wall, but where little chance of external force is expected, they may prove to be space saving. The 2 × 6 or the 2 × 8 wall is usually built where soundproofing is

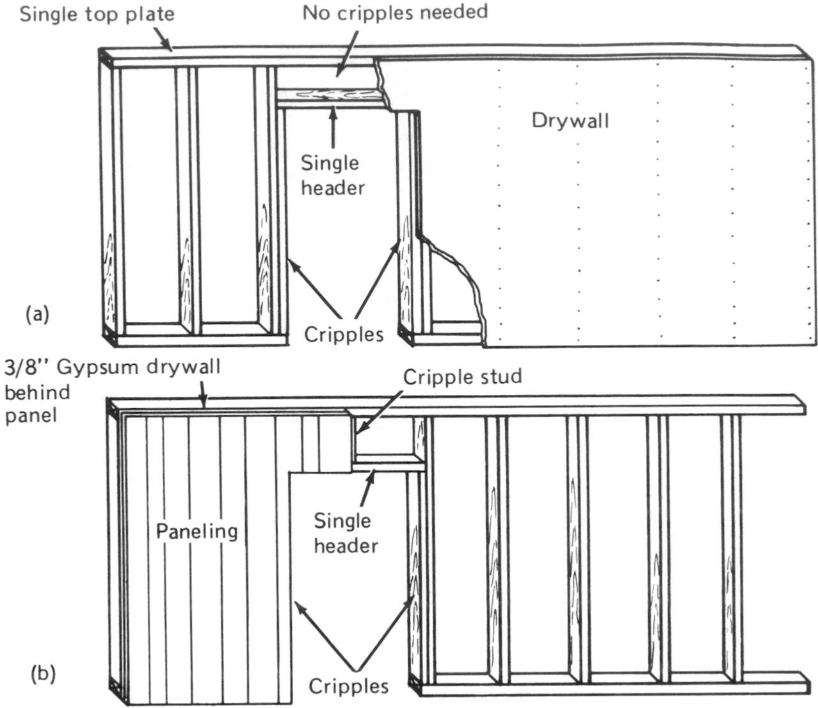

Figure 6–24 Interior partition, non-load-bearing construction details.

needed or where plumbing installations must be accommodated. Figure 6–25 shows a wall with plumbing pipes and helps explain the need for the wider plates and studs.

All the factors mentioned in the preceding paragraphs are equally important in either the conventional 16-in.-o.c. stud wall system or in the 24-in. modular system. Modules are used for partitions just as they are used for exterior walls.

SUMMARY

A wall system is an integration of many products, structural members, and sound design principles. As we point out in this chapter, no one item by itself can adequately support a load or perform a function. Collectively, however, if properly selected, installed, and secured, the products and members are united into a sound design that does all that the requirements call for.

More emphasis has been placed on the fundamental structure of the wall, its studs, header, plates, and bracing than on the products attached to it because the structure is the skeleton and it must be designed correctly. The sheathing, paneling, sidings or interior drywall, lath and plaster, and paneling all contribute to the wall's ability to perform its function. Recall that these may be resistance to wind, earth-

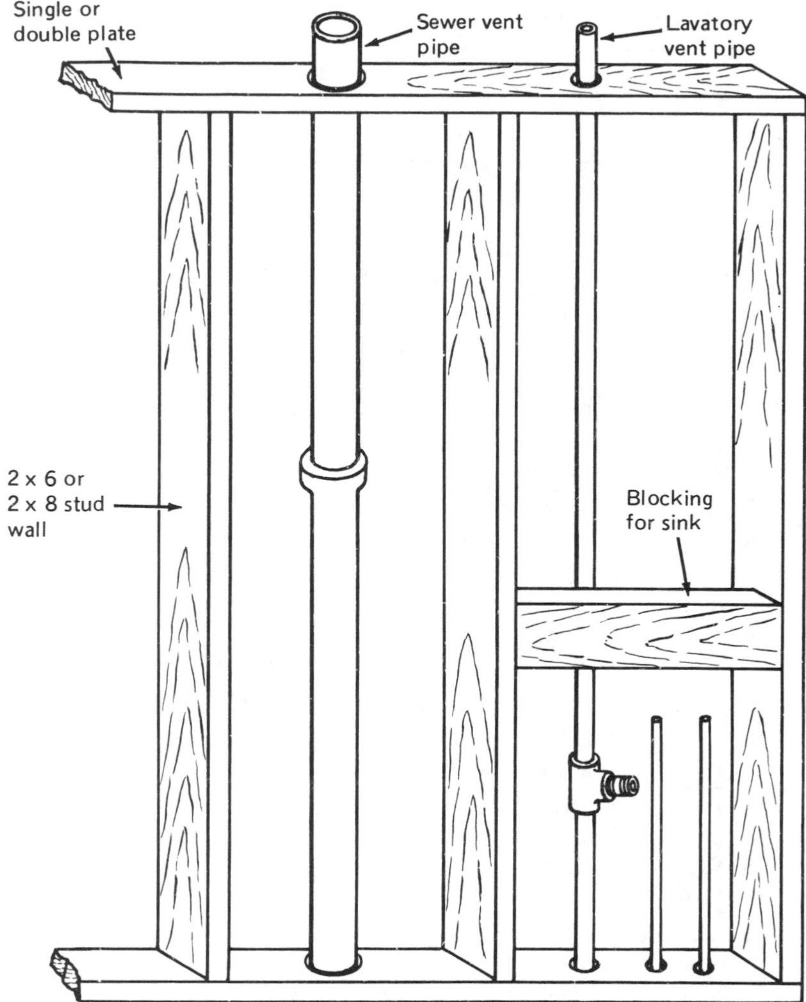

Figure 6–25 6-in. or 8-in. partition housing plumbing vent and water pipes.

quake, sound, heat, or cold. Each problem that must be reckoned with adds to the wall's quality.

Finally, a wall may be constructed conventionally with studs spaced 16 in. o.c., windows and doors located in the most desirable places, and intersecting partitions joining the outer walls wherever desired. Considerably more materials are needed to construct this type of wall than one designed in the 24-in. modular system. The modular concept of wall building reduces the quantity of materials used for its construction, thus saving time and money. It is as sound as a conventional wall design because of the method of construction. Although the conventionally designed wall will be used for many years, wall construction is moving toward the modular 24-in. system.

PROBLEMS AND QUESTIONS

1. What should the exterior wall do for a house?
2. List four purposes for an interior wall.
3. What are the five requirements that a wall must satisfy?
4. What are the two most often used on-center spacings of studs?
   - **a.** 12-in. and 16-in.
   - **b.** 16-in. and 24-in.
   - **c.** 16-in. and 32-in.
   - **d.** 24-in. and 48-in.
5. If you were using the metric system, what size would a stud be?
6. What makes a single stud bend if a section of roof is bearing on it?
7. Why don't studs in a wall bend when loads such as a joist and roof assembly are placed on them?
8. Which grade of stud would deflect the least under a load?
   a. select str **b.** No. 2 **c.** No. 3 **d.** utility
9. Draw a section of wall and label the stud, sole plate, top plate and fire-stop.
10. Can utility studs be used as primary studs in a wall with 24-in.-o.c. spacing?
11. What two purposes does the bonding plate perform?
12. Load-bearing headers must be made of ________ if the span is 62 in.
   a. two 2 × 4s **b.** two 2 × 6s **c.** two 2 × 8s
13. Is there any loss in utility if a header is placed snugly under the wall top plate? Do the jack studs bear any load or is the load passed to the trimmer studs? Explain?
14. What is the use of a trimmer stud in framing an opening for a door or window?
   - **a.** carries the load
   - **b.** reinforces and helps align the jack stud
15. A cripple stud
   - **a.** extends from floor to ceiling
   - **b.** is a short length of stud at stud o.c. spacing
   - **c.** always used over headers
   - **d.** is a full-length wall stud
16. Which bracing of a wall is more effective?
   - **a.** 1 × 6 sheathing
   - **b.** let-in corner boards
17. Which provides the greatest protection against racking forces?
   - **a.** plywood glued to studs
   - **b.** diagonal sheathing
   - **c.** 1 × 4 let-in braces

18. Name four qualities that ½-in. drywall adds to a wall.
19. Describe the main distinctive feature of balloon framing. Are fire-stops used?
20. What are two other names for western framing?
21. Using the plan in Figure 6–20, how many common-length studs would you cut for framing the front wall if they are spaced 16 in. o.c.? How many for exterior and interior corners?
    Studs **a.** 47 **b.** 54 **c.** 60
    Corner studs **a.** 9 **b.** 12 **c.** 16
22. Where would headers be used (Figure 6–20)
23. Where would cripple studs be used? (Figure 6–20)
24. What alteration would you make to the plan in Figure 6–20 to make it fit a model modular 24-in. plan?
25. How many trimmer studs would you save if you placed the three windows and door (Figure 6–20) on-module rather than off-module?
    a. none **b.** 2 **c.** 4 **d.** 6

PROJECT

Using Figure 5–2 and the following additional information, prepare a materials list sufficient to frame the walls of the house in the floor plan. The results of completing this project will prove that you have, first, understood much of the data in Chapter 6, and second, have a clear mental image of the construction details necessary to frame-in the house. Shall we begin?

Dimension of walls (feet)	
Across the back of the house	
Family room	18 × 14
Kitchen	12 × 9
Dining	10 × 9
Master bath	11 × 6
Master bedroom	18 × 16
Left side outside wall	
Family room	14 × 18
Toilet and stairwell	4 × 18
Garage	24 × 18
Across the front	
Garage	18 × 24 (16-ft door opening)
Living room and entrance	22 × 27 (14-ft window and 4-ft door unit)
Third bedroom	13 × 12
Corner bedroom	16 × 16
Right sidewall	
Master bedroom	16 × 18
Corner bedroom	16 × 16
Main baths and closets	11 × 11

Dimensions of related materials

Headers on outside walls are made of 2 × 12s and no cripple studs are used.
Headers on inside walls are 2 × 6s and cripple studs are used.
Single lower and double top plates are used.
Allowance for cripple studs below window sills is 150 ft.

A. Common Studs and Common-Length Trimmer Studs

1. Calculate the common studs at 16 in. o.c. and full-length trimmer studs needed. Since windows up to 4 ft wide require cripple studs, ignore these windows in making your calculations. After defining the wall's length, you may use either formula to calculate the number of common studs:

Formula 1. $\# \text{ common studs} = \dfrac{\text{number of lin. ft wall} \times 3 + 1}{4}$

Formula 2. $\# \text{ common studs} = \dfrac{\text{number of lin. ft} + 1}{1.33}$

a. Back wall __________
b. Left outer wall __________
c. Across the front __________
d. Right outer wall __________
e. Right garage wall and left living room __________
f. Interior garage wall (rear), bath, stair well, and closet __________
g. Interior kitchen walls __________
h. Interior dining room wall (right), and entrance wall (allow 5 ft for closet opening and 3 ft for hall) __________
i. Third bedroom __________
j. Partition between master and corner bedroom __________
k. Bathroom walls and closets (master bedroom) __________

Subtotal __________

2. Calculate the number of exterior and interior corners needed (count only common-stud-length pieces). Since you made allowance for all wall studs above and you know that an exterior corner needs three studs per corner, you should only add one more for these corners. This rule does not apply to inside corners

a. Exterior corner studs__________
b. Interior corner studs __________

Subtotal __________

3. Total number of common and common-length trimmer studs plus 150-ft allowance for cripples in 8-ft lengths __________

B. 2 × 4 Plates

Since the first element of your calculations in Part A required the length of the wall, the amount of plate materials can easily be determined with those data. This house will use a single sole plate and double top plates.

Formula: number of lin. ft plate material = total length of all walls in feet × number of plates to be used

= __________ × __________

= __________

C. Headers

Added information necessary to define the header materials is as follows: All windows except the kitchen, bath, and living room are standard 36-in.-wide double hung. The kitchen and bath are 30-in.-wide double-hung and the living room is a 14-ft window unit with mullions. Doors are 32 in. wide except for the front door unit, which is 4 ft wide. The contractor wants to use 2 × 12s for headers in the outer wall construction and no cripples over the headers. He also plans to use 2 × 6 headers on all interior door openings and closet openings. Remember that all headers are doubled!

1. Calculate the number of feet of header materials for 2 × 12 materials.
   a. Number of windows under 4 ft wide = _____ × length of each in feet _____
      = _____ lin. ft
   b. Garage door header = _____ lin. ft
   c. Living room window = _____ lin. ft
   d. Front door header = _____ lin. ft
      Total 2 × 12s = _____ lin. ft
2. Calculate the number of feet of header materials for 2 × 6 materials.
   a. Number of interior door headers = _____ × length of each in feet _____
      = _____ lin. ft
   b. Number of closet door headers = _____ × length of each in feet _____
      = _____ lin. ft
   c. Master bedroom closet = _____ lin. ft
      Total 2 × 6s = _____ lin. ft

This project was important if you never before calculated the framing materials needed to frame in the walls of a room or house. In doing the project this way you should have been able to recognize how the various dimensions were used in determining the various quantities. However, if you have a full set of floor plans with all dimensions readily available, you can use the more direct route to identifying the number of studs. Simply add up the total number of wall lengths, then divide the sum by 4 and add corner and trimmer studs. Then make allowances for large openings.

Answers:

Part A: **1. a.** 69 ft = 52 common + 10 trimmer = 62
b. 42 ft = 32 common + 2 trimmer = 34
c. 44 ft = 33 common + 7 trimmer = 40
d. 36 ft = 27 common + 4 trimmer = 31
e. 28 ft = 21 common + 0 trimmer = 21
f. 44 ft = 33 common + 6 trimmer = 39
g. 21 ft = 16 common + 2 trimmer = 18
h. 24 ft = 18 common + 1 trimmer = 19
i. 50 ft = 38 common + 4 trimmer = 42
j. 16 ft = 12 common + 0 trimmer = 12
k. 30 ft = 23 common + 5 trimmer = 28
Subtotal 346
2. a. 11 exterior corners (5 on outside walls and 6 on inside walls) = 23
TOTAL: 369

b. 16 interior corners (8 in the outside walls and 8 in interior walls) = 48
Subtotal 71

3. Total number common + trimmers + allowance for cripples at 8-ft lengths (19) = 436

Part B:

404 × 3 = 1212 lin. ft

Part C: **1.** **a.** 16 × average length 3.5 ft = 55 lin. ft, doubled = 110 lin. ft
b. 1 = 18 lin. ft, doubled = 36 lin. ft
c. 1 = 16 lin. ft, doubled = 32 lin. ft
d. 1 = 4.5 lin. ft, doubled = 9 lin. ft
Total 2 × 12s 187 lin. ft

2. **a.** 14 × average length 3 ft = 42 lin. ft, doubled = 84 lin. ft
b. 3 × average length 6 ft = 18 lin. ft, doubled = 36 lin. ft
c. 1 @ = 12 lin. ft, doubled = 24 lin. ft
Total 2 × 6s 144 lin. ft

Chapter 7

The Roof—Part I: The Common Rafter

In this chapter and the next three chapters we discuss the various rafters used in roof construction. The common rafter is defined first so that many of the properties of all rafters may be identified and a sound understanding of the function of a rafter is established.

There are many aspects about a common rafter that are important to know so that they may be used by the carpenter. He or she needs to know where to use a common rafter in a roof assembly versus using a hip or jack rafter. Terms such as pitch, run, rise, and others must be thoroughly understood, as well as the forces the rafters must sustain and the forces placed on walls. Then, too, the carpenter must be able to calculate the rafter length after defining its rise and run from plans. Having obtained all this information, the carpenter then must select the proper rafter grade, size, species, and finally determine how to locate it properly on the wall.

DEFINING A COMMON RAFTER'S LOCATION IN A ROOF ASSEMBLY

Since a roof frame may be made from a variety of different types of rafters, it is important to recognize where common rafters are used. Figures 7-1 through 7-5 illustrate several of the most common roof styles. The shaded area identifies the area in the roof where the common rafter is used.

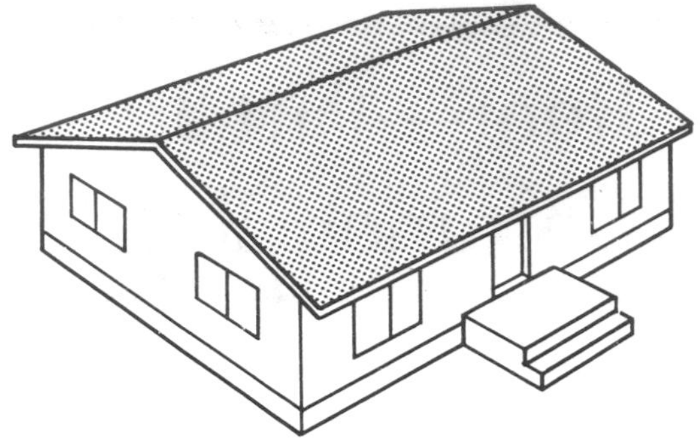

Figure 7–1 Gable roof made entirely with common rafters.

The roof in Figure 7–1 is the common gable roof design. It is constructed entirely with common rafters as the shaded area indicates. The roof in Figure 7–2 is the *shed* or *low-level pitch* roof design. Each rafter is a common rafter, as the shaded area shows. The rafters in this style of roof also act as ceiling joists, since ceiling materials are attached to them.

The *gable* roof shown in Figure 7–3 is constructed with common rafters, as are the dormer roofs. The front dormers use common rafters in the shaded area. The area left white is not constructed from common rafters but uses a type called *jack rafters.*

The *hip-and-valley* roof shown in Figure 7–4 does make use of the common rafter, where the shaded areas show two different locations. They are the areas where rafters are not connected to either hip or valley rafters. This means that the common rafter attaches at the roof peak and rests on the wall plate.

The roof in Figure 7–5 is the *gambrel* roof design. It is constructed with common rafters. One set of common rafters is cut to fit the very steep pitch of the lower section; the other set of common rafters is cut to fit the low pitch of the upper section. The dormer's roof is made with common rafters in shed fashion.

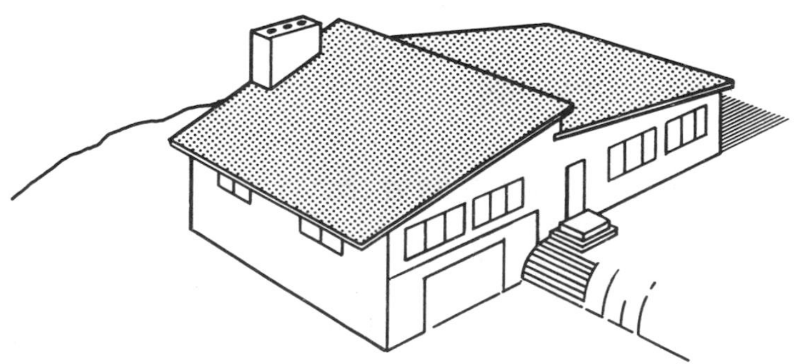

Figure 7–2 Shed or low-level pitch roof, made entirely with common rafters.

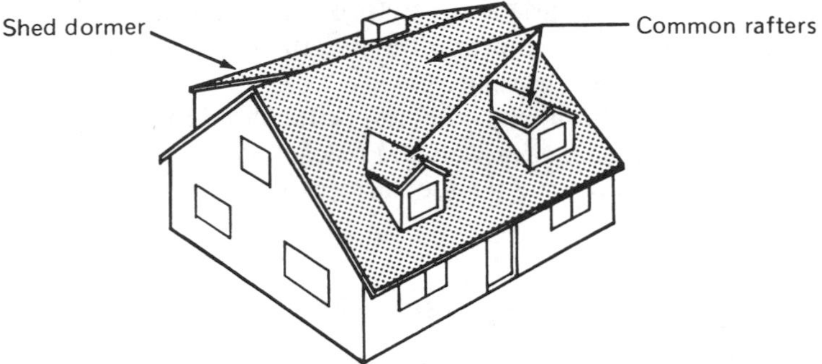

Figure 7–3 Dormer roofs, using common rafters for part of their makeup.

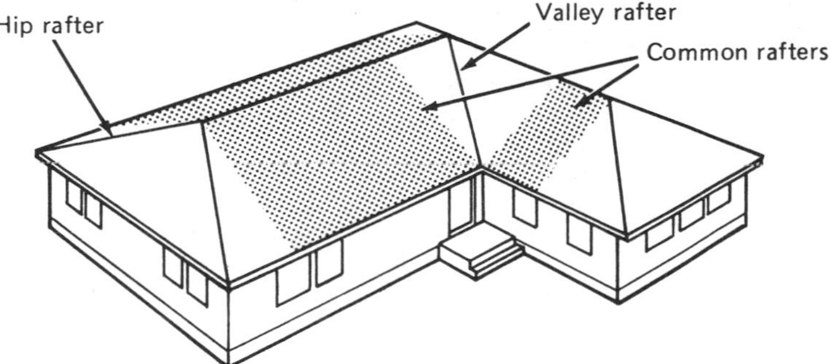

Figure 7–4 Hip and valley roof, using common rafters outside the area of hip or valley.

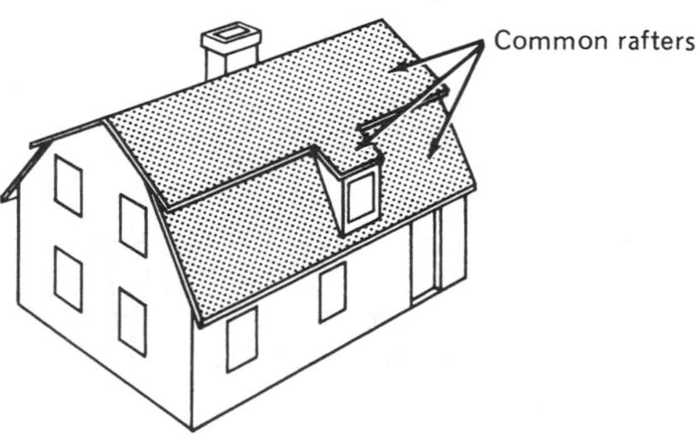

Figure 7–5 Gambrel roof design using common rafters.

There are other varieties of styles of roofs that are really combinations of one or more of the designs shown in the five figures. They, too, may use one or more common rafters in their frame.

A clear definition of a common rafter should now be apparent. It is the one that attaches to the exterior wall and reaches to the ridge or peak of the roof. This definition even holds true for the dormer common rafter.

PITCHES AND HOW THEY ARE DEFINED

The use of a specific pitch stated in fractional value has largely been replaced with the more general terms a *low-pitched* roof, a *high-pitched* roof, or a *steeply-pitched* roof design. However, the term "pitch" is understood by all who refer to pitch in any terms. Therefore, a fundamental understanding must be the common ground. All fractional notations (¼, ⅓, ½, etc.) of pitches of roof design relate to its flatness or steepness in terms of percentage or fractional value of total span versus total rise.

The pitch may be a fractional value of a full pitch as shown in Figure 7–6. When related to a framing square, the pitch is obtained by doubling the 12 in. (shown in Figure 7–6), then dividing the 24 in. into the 4 in.; for example:

$$\tfrac{1}{6}\text{ pitch} = \frac{4\text{ in.}}{24\text{ in.}} = \frac{1}{6}$$

$$\tfrac{1}{2}\text{ pitch} = \text{double 12-in. run} = \text{24-in. span}$$

$$\frac{\text{12-in. rise}}{\text{24-in. span}} = \frac{1}{2}$$

The pitch table (Figure 7–6) is standard; but it, too, is not easily understood. The table indicates, for example, that a ½ pitch has a run and rise of 12. The pictorial just below the table translates the table data. Notice that the rise and run are indeed 12 ft in this case. However, the span is twice the run or 24 ft, and when this value is used, the pitch can be calculated.

$$\frac{\text{12-ft rise}}{\text{24-ft span}} = \frac{1}{2}\text{ pitch}$$

Now look at the ⅓ pitch column in the pitch table. It shows a run of 12 and rise of 8. If the rise is ⅓ of the span and the rise is 8 ft, the span must be 24 ft. The pictorial beneath the pitch table shows this. The rise is indeed one-third the total span. Therefore, the mathematics is correct.

$$\frac{\text{8-ft rise}}{\text{24-ft span}} = \frac{1}{3}\text{ pitch}$$

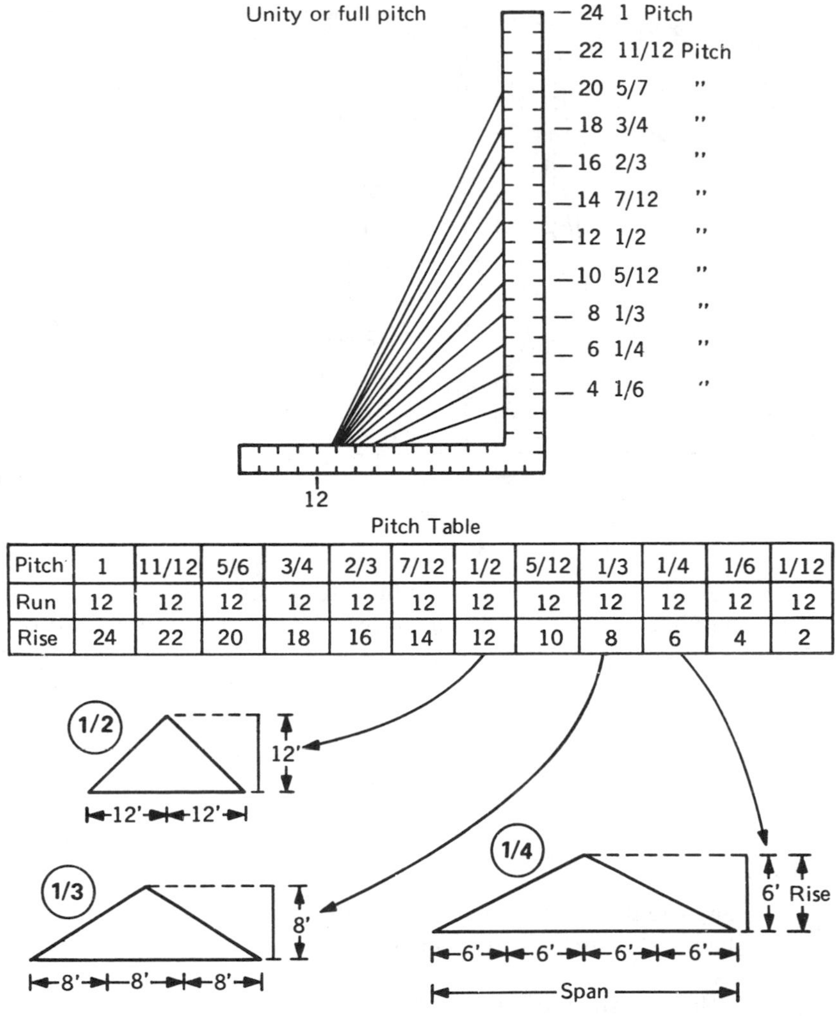

Pitch Table

Pitch	1	11/12	5/6	3/4	2/3	7/12	1/2	5/12	1/3	1/4	1/6	1/12
Run	12	12	12	12	12	12	12	12	12	12	12	12
Rise	24	22	20	18	16	14	12	10	8	6	4	2

Figure 7–6 Pitch table, square and span.

The last picture illustrates the ¼ pitch. Notice that the rise is indeed one-fourth the span.

$$\frac{\text{6-ft rise}}{\text{24-ft span}} = \frac{1}{4} \text{ pitch}$$

Not all pitches need to be a fraction of a full pitch. Figure 7–7 illustrates that a pitch may indeed be a multiple of a unity or full pitch. The number 24 in the figure has been selected to simplify the understanding. A full pitch requires a span

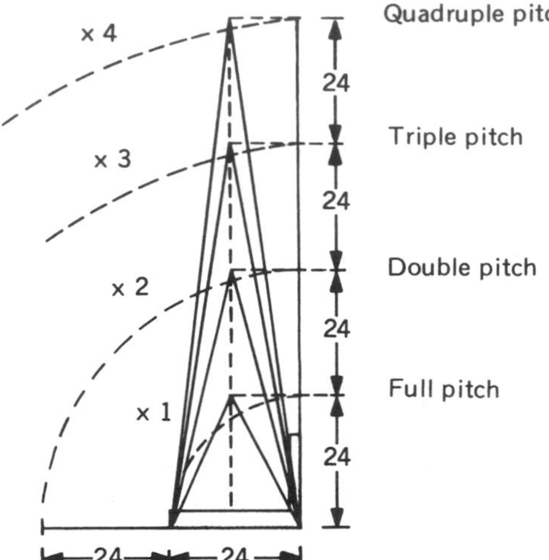

Figure 7–7 Steeple pitches.

and rise equal to each other. A double pitch requires a rise twice the value of a span. If using Figure 7–7 as an example, a rise of 48 ft and span of 24 ft result in a double pitch.

$$\frac{\text{48-ft rise}}{\text{24-ft span}} = 2 \text{ pitch}$$

Example

Suppose that the building span was 10 ft and a double pitch was to be designed and built. What would its rise be in feet?

Solution Transposing the formula,

$$\frac{\text{rise}}{\text{span}} = \text{pitch}$$

to find rise,

$$\text{pitch} \times \text{span} = \text{rise}$$

Thus

$$2 \times 10 \text{ ft} = 20 \text{ ft}$$

The total rise of the roof would be 20 ft. Pictorially, the layout would look as in Figure 7–8. For each foot of run the rise is 4 ft. Since the run is one-half the span (5 ft) the total rise is 20 ft, as the math showed.

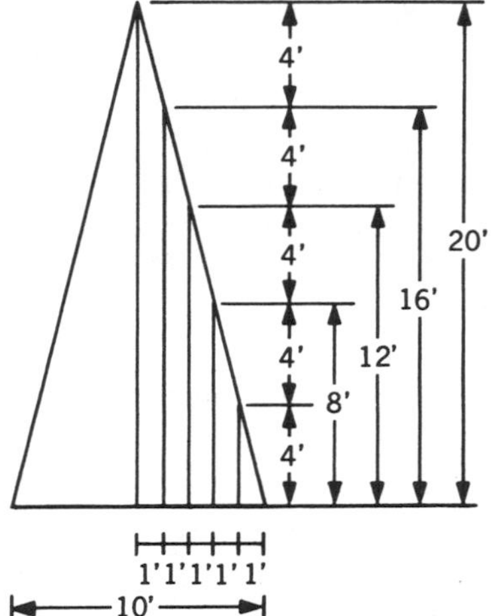

Figure 7-8 Quadruple pitch layout showing run and rise.

Refer once again to Figure 7-7 and notice that a pitch may also be triple or quadruple. These very steep pitches are frequently used on steeples of churches and on the lower section of the gambrel or mansard roof design. The steep incline on contemporary structures such as townhouses is now referred to as a wall whenever its angle approaches 90° from the horizontal.

The current use of pitch in blueprints and elevation drawings is shown in Figure 7-9. It is a triangle located just above one of the sloping rooflines. The one shown indicates that the rise is 4 in. for each 12 in. of run; this is commonly stated as "The roof has a pitch of *4 in 12-in.*" If there is more than one pitch, as in the gambrel roof design, for example, two triangles would be included in the elevation drawings.

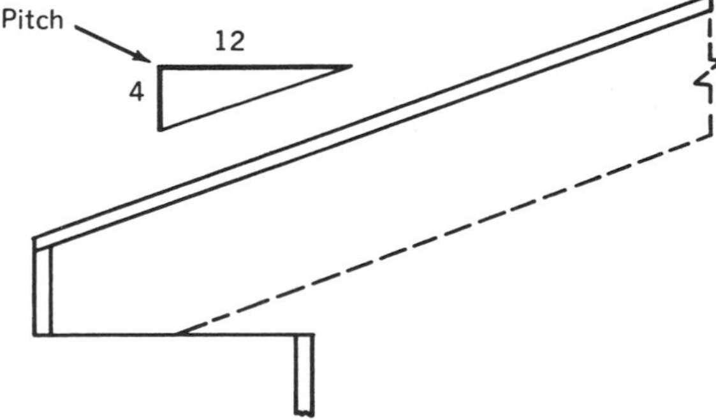

Figure 7-9 Pitch notation on elevation drawings.

TERMS USED WITH THE COMMON RAFTER AND THEIR LOCATIONS

There are several terms used in describing parts or special cuts on a common rafter. Each contributes to understanding some part of the rafter and is part of the terminology of carpentry. They are all shown in Figure 7-10 and defined below.

1. The *overhang* or *rafter tail* is the section of rafter that extends out beyond the exterior wall line. This part of the rafter is used as part of the cornice of the building. The vertical cut (a), and many times the horizontal cut (b), are made on the end of the rafter tail. The rafter tail may project any distance past the wall; however, the most common overhangs are 6, 12, and 24 in.
2. The *bird's mouth* (c) is a right-angle notch that is cut in the underside of a rafter. It is designed to allow the rafter to seat properly on the wall. The horizontal cut (c′), also named the heel, seat, or bottom cut, must be made accurately because the rafter and roof loads bear an the wall through this contact surface.
3. The *crown* (d) in a rafter is the natural bend through its length when sighted along the narrow or 2-in. edge. The use of the crown is always an important part in rafter layout and roof construction. All crowns must be set toward the sky or "crown up" so that the roof dead and live loads are borne better.
4. A *dead load* (e) is not exclusively an element of a rafter. It is the accumulated weight of the rafter plus the roof materials resting on it. These materials are the roof sheathing, tar paper, and shingles.
5. The *live load* (f) is the weight placed on a roof by the workers, snow, ice, or the like resting on the roof for short periods.

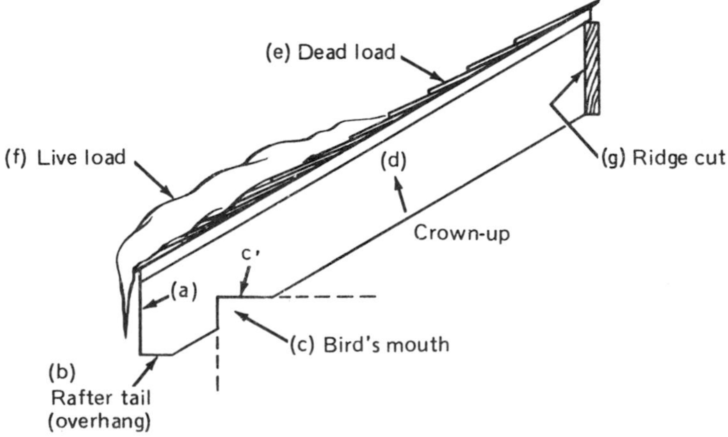

Figure 7-10 Common rafter terms and their locations.

6. The *ridge cut* (g) is a vertical cut at the top of the rafter. This cut must be made so that the entire surface attaches against the ridge board.

FORCE OF RAFTER AND LOADS PUSHING AT THE SUPPORTING WALL

The roof, especially the gable roof, which is made from common rafters, develops considerable force unless proper steps are taken to reduce it. The weight of the rafters and other roofing materials try to settle straight down; but because they are A-shaped and fastened at the plates of the wall the force is directed to the wall. Figure 7–11 illustrates the idea just expressed as well as the final outcome if allowed to happen.

There are three methods that are used to overcome this condition. A *collar beam* or *collar tie* as shown in Figure 7–12a is nailed horizontal to the rafters and about 24 in. down from the ridge. This piece (1 × 4 or 1 × 6) is nailed to every other set of common rafters or a maximum of 48 in. on center. The result is that the ridge is held firmly in place and the loads on the roof are partially absorbed by the collar beam. The remainder of the loads are concentrated on the wall plate.

The second method used to overcome the remaining force is the use of ceiling ties between outer walls. L-shaped assemblies (Figure 7–12b) made from 2 × 4s and 2 × 6s are nailed together and to opposite plates before the rafters are installed.

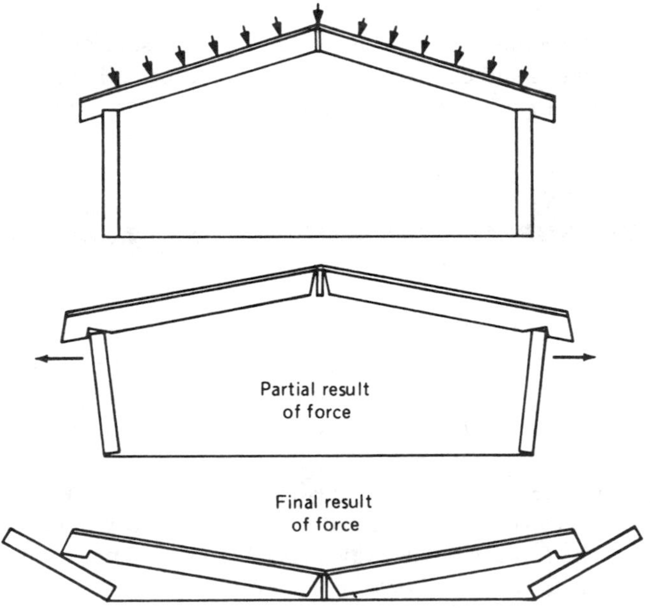

Figure 7–11 Results of forces when roof is improperly braced.

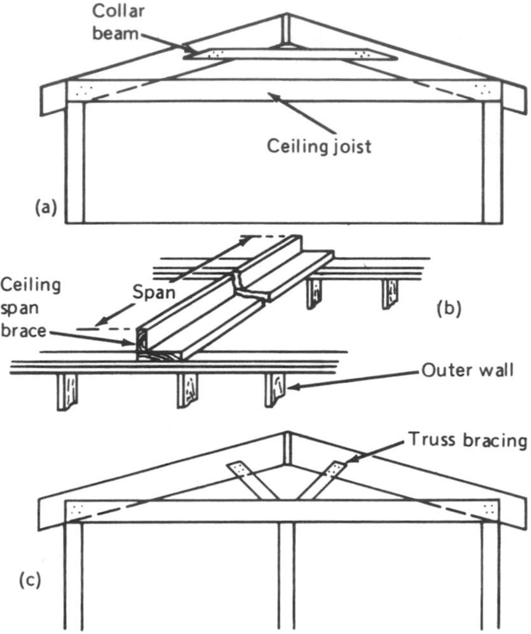

Figure 7–12 Bracing techniques: collarbeam, spanning brace, and truss bracing.

The third method is the use of a truss brace shown in Figure 7–12c. It is installed from bearing wall midway through the building span at a 45° angle to rafters. Most of the time collar beams are sufficient, especially if ceiling joists are well nailed to the plates and each other. However, if especially large rafters are used and heavy deposits of snow are the rule rather than the exception, a combination of the three methods is recommended.

PROPER MEASUREMENT POINTS FOR DETERMINING RAFTER LENGTH

Thus far in this chapter the discussions and descriptions have centered on the roof assembly using the common rafter. This section begins the study on the rafter design. When rafters were beams hewn from logs, the measure of the rafter length was taken through the center of the width. It was understood that this measuring technique provided a more reliable length since edge widths were not always parallel and consistent. Today, machined beams and rafters are uniformly sized. This means that the crown edge of the rafter is the line to use in determining rafter length. The use of the top edge in no way alters the basic principle of determining rafter length.

To illustrate the proper points of measurement, and at the same time verify the comment made in the preceding paragraph, carefully study Figure 7–13. One of the first requirements in determining the rafter length is to determine the *run*. This

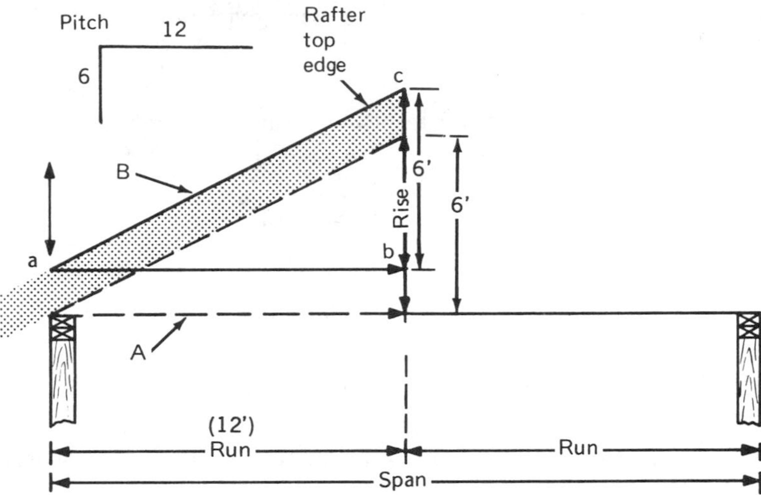

Figure 7–13 Laying out the rafter from its proper points of measure.

is equal to one-half of the *span,* as shown in Figure 7–13 by the dashed line labeled A, beginning at the outer edge of the wall and extending the full run. The other dashed line represents the length of the rafter without considering any rafter tail. The rise is measured as 6 ft. Since the run is 12 ft and rise is 6 ft, these measurements do correspond to the pitch of *6 in 12.*

Now note carefully that if a perpendicular line is drawn from the outer wall line up through the area of the rafter width to its top edge, a second set of lines (solid) can be drawn. These lines parallel the dashed lines. The result is that the rise is not changed; it is still 6 ft.

Two things are revealed: (1) the rafter length can be measured along the rafter's top edge; and (2) the rafter length is the longest line of the *right triangle* and is called the *hypotenuse* in mathematics.

The importance of knowing that the rafter length is equal to the hypotenuse of a right triangle is brought out dramatically during the estimating phase of bidding on the construction job. By sitting at the desk the carpenter-builder is able to calculate the rafter length to determine what size lumber they will be cut from. How does he do this? The formula that must be used was developed as the *Pythagorean theorem* centuries ago. It looks like this:

$$ac = \sqrt{ab^2 + bc^2}$$

Translated to rafter terms, it looks like this:

$$\text{rafter length}^* = \sqrt{\text{run}^2 + \text{rise}^2}$$

*Does not include overhang or rafter tail or reduction for one-half the ridge thickness adjustments.

To calculate the rafter length from values given in Figure 7–13, the problem looks like this:

$$\begin{aligned} \text{rafter length} &= \sqrt{\text{run}^2 + \text{rise}^2} \\ &= \sqrt{12^2 + 6^2} \\ &= \sqrt{144 + 36} \\ &= \sqrt{180} \\ &= 13.4 \text{ ft or } 13 \text{ ft } 3 \text{ in.} \end{aligned}$$

Now here is where a calculator comes in handy.

1. Square the length of the run and store it in memory or write it down.
2. Square the rise after first calculating it.*
3. Add the squared value of the run to the squared value of the rise and use the square-root button ($\sqrt{}$) to find the rafter length.
4. Convert the decimal portion of the result to inches.

The procedure used on most calculators goes like this:

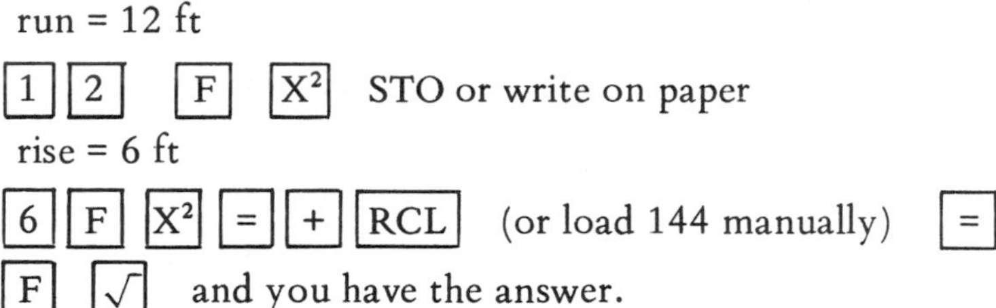

Some calculators that have function buttons for X^2 do not need the step F.

With the rafter length known, any overhang can be added to obtain its full length. This length of the rafter may be retained for use on the job where it can be measured with a tape. However, the carpenter's square is frequently used.

A well-designed carpenter's square provides tables on one of its surfaces for the hypotenuse values per foot of run for many different pitches. Before examining the variety of them, it is important to have an understanding of their origin and accuracy. To help illustrate this subject, Figure 7–14 is provided.

For a rafter run of 10 ft, there is a 5-in. rise per foot, for a total rise of 4 ft 2 in. (50 in.). Mathematically, the hypotenuse of each 12 in. is 13 in.:

*Total rise = rise per foot run × number of full and partial foot runs (e.g., 5 in. × 10 ft 6 in. = 5 in. × 10½ ft = 52½ in. rise).

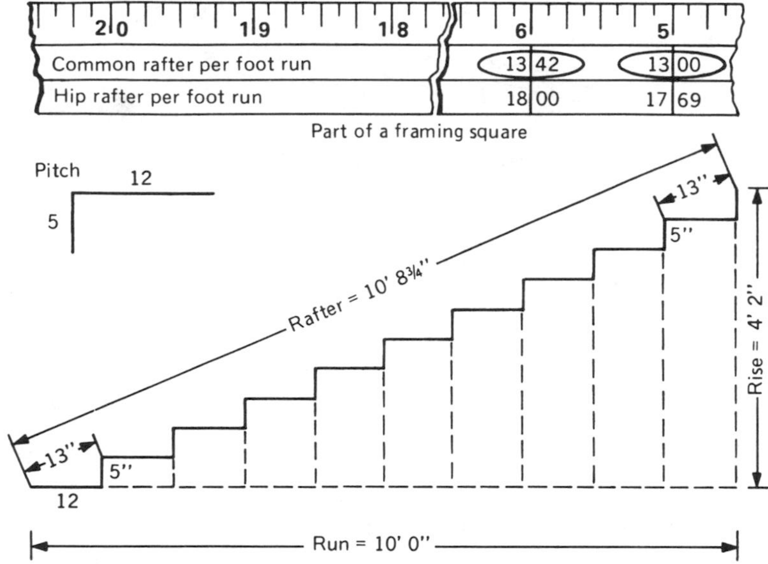

Figure 7–14 Framing square numbers correspond to the length of the rafter for each foot of run.

$$
\begin{aligned}
\text{rafter length per foot of run} &= \sqrt{\text{run}^2 + \text{rise}^2} \\
&= \sqrt{12^2 + 5^2} \\
&= \sqrt{144 + 25} \\
&= \sqrt{169} \\
&= 13.00
\end{aligned}
$$

Notice the partial table found at the top of Figure 7–14. Tables of this kind are found on many different types of framing squares. Under the 5-in. mark is the figure 13.000. This is the value that was found by calculations. If a *6-in-12* pitch were used, the value 13.42 under the 6 is the rafter length for each 12 in. of run of a common rafter.

To assist carpenters and builders with the calculation of rafter lengths, especially during estimating and even on the job, Table 7–1 and the corresponding drawings in Figure 7–15 provide lengths per foot of run of the most common pitches.

TABLE 7–1 LENGTH OF COMMON RAFTERS PER FOOT OF RUN

Rise (in.)	3	4	5	6	7	8
L/ft run (in.)	12.37	12.65	13.00	13.42	13.89	14.42
Rise (in.)	9	10	12	15	18	24
L/ft (in.)	15.00	15.62	16.97	19.21	21.63	26.83

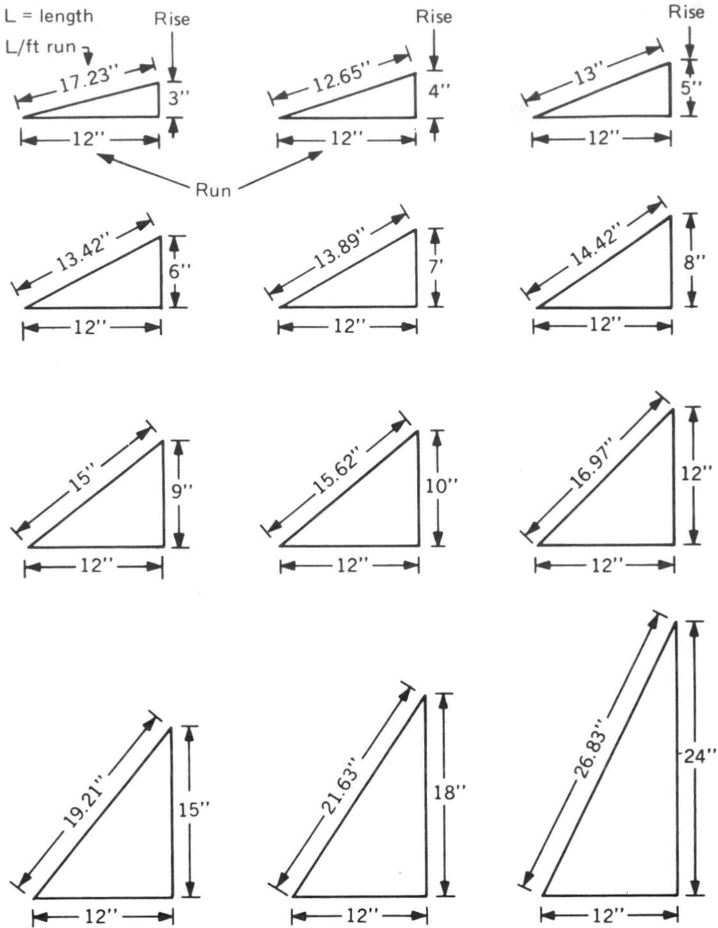

Figure 7-15 Rafter length per foot of run for all commonly used rises.

COMMON RAFTER LAYOUT, PHYSICAL POSITIONING

Review Figure 7-16 until the following data have been found:

1. The building span line
2. The rafter run line
3. The rafter rise line
4. The rafter slope line
5. The 2 × 4 block that represents the wall's plate

Objective: Simulate a rafter's position by using the wall (or driveway) as a reference area and formally lay out and fit a rafter to the layout.

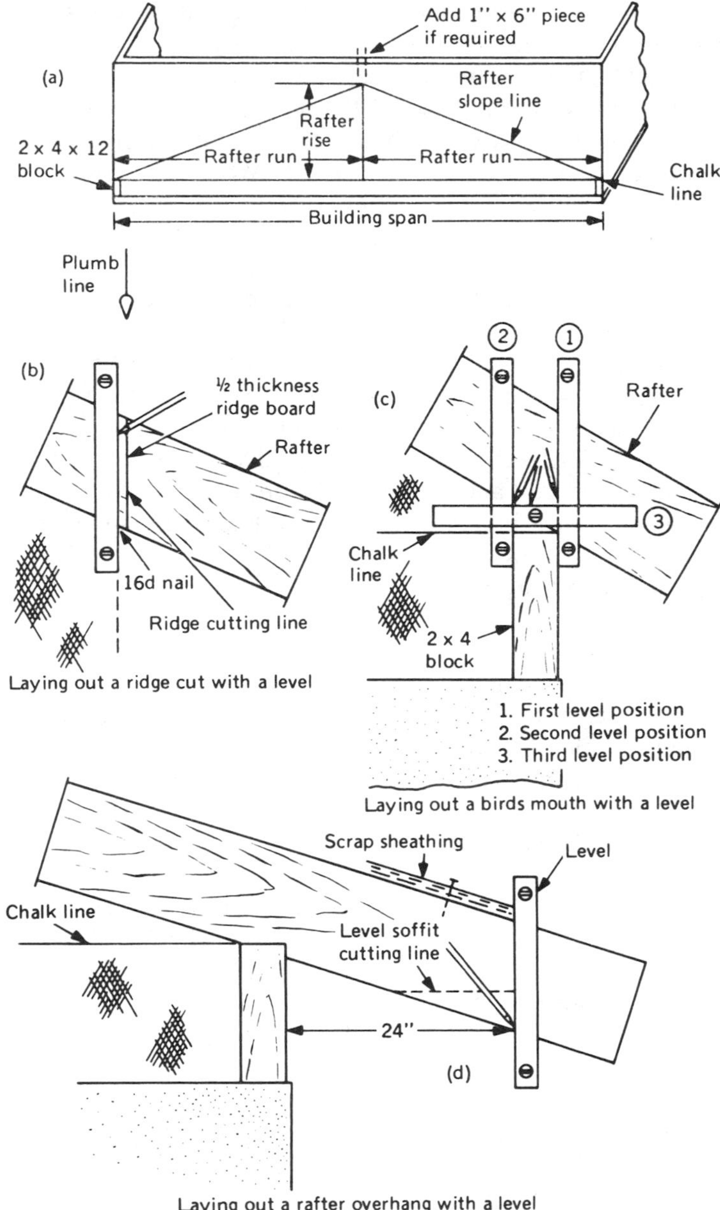

Figure 7–16 Common rafter layout: physical position.

A plate-reference chalkline needs to be snapped from 12 to 18 in. above the sill of the wall. This allows sufficient height for the projection of the portion of the rafter that will overhang the wall line and become the rafter tail.

Calculate the total rise of the rafter by using the formula: rise × rafter run = total rise in inches. The line for the rise must be perpendicular to the baseline, so a plumb bob can be used effectively. Measure the rafter run along the baseline. Suspend the plumb bob above the mark on the baseline and mark the upper end of the plumb bob's line. Snap a line connecting both points. If the total rise is greater than the height of the wall, temporarily nail a 1 × 6 or 2 × 6 approximately centered where the vertical line is to be snapped. Snap the line on the board. Measure *up* from the baseline and mark the total rise. Then snap the slope (hypotenuse) lines between the total rise mark and the outer corners of the baseline. Next, nail a scrap piece of 2 × 4 × 12 in. alongside the corner stud with the block's *top* flush with the base chalk line. Drive a 16d common nail partially into the 1 × 6 or 2 × 6 (if used) at the intersect point of the vertical span (center) line and the slope line. Lay a rafter on the nail and 2 × 4 block and, after ensuring that the ridge end is fully beyond the centerline, tack the rafter in place with an 8d common nail.

Laying Out a Ridge Cut

Lay out the ridge cut, using the level shown in Figure 7–16b. Next, subtract one-half the thickness of the ridge board by measuring in a distance from the vertical line just drawn. Draw another vertical line even with the mark just made and label it "cutting line." While the rafter is in position, the bird's mouth and overhang should be laid out, after which all cuts to the rafter can be made.

Marking a Bird's Mouth

Use a 24-in. level and a pencil to make a bird's-mouth layout as shown in Figure 7–16c. This technique results in a seat equal to the 2 × 4 stud width.

Laying Out an Overhang or Rafter Tail

Only one example is provided of the laying out of an overhang. You must use the principle to fit the specific requirements. Figure 7–16d shows how the principle is employed. Assume that a 24-in. overhang is being laid out. Measure on a horizontal plane (from the 2 × 4 block) 24 in. and make a mark. Use the level in plumb fashion to make a vertical line.

Next, review the plans and determine the width of the fascia board so that the amount of the lower edge of the rafter end that needs to be trimmed is defined. Also note how much is needed and used to retain the soffit if a box cornice is installed. The plan should show this detail, or it may be calculated as follows: Subtract the lip and the soffit's thickness from the total width of the fascia board. Record the remainder.

Lay a piece of roof sheathing material (a scrap piece) on the rafter's top edge, and with a ruler measure down from the plywood's top edge a distance equal to that recorded. Mark along the vertical line previously drawn. Draw a level line from this mark *back* toward the bird's mouth. The rafter-end layout is complete.

Checking the Cuts and Verifying the Fit

Cut out the rafter and another one. Try them for fit on the opposite sides with a scrap piece of stock that represents the ridge inserted. If they fit (which they should), use the rafter as a pattern to lay out and cut all rafters required. *Caution:* Be sure to keep the crown *up* on all rafters.

Common Rafter Layout, Stepping-Off Method

The stepping-off method is a step-and-repeat technique that is frequently used to step-off the rafter length in successive units. Briefly, the process involves a move or step of the framing square for each full foot of rafter run and a partial step for a fraction of a foot of run. The connecting points of framing square to rafter's top edge are the pitch numbers. In Figure 7–17 points A and B are the pitch numbers that are found on the square and used along the rafter's top edge. At point A the 4 from "4 in 12" is used; at point B the 12 is used. Since the run is 6 ft 4 in., six full steps are used. While the square is positioned for the sixth step, a mark should be made on the rafter where 16 in. falls on the square (see section c). The seventh step, a partial step, is made using *4 in 12* but the square is aligned over the 16-in. point marked during the sixth step. The ridge and tail cuts are perpendicular and the bird's mouth is made as shown in Figure 7–17b. The seat cut should always be equal to the 2 × 4 plate width or wider.

The stepping-off method, although used extensively, is not accurate. If each step of the square is just a pencil's thickness off, the rafter's length could be 1⁄16 in. longer or shorter per step. An experienced carpenter is very careful to avoid this situation by using one edge of the pencil mark rather than the entire mark. By using the edge of the pencil mark, greater accuracy is achieved throughout the stepping-off sequence.

Observation

Much time and effort can be saved in estimating rafter needs and laying out rafters if a calculator is used to do the math necessary to determine its length. The formula is very easy to use once it is understood. The carpenter must learn how to change a decimal into a fraction of an inch, but a short table can be made for quick reference. When the construction industry switches to metric measurements, calculations will become very easy because no conversion to fractions is needed.

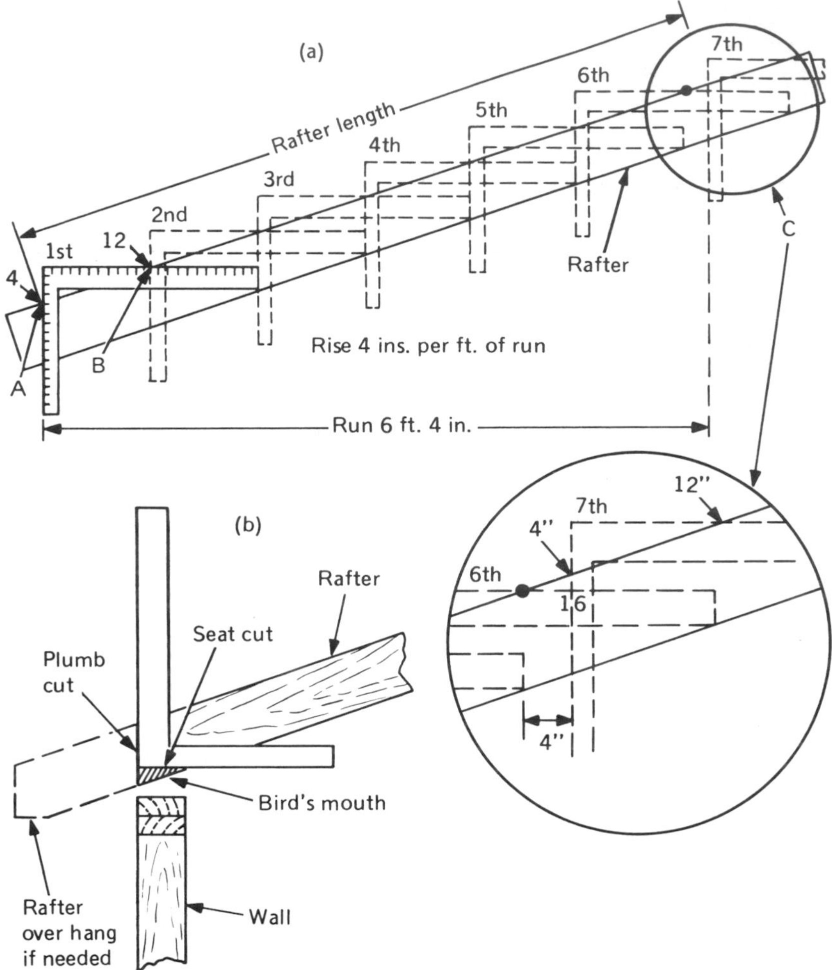

Figure 7–17 Common rafter layout: stepping-off method.

The different cuts needed to be made on a rafter can be done easily by first laying them out with a framing square. These include the ridge cut, bird's-mouth cut, and rafter tail cut. Before making the ridge cut, an allowance for one-half the ridge board thickness must be subtracted from rafter length; then the cut can be made. The rafter tail cut can be laid out and cut initially, that is, when the rafter is on the bench, or it may be done after all rafters are installed. Many experienced carpenters believe that cutting the rafter tails after the rafters are in place makes for a more perfect building line. There is considerable truth in this idea, since frequently there may be variations in top plate alignment.

SELECTING THE PROPER-SIZE RAFTER

The same engineering data that went into selecting the joists (Chapter 5) are used to select rafters. Tables in Appendix C summarize industry standards for rafters and their loads. Modulus of elasticity and fiber in bending factors are again primary for quality control of the lumber grade and species necessary for adequate support. Both live and dead predicted loads must be considered along with the slope of the roof. The usual calculated live loads result from snow deposits and personnel working on the roof. Obviously, snow is more serious. Values of 20 psf, 30 psf, and 40 psf are consistently used as live load values. Dead loads usually are established as 10 to 20 psf. The slope of the roof (pitch) is also considered as a factor. Some tables in the appendix are compiled for pitches of less than 3 in 12, while others are for pitches of high slope, or greater than 3 in 12.

Recall that the three ratings for stiffness are S/360, S/240, S/180. It was stated in Chapter 5 that an S/180 rating would be acceptable for rafters under some conditions. S/360 and S/240 are both acceptable since the load stresses placed on floor joists are usually considerably more than on the roof. A study of the tables in Appendix C shows that certain members (block type) are limited to S/180 and others are limited to their grading. The next two examples aid in understanding the importance of selecting the proper timbers for rafters.

Example 1

Select a rafter to carry a 40-psf load whose pitch is *6 in 12* and will be used on a building with a 36-ft span. Spacing will be 24 in. o.c. and rafters must have a deflection limitation of S/240.

Solution

1. Rafter run = 18 ft or half the building span.
2. At a rise of *6 in 12* the rafter length (not including overhang) is 20 ft.
3. Only 2 × 12 grade Dense Sel Str can be used. See Table 9, Appendix C.
4. Dense Sel Str has E of 1.9 million psi. F_b of 2100 to 2400 psi.
5. Species meeting both grade and load standards are Douglas fir and larch. See Appendix A.

Example 2

Select a rafter by species and grade that will meet the following requirements. The building will have a span of 26 ft and an *8-in-12* rise. The predicted dead load will be 15 psf. The allowable unit stress in bending (F_b) will not exceed 1200 psi under normal conditions, or 1500 psi with a 7-day loading. The rafter spacing equals 24 in. o.c.

Solution

1. Rafter run = 13 ft or half the building span.
2. At a rise of *8 in 12*, the rafter length (not including overhang) is 16.5 ft.

3. Several possible grade choices (see Table 14, Appendix C):
   a. 2 × 8 grade (Sel Str)
   b. 2 × 10 grade (No. 2)
   c. 2 × 12 grade (No. 3)
4. Species data for the three grades (see Appendix A):
   a. For 2 × 8 rafters grade Sel Str, the acceptable species are:
      (1) Douglas fir–larch (F_b normal 2050 psi; 7-day load 2560 psi)
      (2) Douglas fir Larch (north) (F_b normal 2050 psi; 7-day load 2560 psi)
      (3) Southern pine (F_b normal 2000 psi, 7-day load 2500 psi)
   b. For 2 × 10 rafters No. 2, all the previous species are acceptable.
   c. For 2 × 12 rafters grade 3, none of the previous species are acceptable. Typical reason: N/A Douglas fir–larch grade 3 has F_b normal 850 psi and 7-day loading 1060 psi.
5. Stock selection: Acceptable grade and species include: the 2 × 8, grade *select structural* fir or better; or 2 × 10 grade 2 fir, Douglas fir, or southern pine.

The methods used to solve these two examples are based on sound engineering data. All judgments of this kind should be founded with facts of the types listed in the appendixes and taken from the blueprints or specifications. Where a range of species and sizes of materials is adequate to perform the functions of rafter, economics becomes an important factor. Holding costs low is good business sense as long as the quality of the materials for the job is maintained properly.

PROPER LOCATION OF RAFTER ON OUTER WALL

Each rafter in a roof assembly needs to be spaced 16, 24, 32, or 48 in. on-center, depending on the design of the roof. The spacing does not, however, dictate each rafter's position on the wall with respect to the stud position in the wall, although it could. The roof load bears its weight onto the wall through the rafters and it is important that the plates bear as little of this weight as possible. Therefore, the force of the roof's weight should pass through the plates into the studs and to the foundation.

To provide the best opportunity for the transfer of force to be passed to the studs on posts and not be borne by the plates, it is essential that rafters be aligned either directly over the stud or post or aligned adjacent to the stud. If rafters are installed midway between the studs, the on-center spacing will cause the plates to bear the load and the possibility of shearing the plates exists.

Several opportunities exist for the carpenter in the layout of the rafter positions on a wall or each rafter aligned directly over the stud or post in the wall, as Figure 7–18a shows. This assures that the roof load is borne directly by the studs or posts and transferred through the plates with no shear pressure. The ceiling joists are aligned next to the rafters and both are nailed to each other and fastened to the plates. Figure 7–18a is a front view of the isometric drawing of the wall and roof assemblies.

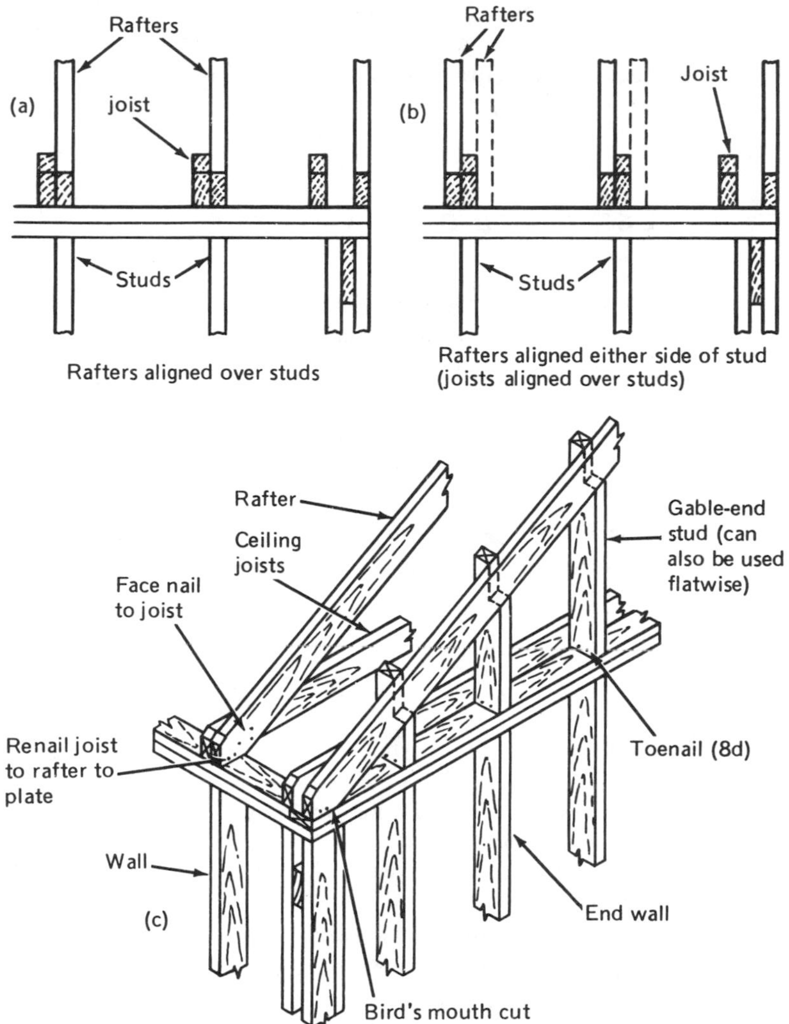

Figure 7–18 Rafter positioning on the outer wall: alignment with studs.

By adopting this plan for rafter positioning, the layout on wall plate and ridge (if used) begins at the same corner as the layout for wall studs and floor joists. This technique places all structural members in alignment.

If the alternative plan is used as Figure 7–18b shows, the ceiling joists are aligned over wall studs. The rafters are then located either side of the ceiling joists. This technique locates all rafters within 2 in. of the wall stud and is acceptable in building codes. There is, of course, some shear force placed on the plates by the roof load. However, the plate stock does have sufficient strength to accept this load because of the close proximity of the stud alignment below it.

If beam-and-post construction is used, the rafter beam is aligned and fastened directly to the supporting post below it. This type of construction is used where exposed beam or cathedral ceilings are used.

PROBLEMS AND QUESTIONS

1. In which of the following roof designs may a common rafter be used?
   a. gable **b.** hip **c.** shed, gambrel **d.** dormer **e.** all of the above
2. Which fractional value indicates the steepest pitched roof?
   a. ¼ **b.** ⅓ **c.** ½
3. When defining a pitch, is the *run* or *span* of a building used along with the rise?
4. What is the pitch of a roof of a building that has a span of 12 ft and rafter rise of 6 ft?
   a. ¼ **b.** ½ **c.** ¾
5. Transpose the formula *rise/span* = *pitch* to find the rise when span and pitch are shown.
6. What does the following symbol mean on an elevation blueprint?

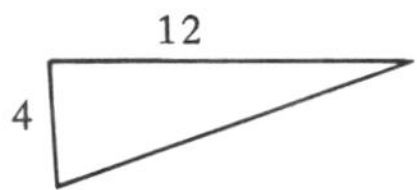

7. How is the notation in the symbol in Question 6 read?
   a. 4 on 12 **b.** 4 in 12
8. What portion of a common rafter would have a *rafter tail?*
   a. the lower portion
   b. the portion near the ridge cut
   c. rafters never have tails
9. What is the purpose of a *bird's mouth?*
10. Define the term *crown* as it applies to a rafter.
11. What parts of a roof account for the dead load on a rafter?
12. Why is the ridge cut always a plumb cut?
13. What member added to a gable roof reduces the forces that are on the plates?
    a. rafter **b.** collar beam
14. List two methods that prevent the roof from causing the walls to bow outward because of forces.
15. Why can a uniformly sized rafter be measured for length along its crown edge?
16. What is the length of a common rafter if its run is 13 ft 6 in. and the rise is 6 in 12?
    a. 13 ft 6 in. **b.** 15 ft ⅛ in. **c.** 17 ft 2½ in.
17. What is the lowest grade and smallest size of lumber appropriate for a rafter that is to run 12 ft, have a rise of 7 in 12, and have a deflection not to exceed S/240?
18. Why is it important to align rafters over wall studs?

PROJECT

A. Description and Specifications

In Chapter 5 we designed a porch floor assembly 20 ft wide by 30 ft long. To be a porch, this floor assembly must have a roof. For the project in this chapter you should be able to identify the common rafters and other roof materials necessary to build that much of the roof. In Chapter 8 we complete the roof design, which includes the valley rafter. Let's make several assumptions. Turn to Figure 7–1. Imagine that the stoop shown is really the porch floor we designed in Chapter 5. Our goal here is to design the porch roof that will integrate with the roof of the house shown. Since the house employs a gable roof design, the porch roof should also have a gable end along the 30-ft span. Also notice that there is a 12-in.-wide overhang on the gable end and across the front and back. Figure P7–1 shows how the porch ceiling assembly was built. Then we can proceed with roof construction. Assume that the depth of the house is 32 ft (a run of 16 ft plus allowance for overhang). Further, assume a pitch of 6 in 12 and that the rafters in the house are 2 × 8s No. 2 KD southern pine. Each set of rafters employs a collar tie. Assume that further work was already done, as Figure P7–1 shows. The exposed view of the house's rafters is shown. Notice that the rafter tails of those within the boundary of the porch have been trimmed flush with the wall plates.

1. What do you notice about the frame for the porch ceiling?
   a. Where does the girder rest?
   b. With what edge does the top of the 2 × 8 frame align?
   c. In what direction will the common rafters of the porch run?
2. Assume that the valley will be an equal-pitch valley. This means that the pitch for the roof will be _______________.

Look at the detail A-A of Figure P7–1 and notice that there are two 2 × 4s nailed to the end of the rafter tail. The fascia and soffit have been cut back far enough for construction of the valley.

B. Common Rafter Roof Design and Material Needs

1. How many linear feet of 2 × 8 ridge will we need for the roof? (See Figure P7–2.) Did you allow for splice blocks on both sides of the joint?
2. Define the length of the rafter.
   a. From Figure 7–15, what is the length per foot of run of the rafters that we must use?
   b. What is the run and length of the basic rafter?
   c. What is the length per foot of run and length of the rafter tail?
   d. Now add up the totals and indicate the total length of each rafter. Round your answer up to the next even-foot length.
3. Define the number of common rafters needed. *Specs:* Space rafters 16 in. o.c. Select a rafter of the same rating as was used on the main roof. (Use Table C–5 in Appendix C.)
   a. What size and grade of rafter should be used?
   b. How many common rafters are needed?
4. Define the materials needed for collar beams or ties for common rafters.
   a. We know that there is one collar tie for every pair of rafters. Therefore, how many are needed?

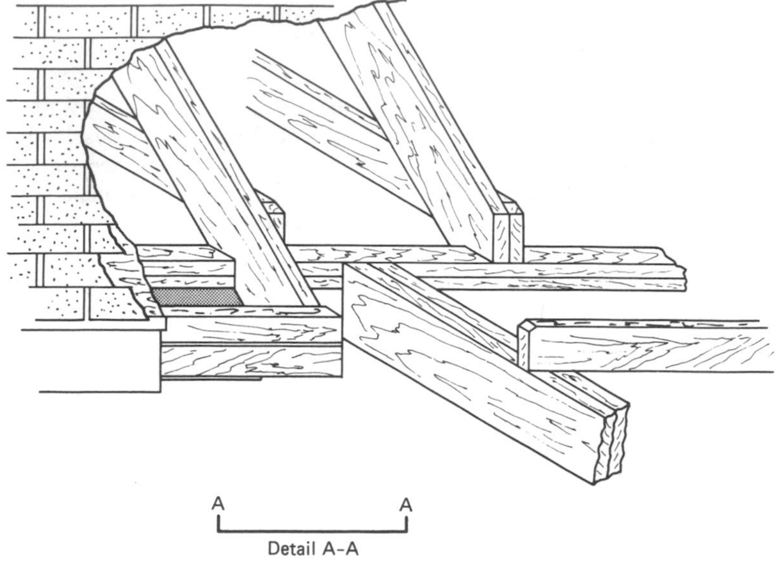

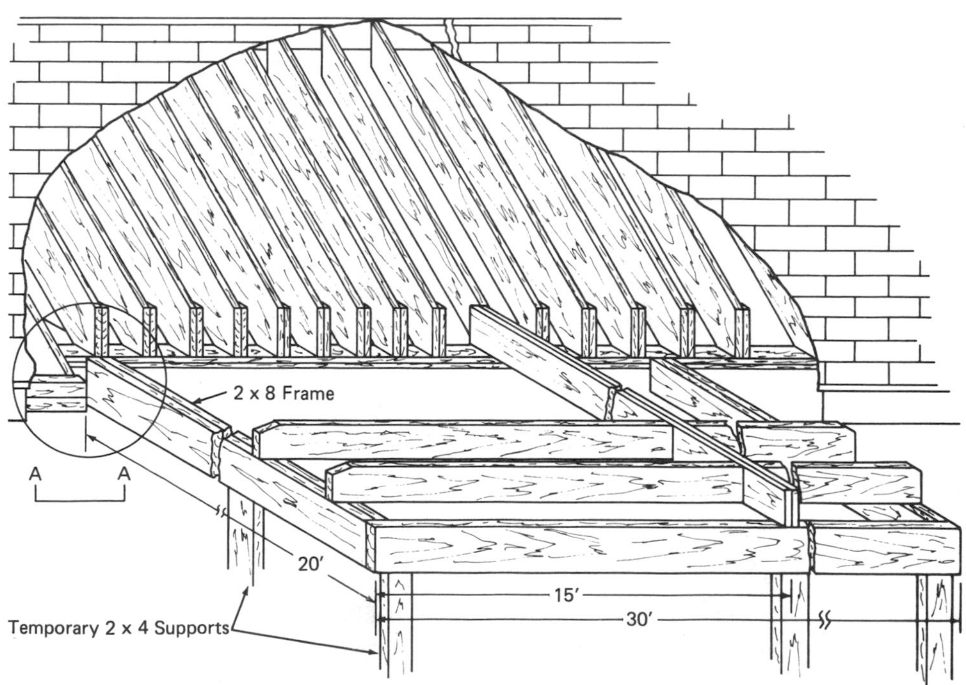

Figure P7–1

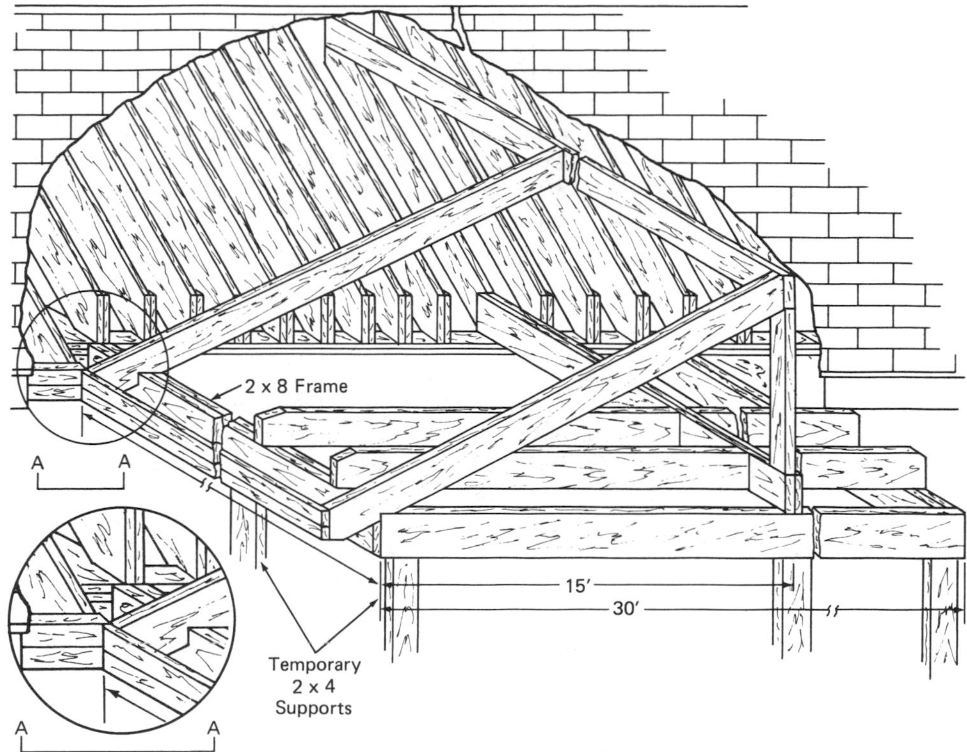

Figure P7–2

b. Consider placing each collar tie at a point equal to three steps of the framing square down from the ridge cut on the rafter. Use Figures 7–14 and 7–15 to help define the length. The length of each will be approximately how many feet long?

c. What size and type of material are used for the collar ties?

5. In summary, the order for common rafters and ridge boards will be:

Quantity	Size and material
________	________________
________	________________
________	________________

Check all of your answers against those that have been provided. If the numbers differ drastically, recheck your work.

Answers:

Part A: **1. a.** On the top of the wall plate of the house, ______________

b. It aligns with the exterior wall plate's top edge, ______________

c. The rafters will run in the same direction as the porch ceiling joists

2. 6 in 12

Part B:

1. Approximately 35 to 36 ft plus 2 blocks of 2 × 8 × 16 in. = 39 ft
2. **a.** 13.42 in.
   b. 15 ft, 16 ft 10 in.
   c. 13.42 in.
   d. approximately 16 ft 10 in. + 13½ in. = 17 ft 11½ in. or 18 ft
3. **a.** 2 × 8s, No. 2 KD or No. 2 southern pine or Douglas fir
   b. providing allowable room for the valley, we should base the number of common rafters on 19 ft of wall:

$$\text{number of rafters} = \frac{19\text{ ft}}{16\text{ in. o.c.}}$$
$$= 14.25 \text{ rafters per slope} \times 2 \text{ slopes}$$
$$= 28 \text{ rafters}$$

4. **a.** 13 (none is needed for the outer rafter pair)
   b. 6 ft (3 steps down on each side × 12 in. per step = 6 ft)
   c. 1 × 6 No. 2 southern pine
5. 28 2 × 8 18 ft No. 2 southern pine (rafters)
   2 2 × 8 18 ft No. 2 southern pine (ridge boards)
   1 2 × 8 4 ft No. 2 southern pine (ridge splice blocks)
   13 1 × 6 6 ft No. 2 southern pine (collar ties).

Chapter 8

The Roof—Part II: Equal-Pitch Hip and Valley Rafters

Hip and valley rafters are more difficult to construct than the common rafter because more variables are involved in their calculations and fitting. Defining their location is relatively easy; however, understanding how to obtain their length requires fundamentals of rise, run, the Pythagorean theorem (right triangles), and adjustments to the top edge of the rafter and bird's mouth.

The method of determining a rafter's length in this chapter uses the approach that the hip or valley is framed in a balanced or equal-pitch design. This means that the left and right portions of the hip or valley are of equal length.

Selecting the proper rafter size is important for several reasons. They are usually made from stock at least 2 in. wider than common rafters so that jack rafter bias cuts can have continuous surface contact. The load-bearing factors also play an important part in the rafter selection process.

DEFINING THE LOCATION OF HIP AND VALLEY RAFTERS

Figure 8–1 illustrates a house that has a roof framed with four hip rafters. Three of the four hips are shown in this figure. Notice that the lower edge of the hip centers over the corner of the building.

In Figure 8–2 a house with a roof framed with both hip and valley rafters is illustrated. As in Figure 8-1, the hip rafters of the house in Figure 8-2 have their lower edges in line with the corners. The hip rafters in this figure are designed to be of equal pitch. There is one valley rafter in this roof, forming an interior corner. Its lower edge aligns with the interior wall corner.

Each dormer on the front of the house shown in Figure 8-3 is framed with two valley rafters. These are located at the intersect point of dormer roof to the main roof. These valley rafters also employ the regular valley design, but may sometimes use the *unequal-pitch* design principle.

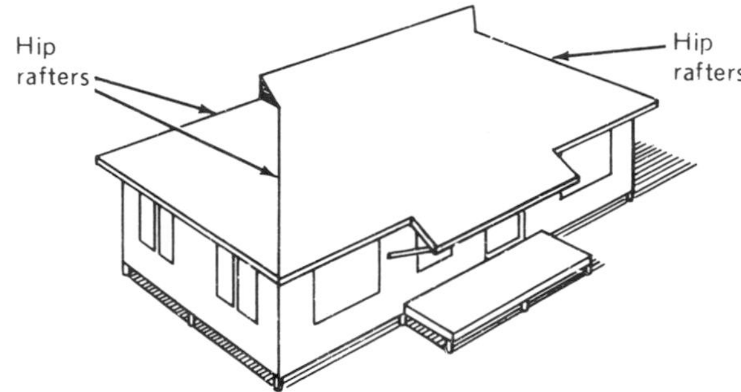

Figure 8-1 Locating the position of hip rafters in a hip roof.

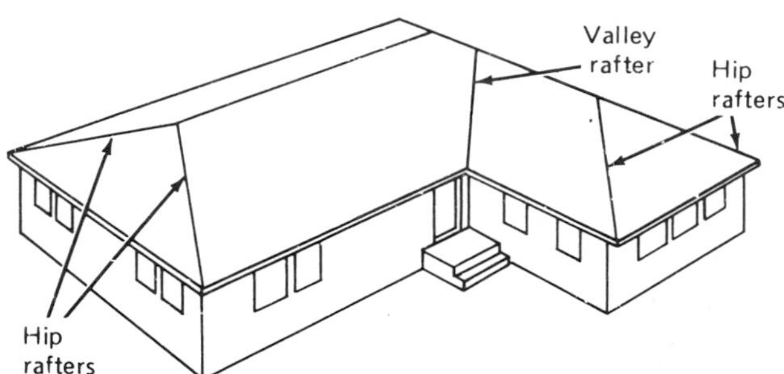

Figure 8-2 Locating a hip and valley rafter in a hip and valley roof.

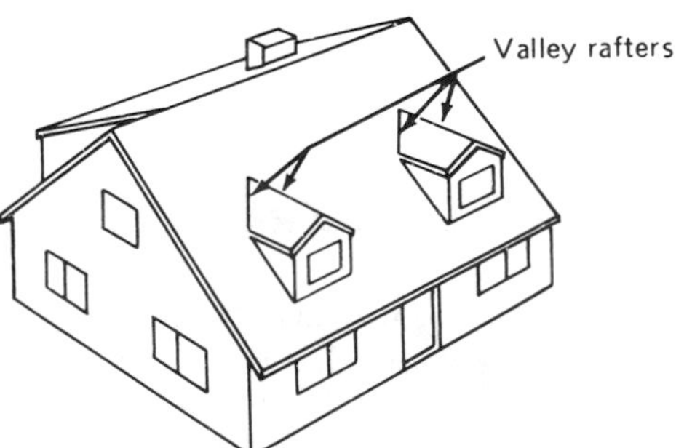

Figure 8-3 Locating valley rafters in a roof and dormer.

TERMS USED WITH HIP AND VALLEY RAFTERS

There are several terms used with hip and valley rafters that are the same as those used with common rafters, and there are several special terms and adaptations of common rafter terms. The terms each have in common are the *rafter rise* (in inches) and the *rafter run* (although the run per foot will not be 12 in.). Calculation of rafter length can be made using the Pythagorean theorem. The hip or valley rafter requires a *bird's-mouth* cut, *ridge* cut, and may need a *tail* or *overhang* with tail cut.

The new and adapted terms are shown in Figure 8–4. The plumb cut of the hip and valley that intersects the ridge and common rafters must be altered with

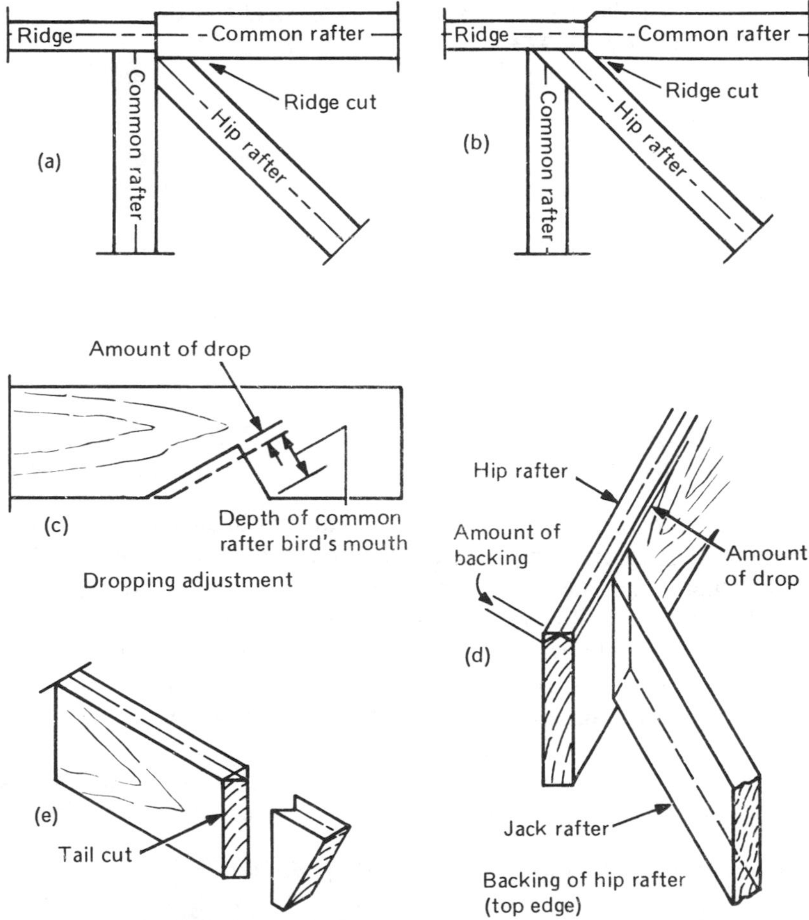

Figure 8–4 Defining terms and their location on a hip rafter (terms also apply to valley rafters).

bevel or side cuts as Figure 8–4a and b show. In part (a) its length is shortened and two angle cuts are made along the plumb cuts. In part (b) an adjustment for length is made and a single angle cut is made so that the hip or valley rafter butts against the ridge. In Figure 8–4c a new term, *drop* or *dropping adjustment,* is shown. This is an adjustment cut made in the bird's mouth to lower the rafter to the level of the common and jack rafters. If dropping adjustments are not used, *rafter backing* must be used, as Figure 8–4d shows. (Valley backing is shown later.) The top or crown side of the entire length of the hip must be backed with a bevel cut so that jack rafters intersect properly and the roof line at the hip is proper. Finally, the *tail cut* shown in Figure 8–4e is plumb cut with two bevel cuts; each cut surface must form an angle that aligns with the tail cut of the jack and common rafters.

DETERMINATION OF HIP OR VALLEY RAFTER PITCH, RISE, RUN, AND LENGTH

Pitch

A review of the elevation plans shows the roofline will quickly identify the pitch of a roof. In the case of a hip or hip and valley roof or just valleys in a valley and gable roof, more than one pitch may be indicated. If only one is shown, all roofs are sloped at the same angle; that is, they all rise to the same height and their runs are equal.

Constructing a roof with a single pitch is easier to understand and build than one having several pitches. An examination of a *regular* or *equal-pitch* roof is used that has a pitch of *6 in 12.*

Definining the Position of a Hip Rafter in an Equal-Pitch Roof

The first requirement is to obtain the outer dimensions of the building, which include its span and length. The span should be a measurement that is shorter than the length. Assume that a building with lines of span and length as shown in Figure 8–5 is to have a hip roof installed. With a building span of 28 ft 0 in. the run is 14 ft 0 in. Since the roof pitches of the span and length are equal (6 in 12), a like distance of 14 ft 0 in. is laid off between points B and D and E and F. A line is drawn connecting points D and F. Midway, or 14 ft 0 in. from either D or F, point A is defined. This establishes the hip rafter positions as AB and AE. Since squares were established, the hip rafter is at a 45° angle to the span and length of the building line. The same method can be used to to define the position of a valley rafter.

Defining the Run of a Hip or Valley Rafter

Using Figure 8–5 again as an example, the hip rafter run is defined as a distance measured from point A to point B on a level line (not up the slope of the rafter). Two methods are used for finding the length of run.

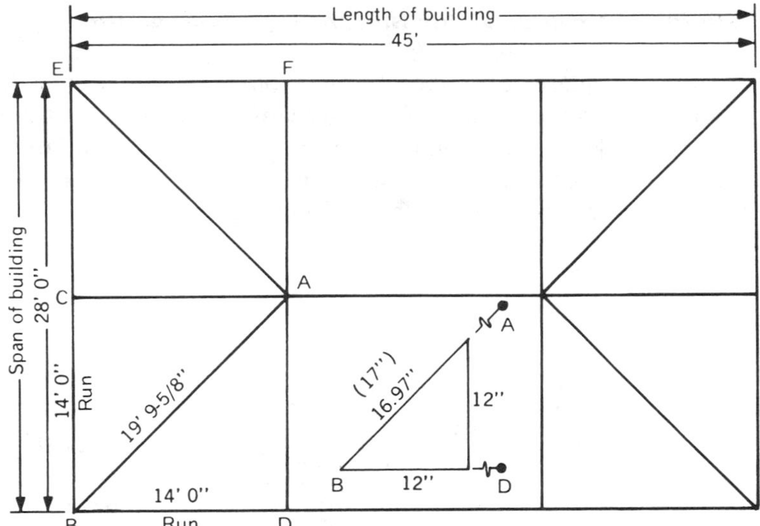

Figure 8–5 Defining the hip rafter run (applies also to the valley rafter run).

Method 1. Use the Pythagorean theorem:

$$\begin{aligned}\text{hip rafter run } (AB) &= \sqrt{BC^2 + BD^2} \\ &= \sqrt{14^2 + 14^2} \\ &= \sqrt{196 + 196} \\ &= \sqrt{392} \\ &= 19.7989 \\ &= 19 \text{ ft } 9\tfrac{5}{8} \text{ in.}\end{aligned}$$

Method 2. Using tabulated data from Figure 7–15, a rise of 12 in. and a run of 12 in. result in a hypotenuse length of 16.97 in., which is the length of hip run per 12 in. of common rafter run. Therefore,

$$\begin{aligned}\text{hip rafter run} &= \text{length per 12 in. times the number of feet of common rafter run} \\ &= 16.97 \text{ in.} \times 14 \\ &= 237.58 \text{ in.} \div 12 \quad \text{(divide by 12 to obtain answer in feet)} \\ &= 19.7989 \text{ ft} = 19 \text{ ft } 9\tfrac{5}{8} \text{ in.}\end{aligned}$$

This means that for each 12 in. of common rafter run the hip rafter (and valley rafter) has a run of 16.97 in. or roughly 17 in.

Defining the Hip or Valley Rafter Projected Length

The framing square has a table imprinted on it that provides the length per foot run of equal-pitch hip or valley rafters for many different rises. It uses a basis of 16.97 in. as a base leg or *run*. Figure 8-6, which corresponds to Table 8-1, provides the length per foot of run for the usual rises of roofs. In the example problem above, the pitch was established as 6 in 12. From Figure 8-6, a rise of *6 in 12* for a hip or

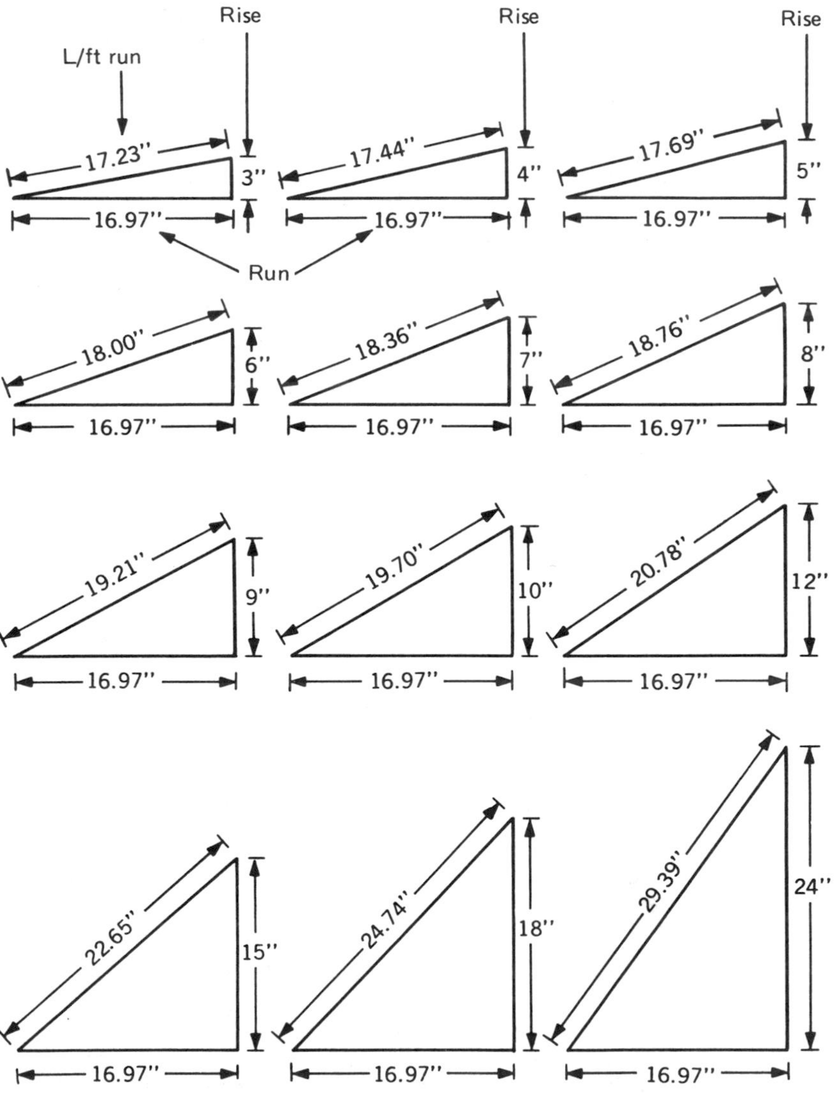

Figure 8-6 Length of hip or valley rafter per foot of run.

TABLE 8–1 LENGTH OF HIP OR VALLEY RAFTER PER FOOT OF RUN

Rise (in.)	3	4	5	6	7	8
L/ft run (in.)	17.23	17.44	17.69	18.00	18.36	18.76
Rise (in.)	9	10	12	15	18	24
L/ft run (in.)	19.21	19.70	20.78	22.65	24.74	29.39

valley rafter has a length per foot of run of 18 in. Therefore, the rafter length may be calculated:

$$\begin{aligned} \text{hip rafter length} &= \text{length per foot of run} \times \text{number of feet of common rafter run} \\ &= 18 \text{ in.} \times 44 \\ &= 252 \text{ in.} \div 12 \quad \text{(to obtain the number of feet)} \\ &= 21 \text{ ft} \end{aligned}$$

With the rafter calculated it becomes important that its measurement be properly established on the rafter itself. Recall that dimensions for rafters may be made through the center of the rafter or in the case of sized members along their top edge. When using the top edge, a perpendicular line (plumb line) must be used to ensure that the rafter length is not changed (see Figure 7–13).

The rafter is laid out as shown in Figure 8–7. Notice that the ridge (plumb) cut is made using 17 in. and 6 in. on the square, as is the bird's mouth.

Up to this point the only difference between layout on a common rafter and that on a hip or valley rafter is the use of 17 in. instead of 12 in. on the square to obtain the plumb and horizontal cuts. However, the remaining cuts need to be made from a new reference point—the center of the top edge of the rafter. Notice in the top drawing in Figure 8–7 that hip rafter's length is from the outer corner of the wall to the midpoint (intersect) of the junction of the common rafters and ridge. This means that some special cuts must be made to each end of the rafter as well as an allowance for length near the ridge.

The lower section of Figure 8–7 illustrates the side view of the hip rafter with its primary cuts for ridge, bird's mouth, and lower end. These may all be laid out with the framing square or with the use of a miter or bevel square (this method is shown later). Also notice the top view of the rafter, with the dashed line bisecting the rafter thickness. This is the centerline that must be used as a reference for the remaining cuts.

Rafter Cuts and Adjustments

Lower-edge bevels. The lower-edge bevels are made at the angle back from the centerline and are cut accordingly. These surfaces then align with the wall plates or sheathing. In the lower end of the valley rafter, the bevels are made opposite

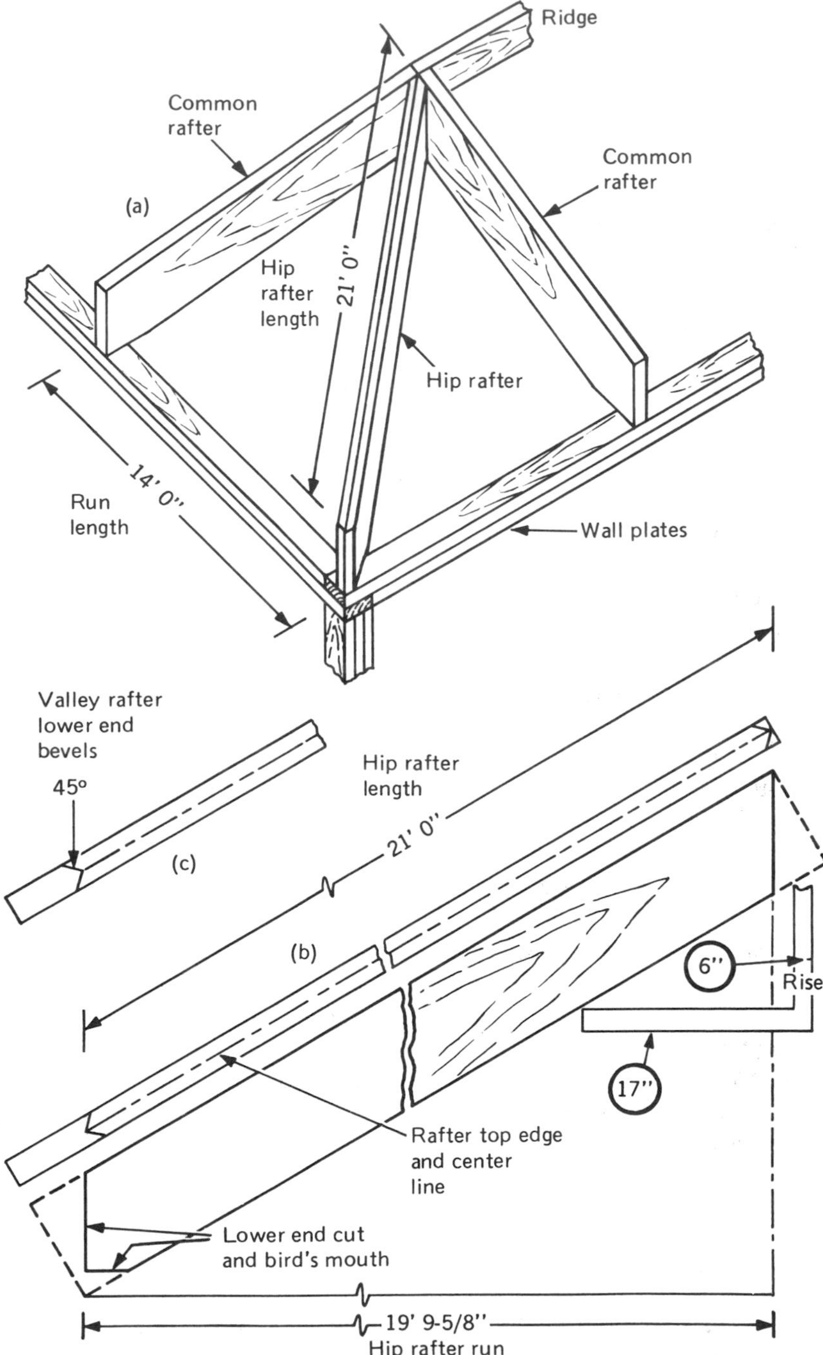

Figure 8-7 Laying out the hip rafter from the rafter centerline.

those of the hip, as Figure 8–7c shows. The layout point is still the centerline of the rafter. These bevel cuts also align with the wall plates. It would appear that simple 45° angles should be cut beginning at the centerline. If the rafter were laid on the horizontal plane, this idea would be accurate; especially so since the layout of the 45° lines are made on the top surface of the rafter. However, when the rafter is tilted up as each rise predicts it will be, these angles must be cut differently so that their cut surfaces still remain in alignment with the wall plates.

The concept just stated is probably one of the most difficult to understand. Figure 8–8 may provide the visual reinforcement needed to understand why the layout and side cuts of the hip and valley are not 45° angles. To begin, the combination square is held at a horizontal plane with the plates (see X). In this position the beveled edges of the hip rafter form a 90° corner, 45° on each half. However, the miter or bevel square shown resting on the top surface of the rafter is not on a horizontal plane with the plates since the hip rafter is shown at some rise. Because of the position of the bevel square and hip (or valley) rafter the angle of the bevel square is *always* less than 45° when positioned as shown.

Most of the better-quality framing squares have a table on the blade that gives the setting for hip and valley side cuts for many different rises. Once understood, they may be applied as needed. However, there are several other methods that may be used. One is to use the bevel or miter square. Setting it for the proper angle is easy. Figure 8–9a shows how it is used to make a single side cut for hip or valley where the rafter ties into the ridge or other valley. In Figure 8–9b side cuts for ridge and lower-end tail cuts are shown easily laid out with miter square. Finally, in Figure 8–9c inside bevels are laid out with miter square for the lower end of the valley rafter. These illustrations lead to the obvious question: How does one obtain the angle needed for a particular rise?

Table 8–2 and Figure 8–10 combine to provide visual understanding as well as a standard of values to lay out side cuts for almost all different rises. Table 8–2 provides the following data:

1. Rises begin at *3 in 17* up to *18 in 17.*
2. Degrees of side cut for each different rise.
3. A conversion from degrees to a numerical value stated in inches. The 2 stands for the framing square's 2-in. blade. the 2⅟₁₆L across from the 3-in. rise stands for a base leg of just under 2⅟₁₆ in. (The L stands for light or shy measurement.)
4. The last column provides bevel square settings using the framing square's blade as a basis.

Now we will apply the tabulated data to a roof with a 6-in. rise. As Figure 8–10 shows, the bevel square is set so that its wooden surface is against the outer surface of the framing square. Its metal blade is set to the values listed in the last column of Table 8–2, "1 in. out, 1⅝ in. inside." When the two values are in alignment with the bevel square's blade, lock the blade in position.

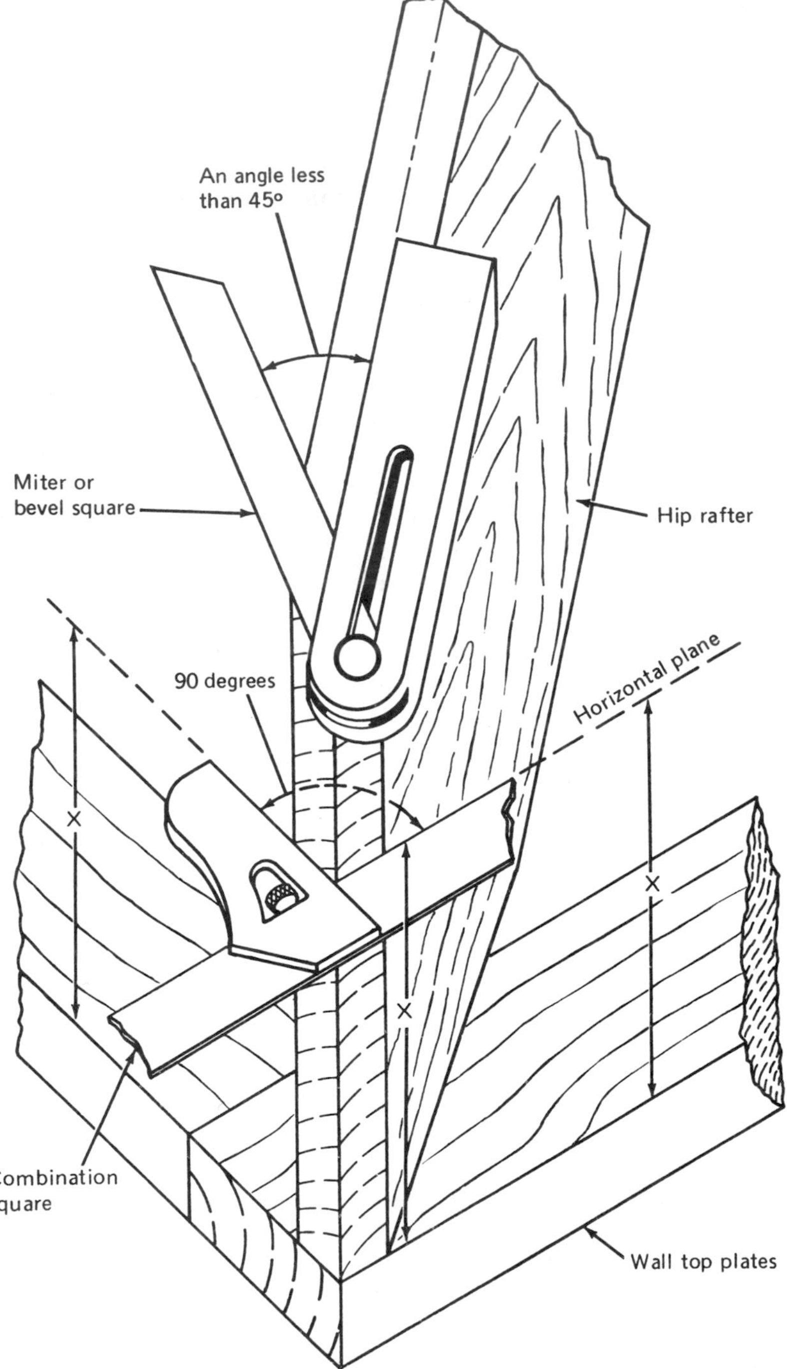

Figure 8-8 Bevel meter cuts align with wall plates; miter square angle is less than 45°, to achieve the miter needed.

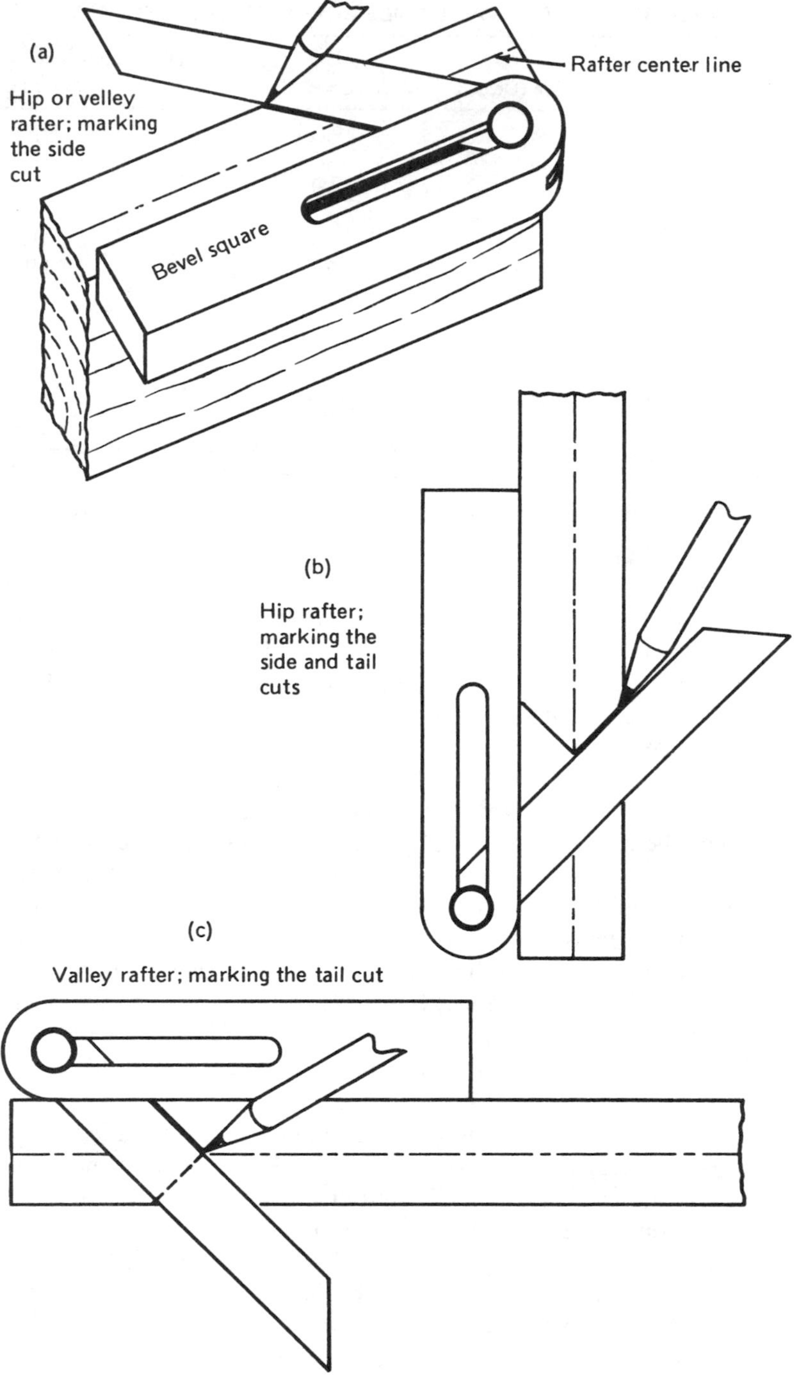

Figure 8–9 Marking hip and valley side cuts and tail cuts with properly set miter square.

TABLE 8-2 HIP AND VALLEY SIDE CUTS WITH BEVEL SQUARE SETTINGS

Rise	Angle (Deg)	× (in.) = 2/tan	Bevel square setting on framing square (see Fig. 8-10)[a]
3	44.5	2 1/16L	1 9/16L
4	44.2	2 1/16	1 9/16
5	43.8	2 1/16F	1 9/16F
6	43.3	2 1/8	1 5/8
7	42.7	2 3/16L	1 11/16L
8	42.2	2 3/16F	1 11/16F
9	41.5	2 1/4	1 3/4
10	40.8	2 5/16	1 13/16
11	40.2	2 3/8	1 7/8
12	39.5	2 7/16	1 15/16
15	37.0	2 5/8F	2 1/8F
18	34.5	2 15/16	2 7/16

Note: L behind number stands for *light* and F for *full.*

[a]All settings are for 1 in. out.

The lower portion of Figure 8-10 shows an expanded inch scale with each of the rises from Table 8-2 located for ease of reading. Any serious carpenter who uses this method of laying out side cuts on hips and valleys could easily duplicate this expanded segment on a small card and carry it in his or her wallet for quick, reliable reference.

Ridge adjustment and bevels. Depending on the technique of construction used, the ridge cut on a hip (or valley) rafter may need to be adjusted for length. Figure 8-11a and a′ show two techniques for making the ridge cut on hip rafters. In a, the rafter length must be shortened equal to the distance between points A and B. This is necessary for several reasons. First, the rafter length was originally calculated from point A on the ridge to the outer corner of the wall. (Recall that this was 21 ft 0 in. in the example.) Since the common rafters both center on point A and are installed as shown, a portion of their surface accounts for some reduction of hip rafter length. Finally, a portion of the ridge board's thickness must be subtracted from the rafter's length (pp. 182–83).

Two practical methods of defining the adjustment length are (1) measuring the distance from point A to point B where it intersects the corner of the common rafters at the ridge, and (2) lay out a full-scale model of the ridge and common rafters on a scrap board or cardboard and then measure the distance from point A to point B. After determining the adjustment length, mark the hip rafter accordingly along its top centerline; then lay out the required bevels and cut along the markings.

An alternative technique of fitting the hip ridge cut is shown in Figure 8-11a′. In this technique the hip rafter is butted to the ridge rather than the common rafters. The adjustment for length is made on the rafter as shown; however, the bevel cuts

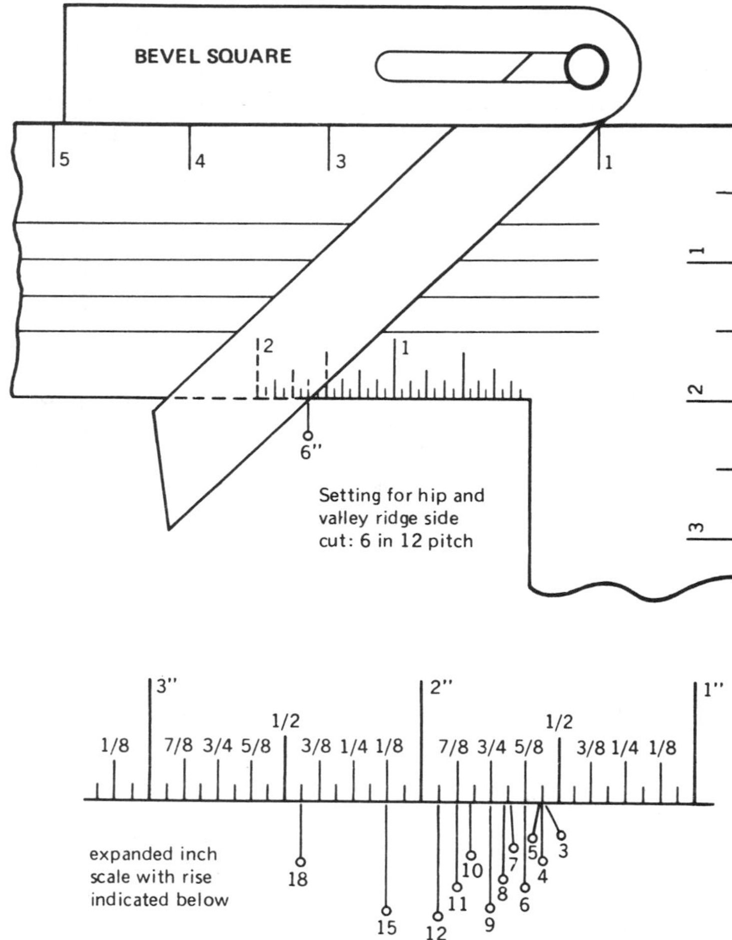

Figure 8–10 Miter square settings for various side and tail cuts for various pitches.

result in one side cut across the full hip rafter with point B as the intersect point of the rafter's centerline and ridge board edge.

While studying this technique, notice that several adjustements must be made to the common rafters. Common rafter 1 must have both of its ridge cut sides beveled, and common rafter 2 must have its ridge cut made at an angle equal to that of a hip jack rafter (covered in Chapter 9).

The valley rafters shown in Figure 8–11b and b′ also require adjustment of their initial lengths to account for ridge board thicknesses. In part (b), where both ridges of the valley are of equal height, the adjustment is made from point C to D. Then bevel cuts are made from the adjustment point D when it intersects the middle of the rafter thickness. Figure 8–11b′ shows that one valley rafter (left) intersects

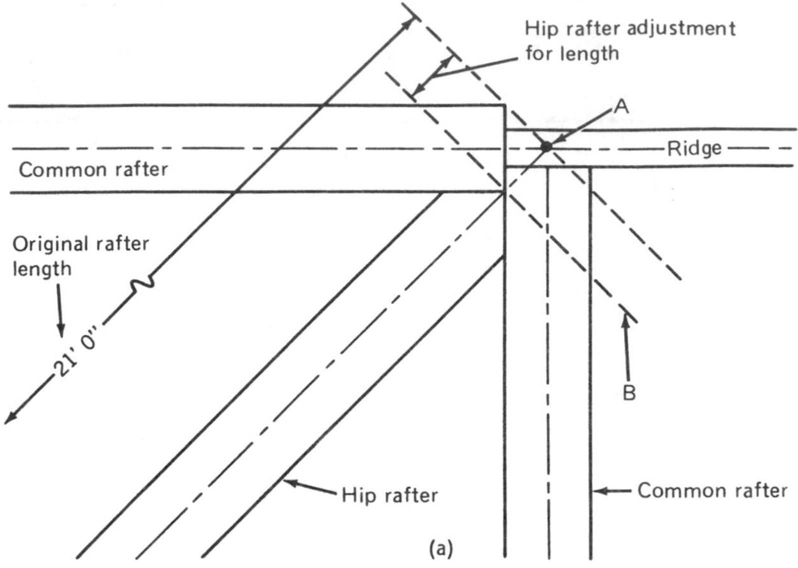

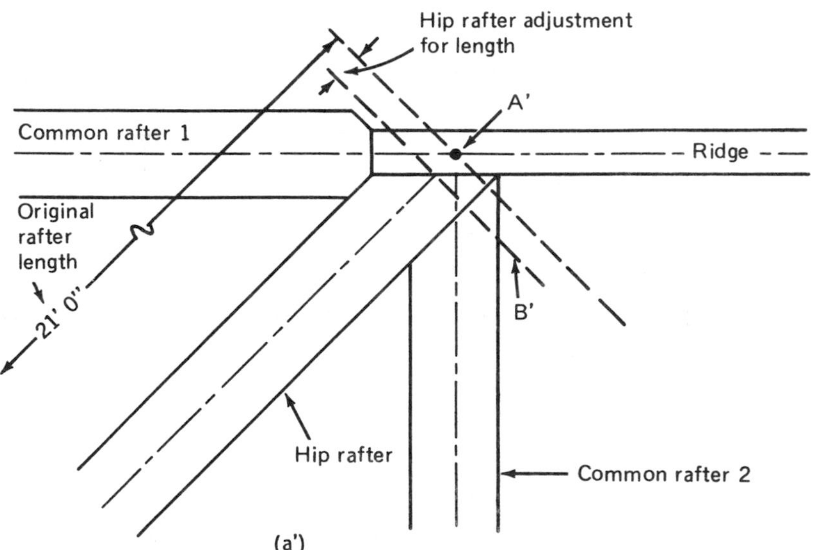

Figure 8–11 Ridge adjustment layout on hip and valley rafter and cornice.

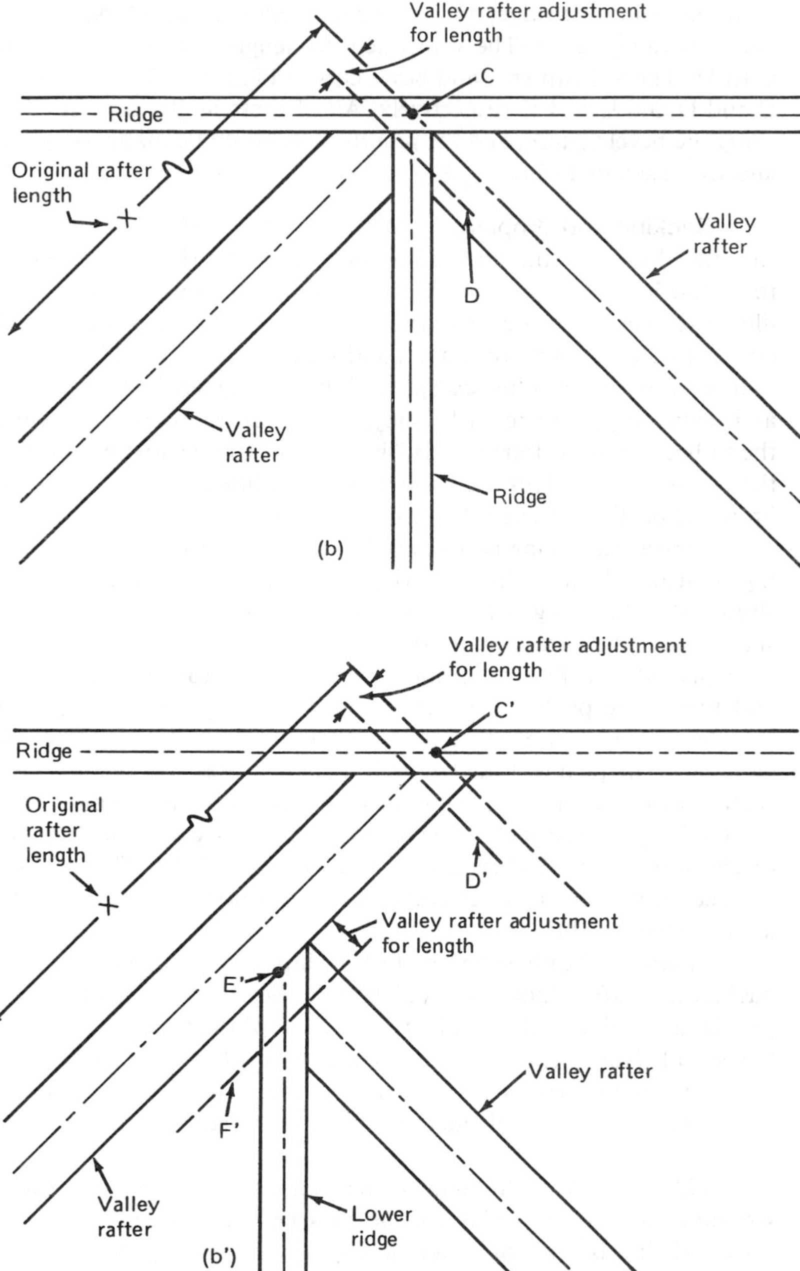

Figure 8-11 (*continued*)

with the main roof ridge, and the second valley rafter intersects with the lower ridge and first valley rafter. The adjustment for length for one valley rafter is the distance C to D. The adjustments and bevel cuts in Figure 8–11b′ are taken between points C and D and E and F, respectively. All of these angles for side cuts can be laid out using the bevel square, and the plumb cuts should be made using the framing square and the rise and 17 in.

Backing and dropping. With the rafter adjusted for length and beveled to fit either between common rafters or against the ridge, it would seem appropriate to install it. Not so, a *backing cut* or *drop cut* must be made first. Figure 8–12a illustrates backing on a hip rafter. Imagine that a sighting is made along the top corner of all the common rafters and jack rafters to the hip rafter. All surfaces are in line along the level line except the hip; it is higher. Notice also that accurately cut and installed jack rafters (the subject of Chapter 9) are short when placed against the unbacked hip rafter; that is, they appear to be short because they do not reach the lip of the hip. This area above the level line and from jack rafter to hip rafter lip is the portion of the rafter to be backed (page 185).

Before examining how to do the backing, consider what happens when sheathing is installed on a hip roof where the hip rafter is unbacked. As Figure 8–12b shows, the sheathing is lifted and only makes contact with the lip of the hip. The space under the sheathing on the hip is of little consequence since nails may easily be toenailed at a slight angle into the hip rafter. However, the roofline is not proper and this is the problem. Cornice construction as well as roof shingling are both affected. Cornice construction as well as roof shingling are both affected. Backing eliminates the problem, as Figure 8–12c illustrates. Notice that one-half of the top surface of the hip rafter is now flush with the level line. Notice also that because of the backing cut the jack rafter's top edge intersects flush with the backed surface of the hip rafter. If sheathing is installed now it lies flat and has a flat contact surface on the hip. Cornice construction can now be made accurately and shingles lay properly on the roof.

Figure 8–12 only shows half of the hip rafter backed surface; what does a fully backed hip rafter look like, and how is the backing cut laid out? Figure 8–13 is provided to aid in understanding. At the left of the figure is an unbacked hip rafter's lower end. Notice that the plumb cut is beveled at horizontal 45° angle back from the center point. To its right is the same hip rafter after backing has been made. Notice the correlation of the right-side backed edge to the level line used in Figure 8–12.

Finally, look at the right of the figure and observe the framing square and a section of hip rafter. To lay out the backing place the square as shown with 17 (the run) and the rise (in this example 6 in.) along the lip of the rafter's top edge. If the hip rafter is of nominal 2-in. stock (actual thickness 1½ in.), divide by 2 and use the results of division as shown to obtain the dimension of backing. One-half of 1½ in. is ¾ in.; therefore, the backing line is established at the 16¼-in. mark on the framing square. For a 2 × __ hip rafter with a 6-in. rise, the backing cutting line is ¼ in. from the lip of the rafter.

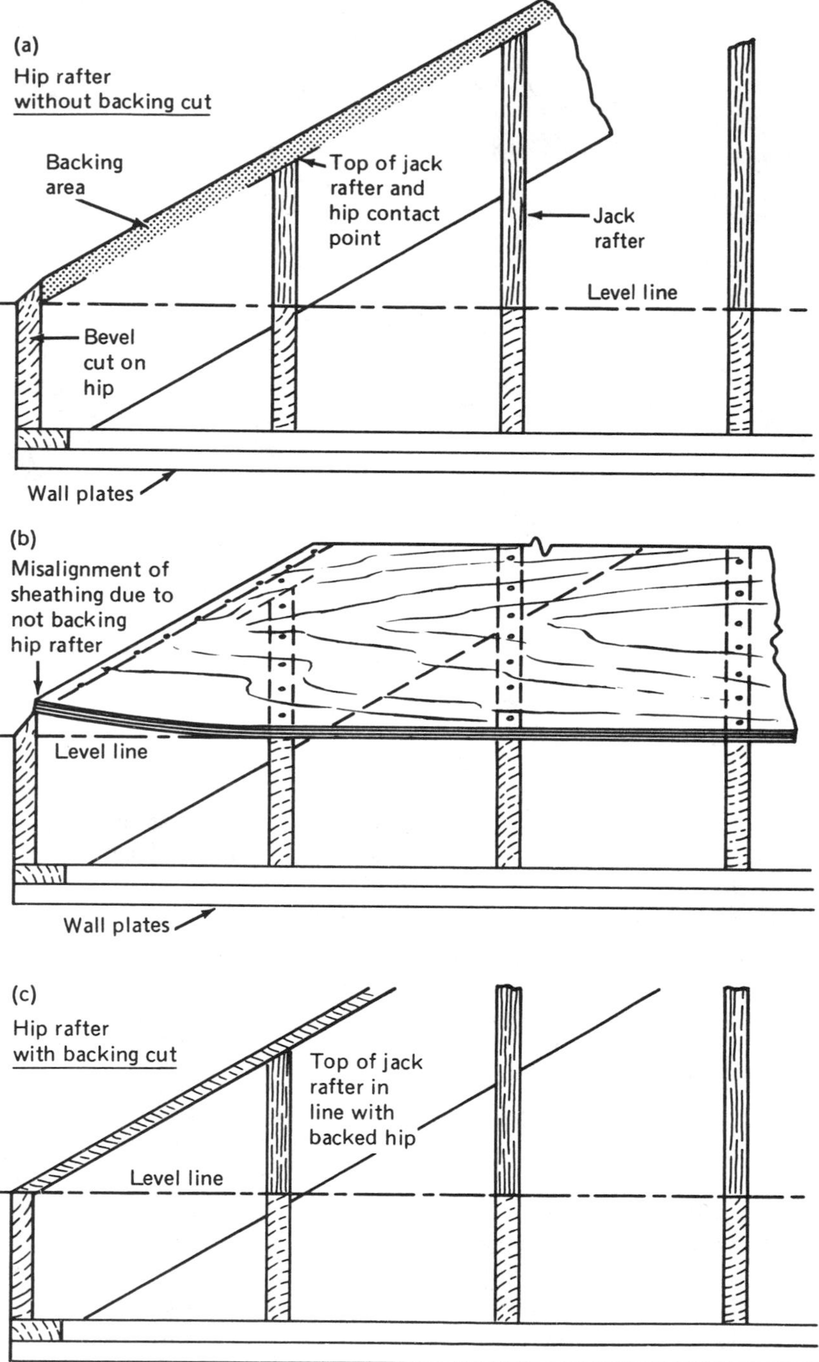

Figure 8–12 Backing the hip rafter to allow proper alignment of sheathing.

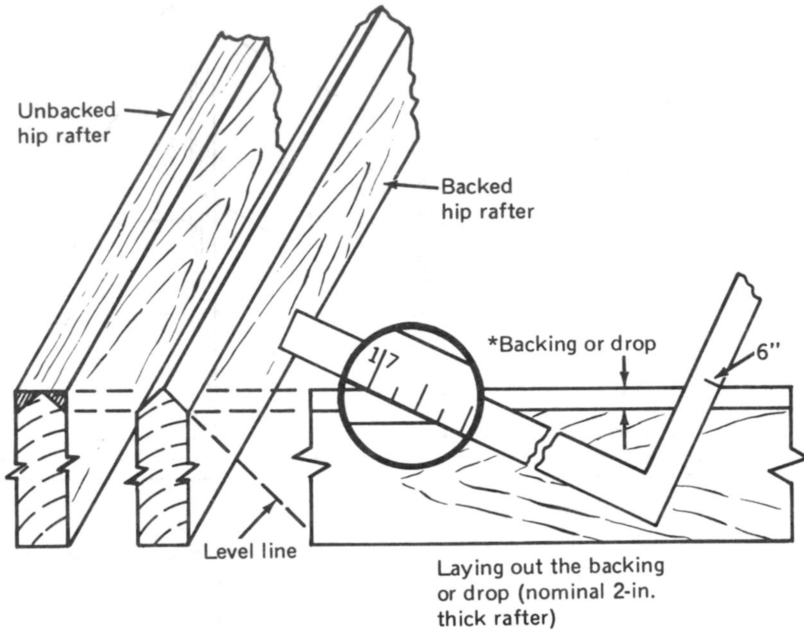

Figure 8–13 Backing the layout with a framing square.

This method, once learned, is easily repeated whenever the need arises, but a reference to the amount of backing needed for all rises reduces the work involved. Table 8–3 provides a quick reference for the amount of backing needed for the usual rises of roofs. Those given are for a nominal 2-in.-thick hip rafter. Note, however, that if a nominal 1-in.-thick hip rafter is used, all values in the table should be reduced by one-half.

Setting a power saw to cut the backing after determining and laying it out is time consuming but correct. Frequently, where roofs are pitched up to *8 in 12* and sometimes even *10 in 12*, a *drop* is made in lieu of the backing. It is quicker to make. Both Figure 8–13 and Table 8–3 indicate that the amount of backing is equal to the amount of drop. Where, then, is the drop cut made?

A *drop cut* is made on the seat cut of the bird's mouth as in Figure 8–14. If, for example, a drop cut is used on a hip (or valley) rafter, with a pitch of *4 in 12* the amount of drop is 3/16 in., as Table 8–3 indicates. A line is drawn parallel 3/16 in. from the bird's-mouth common rafter cut, and then this stock (blackened area) is cut away. When the rafter is installed, it is lowered 3/16 in., thereby bringing the top lip level with the jack and common rafters. A small gap will exist between sheathing and rafter top surface, but this is of little consequence.

A different set of conditions exists on the *valley rafter* that requires the use of backing or a *drop cut*. Figure 8–15a shows the frontal view of the lower end and

TABLE 8–3 BACKING OR DROP ADJUSTMENT VALUE FOR NOMINAL 2-IN. HIP OR VALLEY RAFTER

Rise (in.)	Backing or drop (in.)
2	1/8
3	5/32
4	3/16
5	7/32
6	1/4
7	9/32
8	5/16
9	11/32
10	3/8
12	7/16
15	1/2
18	9/16
24	5/8

Note: If a nominal 1-in. hip rafter is used, use one-half the value listed under the backing or drop column for the rise of the roof.

top surface of an *unbacked* valley rafter. Notice that the level line is lower than the unbacked surface. Figure 8–15b is an illustration of a properly backed valley rafter. Notice that the lowest point of the V groove is in line with the level line.

If a valley rafter is used without a backing or drop cut, several conditions surface that create problems, as seen in Figures 8–16 and 8–17. In Figure 8–16 sheathing installed on the roof and intersecting in the valley *does not* bear on the valley. Notice that if sheathing is square cut (not beveled), its lower lip barely touches the valley. In many cases no contact is made. This condition is unsatisfactory. In Figure 8–17, plywood or other sheathing is caused to lift away from the end of the jack rafter even though the sheathing is beveled. This allows sheathing full

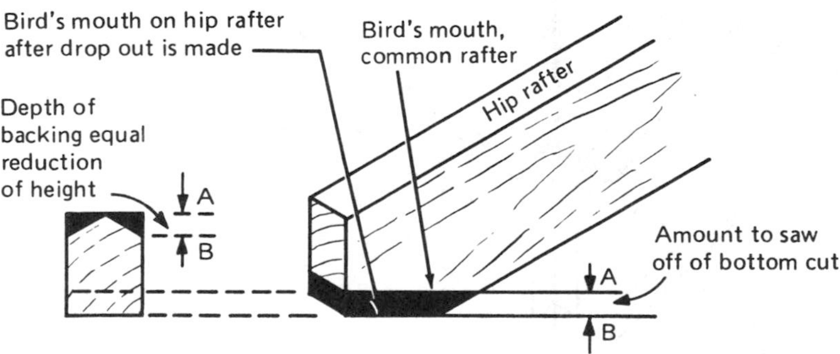

Figure 8–14 Drop cut on bird's-mouth seat.

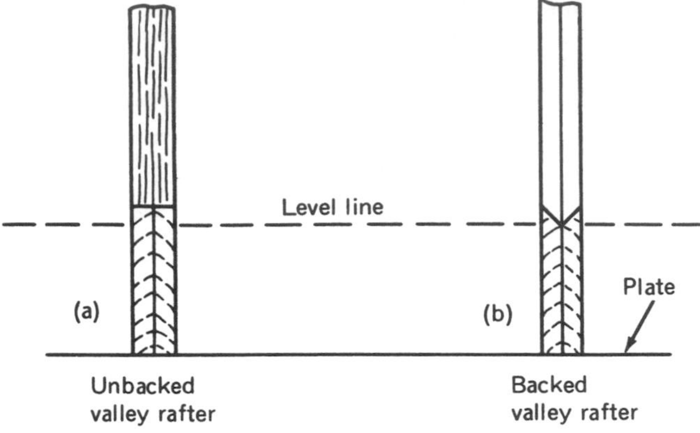

Figure 8–15 Backing on a valley rafter.

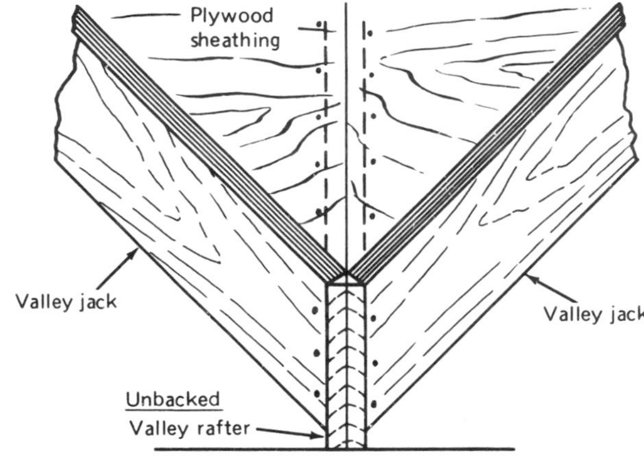

Figure 8–16 Unbacked valley rafter; sheathing does not make contact.

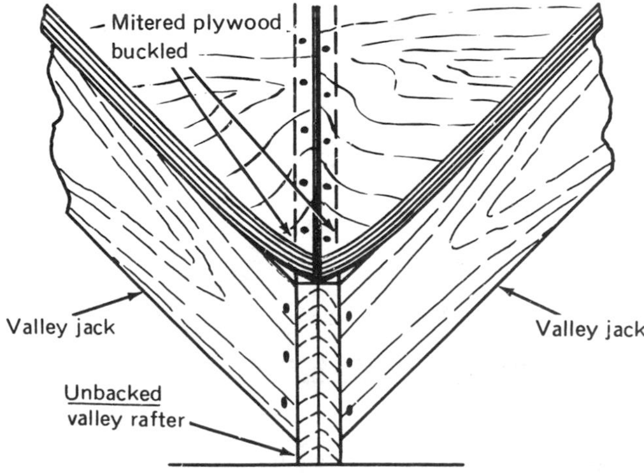

Figure 8–17 Unbacked valley rafter; sheathing mitered but buckled.

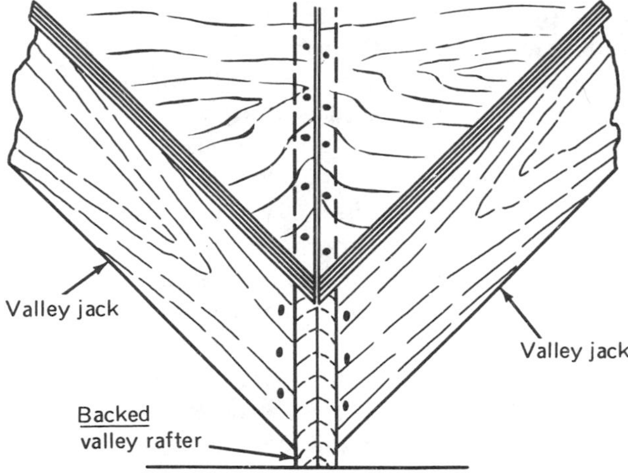

Figure 8–18 Backed valley rafter; properly seated and cut sheathing.

contact on the valley, which is good; however, its lifting away from the jack rafter surface is not. In addition, this lifting causes some problems when cornices are constructed.

If the problems cited above are to be avoided, backing as in Figure 8–18 is one solution. Table 8–3 can be used to obtain the amount of backing (depth of V groove). Once it is defined and laid out on the valley rafter, a power saw is used to make the two necessary cuts. Look at Figure 8–18 and notice that a continuous line extends along the top lip of the valley jack rafter to the lowest point of the backing cut. Also notice that beveled sheathing fits snugly into the V groove and flush with the valley jack.

If a *drop cut* is used in lieu of the backing cut as in Figure 8–19, several adjustments must be made. One adjustment is made to the valley rafter, and another to each jack rafter. In Figure 8–19 an amount of drop is chosen from Table 8–3 according to the pitch (rise) of the roof and is shown as the blackened area (G to H).

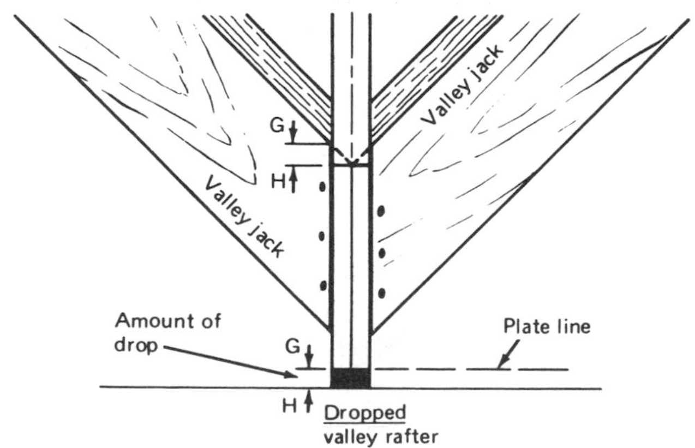

Figure 8–19 Dropped valley rafter and jack rafter alignment.

Since the valley is lowered by the drop cut, valley jack rafters must be installed properly; that is, their top edges must align with the centerline of the valley rafter. Figure 8–20 shows this idea more clearly. Each jack rafter is raised above the top surface of the valley so that the straightedge that lies along its surface touches the valley rafter as shown.

Backing or a drop cut is a very important detail that must be included in the planning and cutting of either a hip or valley rafter, as is the adjustment that must be made during planning and cutting for rafter overhang.

The overhang of the hip and valley rafter performs the same function as on a common rafter; it may also be called *rafter projection* and *rafter tail.* However, this is where the similarity to the common rafter ends. Figure 8–21 provides the reference of the overhang necessary for this study. In Figure 8–21a the hip rafter is shown extending beyond the wall line, and in Figure 8–21b the valley rafter is shown extending beyond the wall plate also. While observing these illustrations, notice that the end cuts on the overhang are just like the ones on the flush rafter studied earlier. Because of the slightly different approach needed to explain and understand the overhang on each type of rafter, the hip is studied first, followed by the valley rafter.

When the overhang is used on a residential house, it is usually made to project horizontally from the wall. The three most common dimensions of projection are the 6 in., 12 in., and 24 in. Figure 8–22a shows these as *run of common rafter projection.* The heavy black line represents the building line. It thus follows that on a hip roof the overhang extends on all walls, as the illustration shows.

On an equal-pitch hip roof the horizontal run of the hip roof is calculated as the hypotenuse of the right triangle whose sides are either 6, 12, or 24 in. With the figure on hand, the next step is to calculate the hip rafter *projected length.* For this

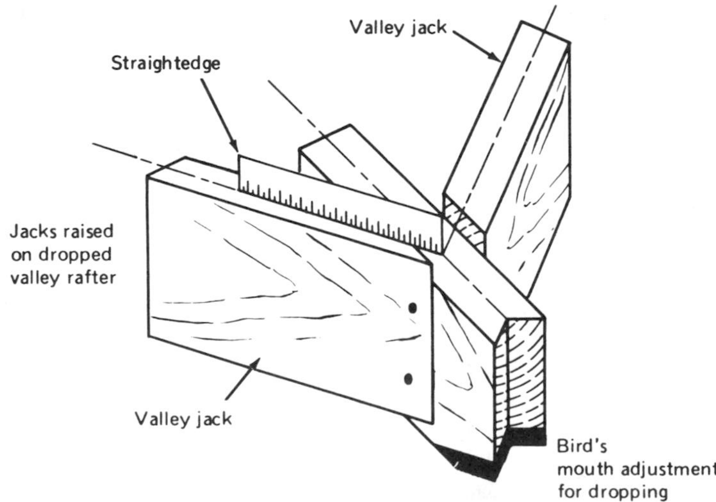

Figure 8–20 Aligning jacks on a dropped valley rafter.

Overhang

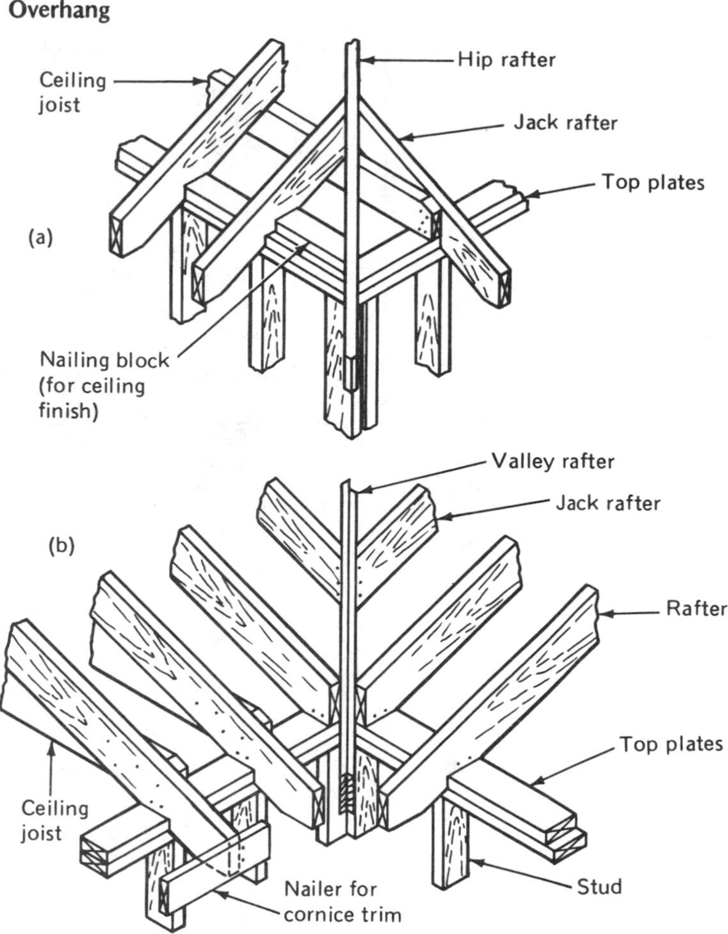

Figure 8–21 Hip and valley rafter tail (overhang).

the rise is used as the leg opposite the right angle. If these values are inserted into the Pythagorean formula, the projected length is defined.

A simple on-site method is usually used by the carpenter. He or she simply steps off the framing square for the projection needed. However, a simpler method is available. All the calculations necessary for any of the three projections are found in Tables 8–4, 8–5, and 8–6. To illustrate their ease of use and effectiveness, consider the following three examples.

Example 1

Determine the projected *hip rafter length* on a roof with a 5-in. pitch and planned 6-in. overhang.

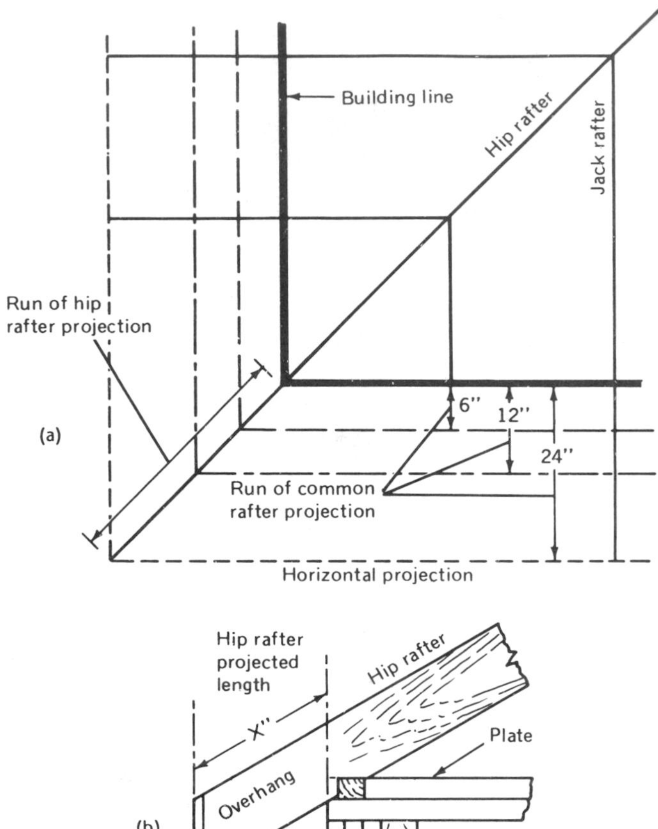

Figure 8–22 Hip rafter projected length.

Solution

1. Figure 8–22b indicates that the projected hip rafter length should extend from the corner of the plate/wall line a distance of *X*.
2. From Table 8–4, for a 5-in. rise, the length is 9 13/16 in.
3. The hip rafter must be extended a length of 9 13/16 in. beyond the corner or original line of measurement.

Example 2

A building is to have a 12-in. overhang. It has a rise of *8 in 12* on an equal-pitch hip roof. Determine the *hip rafter projected length.*

TABLE 8–4 HIP AND VALLEY RAFTER OVERHANG ON AN EQUAL-PITCH ROOF DESIGN: 6-IN. OVERHANG

Rise (in.)	Length (in.)
2	8 11/16
3	9
4	9 3/8
5	9 13/16
6	10 3/8
7	11
8	11 5/8
9	12 3/8
10	13
12	14 5/8
15	17 1/4
18	19 7/8
24	25 7/16

Note: Lengths have been rounded to the nearest 1/16 in.

TABLE 8–5 HIP AND VALLEY RAFTER OVERHANG ON AN EQUAL-PITCH ROOF DESIGN: 12-IN. OVERHANG

Rise (in.)	Length (in.)
2	17 1/16
3	17 1/4
4	17 7/16
5	17 11/16
6	18
7	18 3/8
8	18 3/4
9	19 3/16
10	19 11/16
12	20 3/4
15	22 5/8
18	24 3/4
24	29 3/8

Note: Lengths have been rounded to the nearest 1/16 in.

Solution

1. Again, referring to Figure 8–22b, the projected length of the hip rafter is defined as X.
2. From Table 8–5, for an 8-in. rise the length is 18 3/4 in.
3. The hip rafter projected length for a 12-in. overhang is 18 3/4 in. long.

TABLE 8-6 HIP AND VALLEY RAFTER OVERHANG ON AN EQUAL-PITCH ROOF DESIGN: 24-IN. OVERHANG

Rise (in.)	Length (in.)
2	34
3	34 1/16
4	34 3/16
5	34 5/16
6	34 7/16
7	34 11/16
8	34 7/8
9	35 1/4
10	35 3/8
12	36
15	37 1/8
18	38 7/16
24	41 9/16

Note: Lengths have been rounded to the nearest 1/16 in.

Example 3

An overhang of 24 in. is called for on the blueprints. The roof design is an equal-pitch hip roof whose pitch is *6 in 12*. Find the hip rafter projected length (X).

Solution

1. For the third time, Figure 8–22b indicates that the projection needed is X.
2. From Table 8–6, for a 6-in. rise the length is 34 7/16 in.
3. This indicates that the hip rafter projected length is 34 7/16 in. long.

Each of the tables provides at a glance the required hip rafter projected length for all of the most common pitches of residential roofs.

The materials estimating and planning phases for a hip rafter (and valley rafters) should be done in the office. In earlier paragraphs, explanations of how to obtain the basic hip or valley rafter lengths were given. It then becomes simple to refer to Table 8–4, 8–5, or 8–6 to find the value of the projected length and add this length to the previous findings to determine the full rafter length. It is even possible to use adjusted values from these tables for estimating overhang lengths, where, for example, the run of common rafter projection is to be 8 in.

Finding the *valley rafter projected* length is no more difficult than finding the hip rafter projected length when the run of common rafter projection or overhang is 6, 12, or 24 in. Tables 8–4, 8–5, and 8–6 may be used. Figure 8–23a provides details that illustrate the run of valley rafter projection on a horizontal plane, which is the hypotenuse of the right triangle whose sides are either 6, 12, or 24 in. The value of the hypotenuse then becomes the base leg, and the rise is the opposite leg

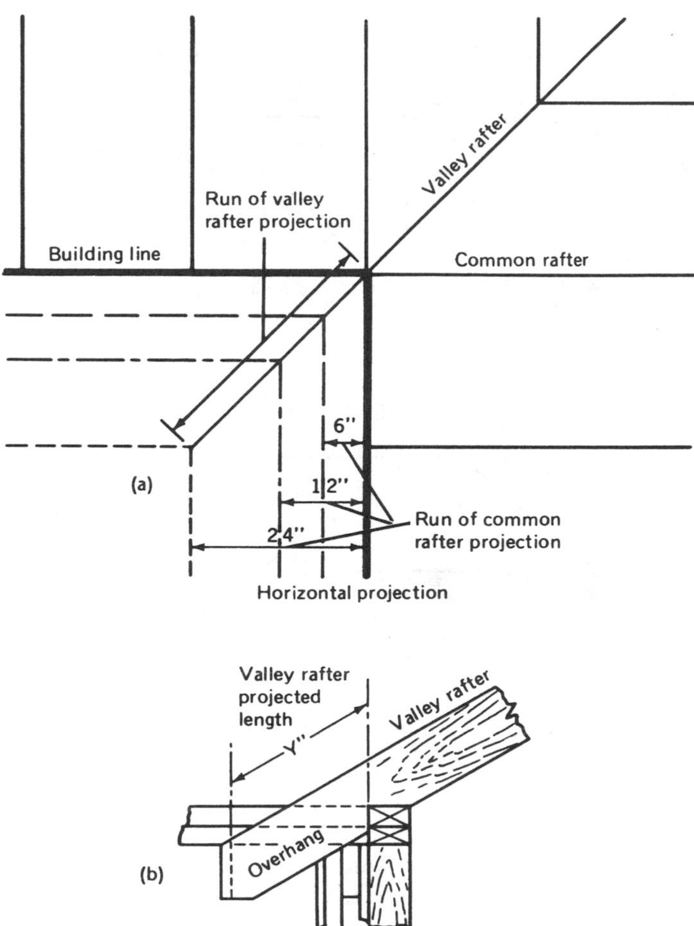

Figure 8–23 Valley rafter projected length.

of a triangle that defines the length Y in Figure 8–23b. However, the value of Y can be selected from one of the three tables mentioned earlier (see examples for finding X and use the same approach for finding Y).

Bird's-Mouth Adjustment When Hip and Valley Rafter Have a Tail

Hip and valley rafters require different bird's-mouth cuts because of the tail projection. Figure 8–24 illustrates the requirements where a miter cut is needed in each bird's mouth. Consider the hip rafter first where a corner miter is made to fit around the plates of the wall. The deepest point of the miter (corner) is the *original rafter*

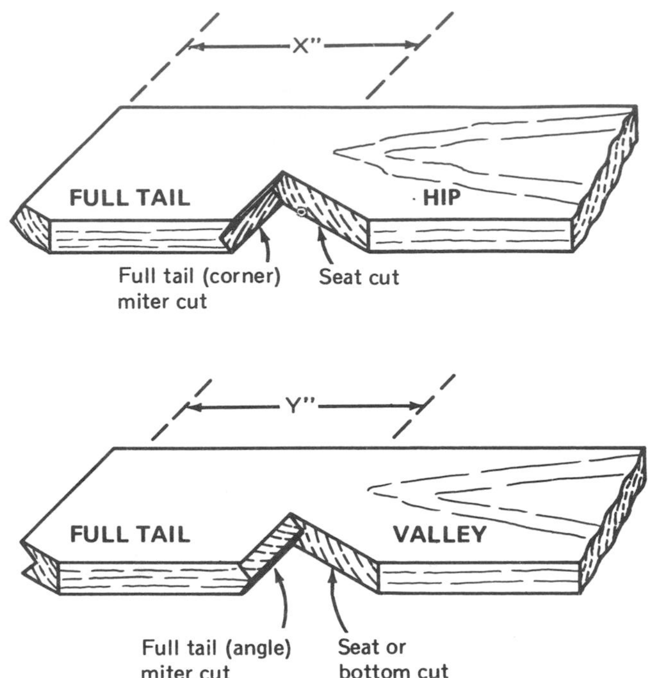

Figure 8–24 Bird's-mouth miters and tail cuts for hip and valley rafters.

length. The tail length *X* projects from this point to the tail end. At the tail end angle miters are made as explained earlier.

If mitering is not used in the bird's mouth as shown, a square cut at the corner must be made. Quite frequently this miter cut is omitted because of the time it takes to cut it (mostly by hand with hammer and chisel). This attitude is prevalent where boxed cornices are employed. However, if open beam construction is employed, the miters must be cut.

Constrast the bird's-mouth cut in the valley rafter with the hip and observe that a miter is made whose angles are easily cut with a hand saw. No other option is available here; they must be cut so that the rafter seats properly into the corner of the wall and tightly against the plates. The end of the tail must have the internal miter as shown.

SELECTING THE PROPER SIZE AND GRADE OF RAFTER

The hip rafter must support live and dead loads as well as resist wind forces. It is a beam whose length is customarily long and all hip jack rafters must be nailed to it securely. Therefore, the selection of one by species and grade to accept and contain the loads distributed over its length must meet strength requirements as well as stiffness requirements of $S/180$, or $S/240$.

Additional strength and stiffness are automatically incorporated into a hip or valley rafter if the one used is 2 in. wider than the common rafter. This procedure is often elected to ensure that the side-bevel cut of the jack or cripple rafters have full nailing and contact surface. This means that:

1. If common rafters are 2 × 6s, use a 2 × 8 for hip or valley rafter.
2. If common rafters are 2 × 8s, use a 2 × 10 for hip or valley rafter.
3. If common rafters are 2 × 10s, use a 2 × 12 for hip or valley rafter.

On occasion the hip or valley rafter is so long that a splice must be made. When this is done the splice must fall approximately midway between jack rafter on-center spacings as Figure 8–25 shows. A short block of rafter stock or plywood must be nailed securely to each side of the splice. When such an arrangement is used on a hip rafter, there seldom is any problem since the rafter is in compression. However, if such an arrangement is used on a valley rafter, additional support in the form of a truss brace should be installed from the rafter to a bearing partition. The valley rafter is in tension, and it supports the loads of the jack rafters, sheathing, and live loads. These forces are considerably more than the hip must support.

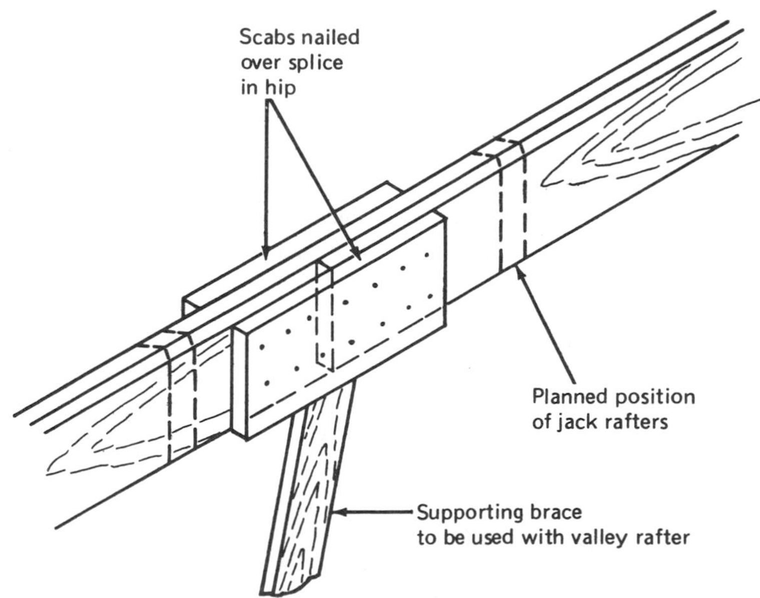

Figure 8–25 Splicing a hip or valley rafter with blocks of rafter stock.

SUMMARY

Many different approaches to laying out and fitting hip and valley rafters are available from numerous sources besides this chapter. However, most of the other presentations employ only the framing square and the practical on-site method used in hip and valley layout. This chapter explains the math that is needed to understand why and how a hip and valley are laid out. It provides tables that at a glance provide values for hip and valley rafters on run, sidecuts, backing, and drop cuts and for the three most commonly used overhangs. Remember that this chapter deals *only* with the *equal-pitch* hip or valley rafter design.

PROBLEMS AND QUESTIONS

1. What is an equal-pitch roof design in a hip or valley?
2. How many hips are shown in Figure 8-1?
3. What type of rafter is used in dormer framing?
   a. hip **b.** valley
4. In addition to a plumb cut at the ridge, a hip or valley rafter probably uses a ________ cut.
   a. end **b.** side
5. In an equal-pitch roof the hip or valley rafter is always at an angle of________from the exterior wall plates.
   a. 30° **b.** 45° **c.** 60° **d.** 90°
6. Hip or valley run per foot of run of common rafter is
   a. 12 in. **b.** 17 in.
7. If a common rafter run is 10 ft and the rise is 4 in 12, what is
   a. The run length of a hip rafter? **b.** The length of the hip rafter?
8. Write a short paper explaining why a bevel square is set for angles less than 45° when marking side-cut bevels on hip or valley rafters.
9. From your study of Figure 8-8, at what angle would you set a portable circular saw table to make the side cut *if* the plumb cut line were already drawn?
   a. cannot determine
   b. an angle of 45°
   c. an angle of 90°
10. Why must an adjustment for length be made to the ridge end of a hip rafter?
11. Explain why backing cuts need to be made to the top edge of each hip rafter.
12. What is dropping? How is the distance of dropping determined?
13. What would the estimated rafter tail length be for a hip that had a 5-in-12 pitch and planned projection of 12 in.?
    a. 12 in. **b.** 18 in. **c.** 24 in.
14. What size of hip or valley rafter would you use if the common rafters were 2 × 8s?
    a. 2 × 8 **b.** 2 × 10 **c.** 2 × 12

PROJECT

We ended the project in Chapter 7 with the identification of the materials necessary for the common rafters, ridge, and collar ties. In this project we continue to build the porch roof. Our goal is to determine the valley rafters needed and the associated work necessary to install the rafters.

A. Getting Ready to Install the Valley Rafter

1. First we define the position of the valley rafter and mark existing common rafters for cutting.
   a. Our first job is to prepare the roof to define the position of each valley rafter. We will make some assumptions and further we will spend time with only one side (valley) since we know that the other is identical but opposite (see Figure P8–1).
      (1) The valley rafter will be made from nominal 2-in. stock and will be 2 in. wider than the common 2 × 8 rafters. Therefore, each valley rafter will be a minimum of a ____________ .
      (2) Since the run of the house is 16 ft, the run of the porch is 15 ft, and the pitch is *6 in 12,* approximately how far below the house ridge will the porch ridge intersect?

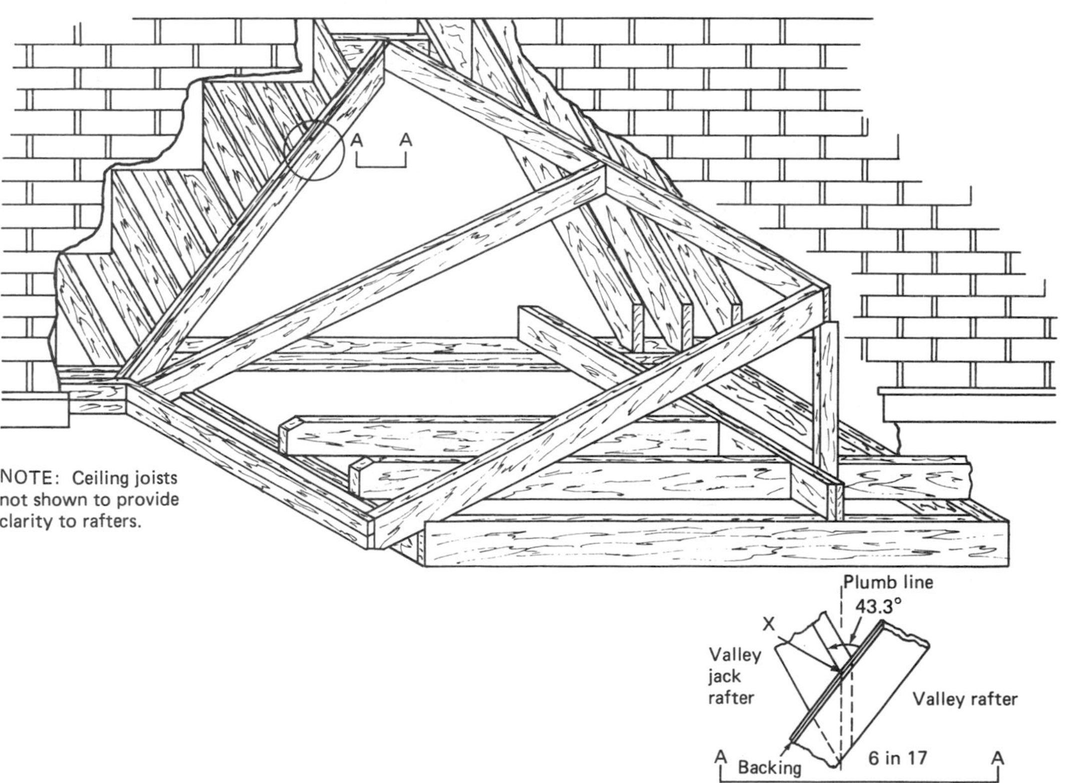

Figure P8–1

(3) Since the valley is an equal-pitch valley, approximately how far from the house's ridge toward the porch will the intersect point of the valley rafters and porch roof ridge be?

(4) As shown in Figure P8–1, a block of 2 × 8 needed to be installed to span the distance of at least two common rafters. How do we make room to install this piece?

(5) After the block is installed, the next thing to do is to install the remainder of the porch ridge.

(6) On the rafter tails of the porch we should install the 2 × 4 pieces onto the vertical cuts, thereby forming the inside corner of the valley.

b. Defining the valley rafter can now be done. Further, other associated tasks can be identified.

(1) The center of the valley must be measured from two intersect points (see Figures 8–11 and Figure 8–21). Where are these points?

(2) Since this line represents the center of the rafter, we must make allowances for one-half its thickness to the left of the rafter placement for the side shown in Figure P8–1. How much should this be?

(3) Now we would snap a chalk line from the two points marked in the previous step across the tops of the common rafters. What tool would you use to mark the plumb line or the surface of one of the common rafters? (*See* Figure P8–1, inset A–A, point *x*.)

(4) With that rafter marked and ready for cutting, what tool would you use to mark the remainder of common rafters that have chalk line marks on them?

(5) What must we do as a temporary measure before we cut each common rafter?

(6) Since we know that the valley is an equal-pitch design, we know that each jack rafter is 45° from the valley. We make a cut in each rafter with the portable saw set to what angle?

(7) As each common rafter is cut, its lower section is removed from the roof and saved. When you have finished, your common rafters should look like those in Figure 9–3.

c. Define the valley rafter's length and cuts. Refer to Figure 8–5. Imagine that points A (the ridge point) and B (the lower center point) are the centerline of the valley. The simplest way to measure the length is with a tape measure. I would certainly recommend this method.

(1) But given that this is a paper-and-pencil project, we will need to calculate the valley's length using other techniques. We can see an inset in Figure 8–5 that provides us with a clue to the horizontal length per foot run for valley rafters. What is this length?

(2) Now let's calculate the basic valley rafter length using data from Figure 8–6 and Table 8–1.

Formula: Rafter length = number feet common rafter run ×

$$\frac{\text{L/foot run (in.)}}{12\text{ in.}}$$

$$= \frac{\underline{\qquad} \times \underline{\qquad}}{12}$$

$$= \frac{\underline{\qquad}}{12}$$

$$= \underline{\qquad}\text{ ft}$$

(3) Now let's add to the basic length enough material to allow for the rafter tail. Since we have a planned 12-in. overhang, a look at Table 8–1 under 6 in. (for a 6-in-12 pitch) gives us the exact length needed. This is ____________ ft.

(4) The total length needed is the sum of the basic and tail lengths, or ____________ ft.

2. Determine the cuts to be made on the valley rafter.

a. What four cuts should we make on the rafter to make it fit into place?

b. To make the ridge plumb cut you would follow Figure 8–11b by using a framing square and the numbers ________ and ________ to mark the plumb cut. To know where to make your mark, you should first mark the center of the clown edge with a pencil line. Then you should set your bevel square for a 6-in. rise, as shown in Figure 8–10. Mark a line from the centerline on the rafter to the edge with the set bevel square. Then using the framing square, make your cutting line along the flat side of the rafter. Repeat the process on the opposite side and cut both sides with the power saw set at 45°.

c. Since you know the actual length, can you use the same tools and setting to make the rafter tail cut? How must you adjust the saw setting?

d. To make the bird's-mouth cut, you would follow the guidance shown in Figure 8–24. The framing square numbers to use are ________ and ________. The position of the bird's mouth can be obtained three ways: (1) place the cut rafter into its actual position and mark the place for the bird's mouth; (2) lay it out by stepping off the location with the framing square; or (3) use the data from Table 8–5, 12-in. overhang. From the interior point of the rafter tail cut, you would measure up to a point ________ in.

e. Finally, you need to make a backing or drop adjustment onto the rafter. From Table 8–3, for the backing or drop adjustment value for a nominal 2-in. hip or valley rafter, what amount will you be using?

B. Installation

1. What is the sequence of installation of the valley rafter? (Order the following steps correctly.)

a. Nail the rafter to the ridge support. ________

b. Nail the 2 × 4s to the rafter tail. ________

c. Insert the rafter into its position. ________

d. Nail each previously cut-off common rafter to the valley rafter. ________

e. Nail the valley to the wall plate. ________

Answers:

Part A: **1. a. (1)** 2 × 10

(2) approximately 6 in.

(3) about 12 in.

(4) remove one or two common rafters that are in the way

b. (1) center intersect point at the ridge and intersect point when the lower edge 2 × 4s join

(2) ¾ in.

(3) level

(4) a bevel square or continue to use the level

(5) brace it with a 2 × 4 to a ceiling joist below (on the part that will be retained)

(7) 45°

c. (1) 17 in.

(2) (15 ft × 18 in.) = 270 ÷ 12 = 22½ ft.

(3) 18 in. or 1 ft 6 in.

(4) 24 ft

2. a. ridge plumb cut, tail plumb cut, bird's mouth, and ridge backing or drop cut

b. 6 and 17

c. yes, the depth of the cut must be exactly correct to form the groove needed (see Figure 8–20)

d. 18 in.

e. ¼ in.

Part B: Order: c, a, e, d, b.

Chapter 9

The Roof—Part III: Jack and Cripple Rafters

In each roof constructed with hip and valley rafters there is a need for jack rafters and sometimes cripple rafters. These are specially cut rafters that are actually modified common rafters. This chapter provides explanations and illustrations about both types of rafter. Definitions are given and comparisons are made between the hip, jack, and common rafters and the jack and cripple rafters. Specialized data are provided for the determination of rafter length and the different cuts.

DEFINING THE LOCATION OF A HIP, VALLEY, AND CRIPPLE JACK RAFTER

Figure 9–1 illustrates two roofs that include the rafters we will study in this chapter. Both roofs are defined as hip and valley roofs. In Figure 9–1a all ridges are at the same height, whereas in Figure 9–1b one ridge is much lower than the other. Even so, the principles of jack rafter identification remain the same.

On the left of Figure 9–1a *hip jack* rafters extend from the building plates to the hip rafter. This illustration shows six hip jacks, four on the end and two on the side. Notice that the upper end of each hip jack is connected to the hip rafter.

The right of Figure 9–1a illustrates a valley and *valley jacks.* These are rafters that are connected to the ridge and extend downward to the valley rafters, where

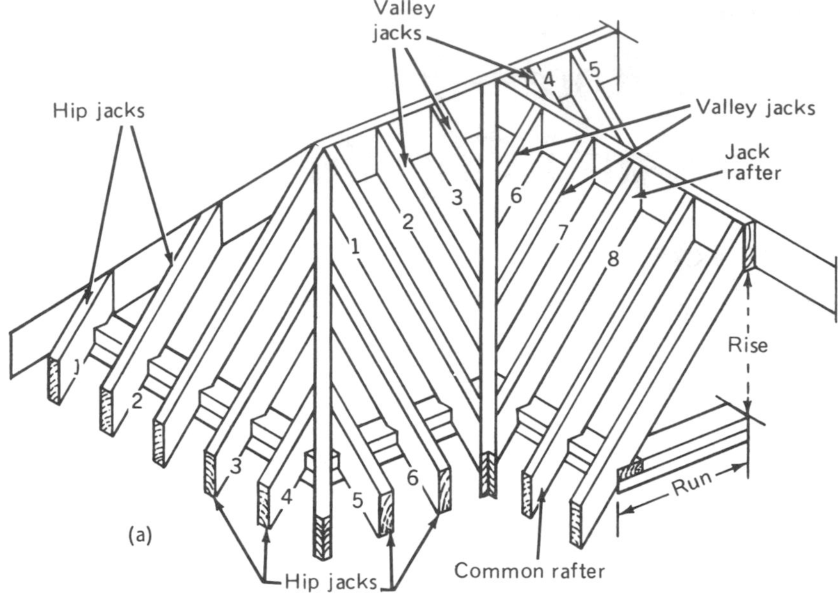

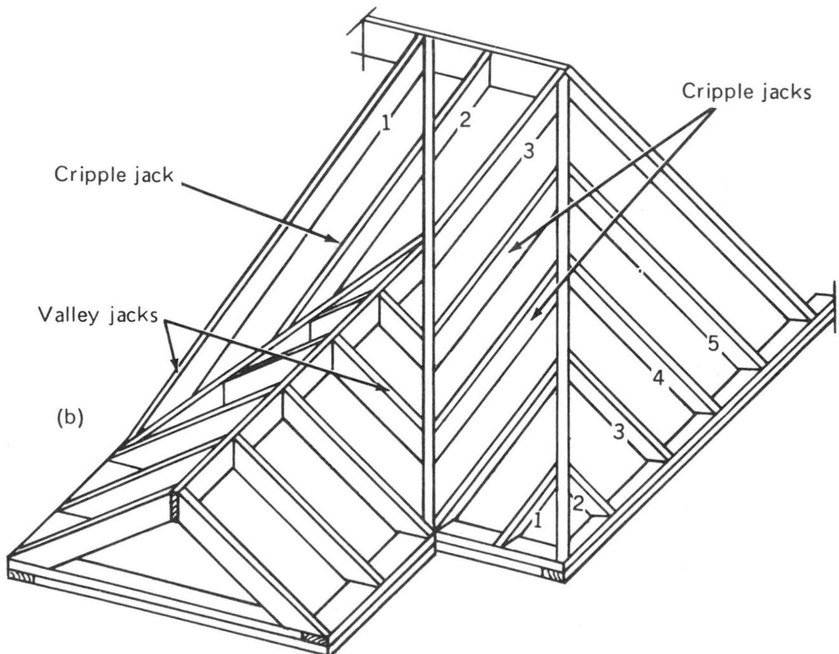

Figure 9-1 Framed hip and valley roof design using jack and cripple rafters.

their lower ends connect. The illustration shows eight valley jacks. Figure 9–1a does not show any cripple jacks.

Figure 9–1b illustrates a roof containing all three types of jack rafters. On the left, the lower roof uses four valley jacks, and the higher roof uses three valley jacks. The lower roof does not use any hip jacks, but the higher one uses five hip jacks. Between the valley and hip on the higher roof are several rafters that do not touch either the wall plate or ridge. These rafters are defined as cripple jacks. Notice that there are three between valley and hip. Also notice that there is one other cripple jack connected between both valley rafters (left of longer valley rafter). It is classified as a cripple jack because it does not connect to either the plate or ridge.

These definitions of jack rafters are always true regardless of the roof design. But even though the hip jacks rest on plates and valley jacks are connected to the ridge, could there be correlation between these rafters and the common rafter to make understanding them easier?

DETERMINE THE COMMON AREAS OF JACK AND CRIPPLE RAFTERS WITH COMMON RAFTERS

Rise and Run Values

The rise and run values of a hip jack and valley jack are the same as those of the common rafter in the same roof. This means that if, for example, a roof has a 4-in-12 pitch, the hip and valley jacks also use the *4 in 12*. How and where are the rise and run or pitch used? These values are used for laying out the bird's-mouth and rafter tail cuts on the hip jack, and to lay out the ridge cut on the valley jack.

Figure 9–2 shows a side view of a hip jack as though looking through the hip so that the correlation of rise and run are understood. Notice that all the hip jacks are in alignment with common rafters. This is as it should be. If a similar view were prepared for the valley jack, their alignment with the common rafter would agree. However, the common connecting point would be at the ridge.

Bird's Mouth and Rafter Tail

Figure 9–2 shows that the bird's mouth and rafter tail of each hip jack rafter are identical to the common rafter's bird's mouth and rafter tail. The pattern that was designed to lay out this portion of the common rafter should be used when laying out the hip jack rafter. Since the length of the jack will be shorter than the common rafter, it is a good idea to draw a plumb line to the rafter's top edge as the figure shows. The rafter's length can then be measured along the top line.

Ridge Cut

Each valley jack rafter needs to have a common rafter ridge cut. This cut is made using the roof pitch and framing square layout, followed by cutting. Normally one-

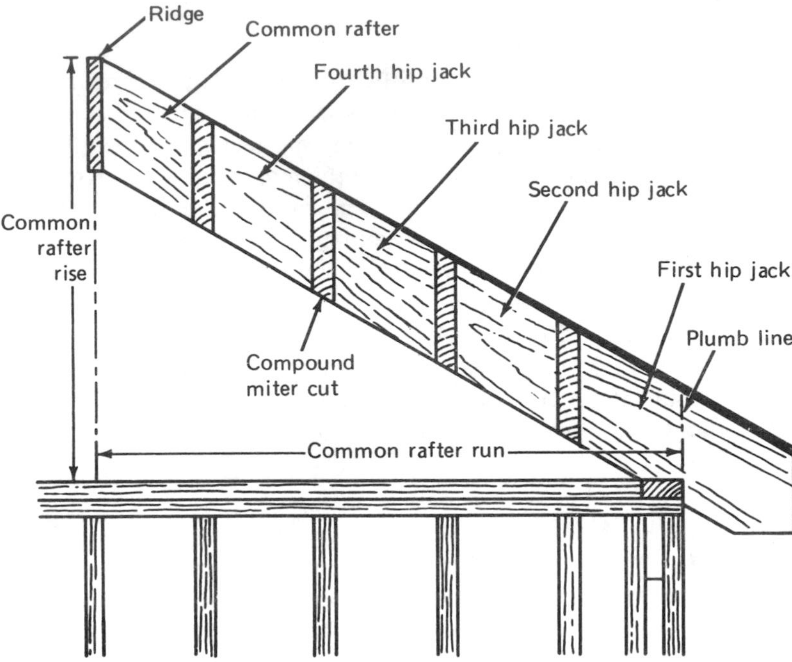

Figure 9–2 Comparing the hip jack rafters to the common rafter.

half the ridge board thickness is subtracted from the common rafter's length before the ridge cut is completed. This allowance is necessary when valley jacks are cut. However, rather than cutting it from the ridge cut, the allowance can be made a part of the compound miter cut at the lower end of the rafter. Figure 9–3 shows four valley jack rafters spaced evenly as they appear with the valley rafter removed. Carefully note that the lower end of each jack has a compound miter cut. One part of the cut is the pitch; the other part is the bevel needed for the rafter to join the valley rafter.

Cripple Rafter and Jack Rafter

The cripple jack rafter does not touch either the plate or ridge, yet its pitch is the same as the common rafter and jack rafter. It therefore must fall on a plane with these other rafters, as Figure 9–4 shows. Since it touches neither plate nor ridge, two compound miter cuts are needed. One cut is made so that the cripple seats properly against the hip rafter; the other cut is made so it seats against the valley rafter. (View Figure 9–1b if reorientation of the cripple jack is needed.) It may also be that the compound miter is made to fit the rafter between two valley rafters. In any event the plumb cut for the common roof pitch is one of the two bevels used in the compound miter.

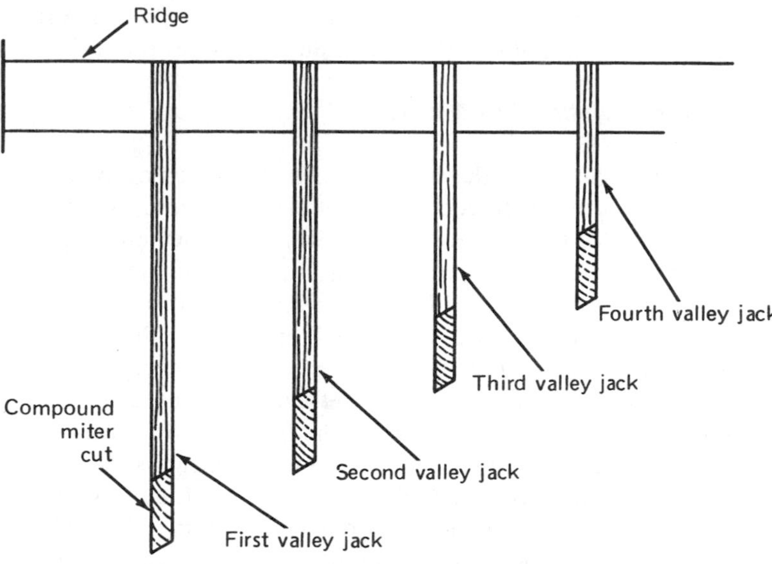

Figure 9-3 Valley jack rafter spaced properly, fastened to ridge, and revealing the compound miter cut needed to seat against the valley rafter.

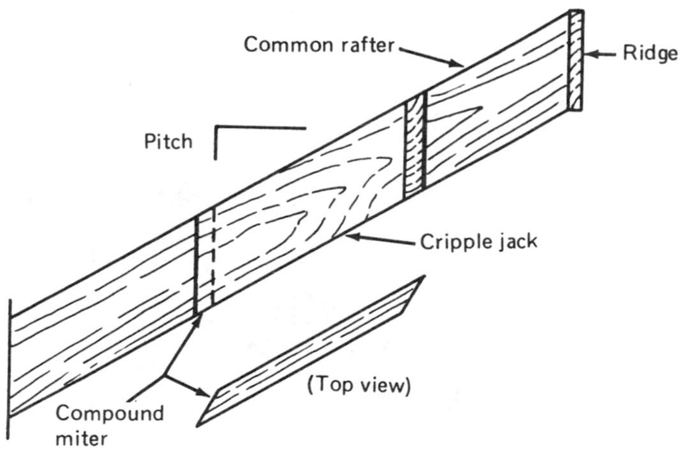

Figure 9-4 Comparing the cripple jack to the common rafter.

DETERMINING THE JACK RAFTER LENGTH AND MITER CUTS

There are several ways to determine the length of a jack rafter. One is by physically measuring its length on the roof assembly, another is by using the jack rafter tables on a framing square, and a third is by calculating the length. The two most practical are the measuring techniques and using the rafter table on a framing square. Sometimes a combination of these techniques simplifies the task.

Practical Measuring Method

Figure 9-5 illustrates the practical measuring method of finding the length of the first hip jack when the roof layout of rafters is from right to left. The last common rafter installed is the reference from which all calculations begin. On-center spacing is marked at the top edge of the hip (in this example, 16 in.). The plate is already marked. However, a level line (chalk line) should be established as shown and the rafter length is measured with a steel tape. Notice that the length obtained is from the *longest* point of the compound miter and does *not* include any rafter tail.

Figure 9-6 shows that this method can be used for finding the length of the shortest hip jack. In this situation the layout of on-center spacing is considered to be left to right. Once again the jack length is from its longest point to the level line.

A handy aid that is easily made on site is the *marking* stick, shown in Figure 9-6. A 1 × 2 or 1 × 4 laid off at on-center spacings from plate markings can easily be tacked to common rafters and hip rafter. *A word of caution:* The marking stick must be parallel to the level line for accurate jack lengths to be determined.

The methods used in finding the hip jack length are easily adapted to finding the valley jack lengths. Figure 9-7 shows that the longest valley jack's length is found with square and tape measure. The square provides a true on-center point on

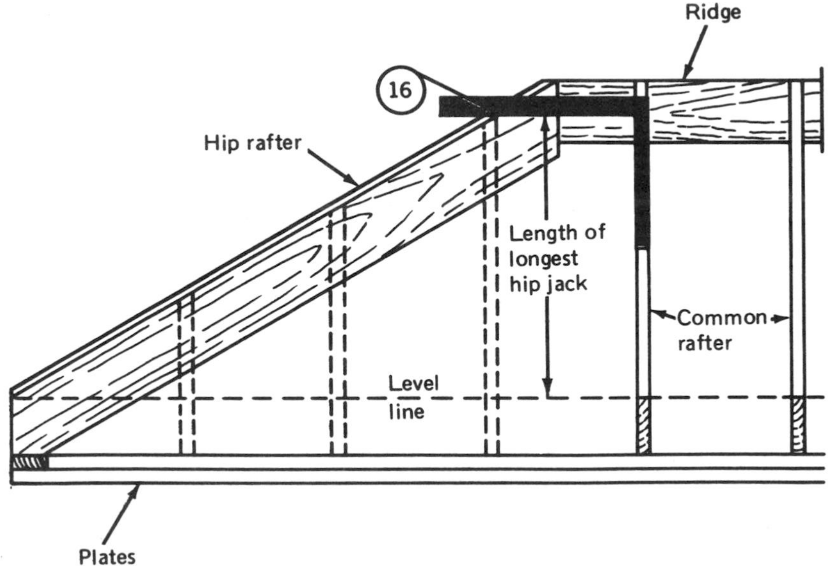

Figure 9-5 Practical method of defining the longest hip jack rafter.

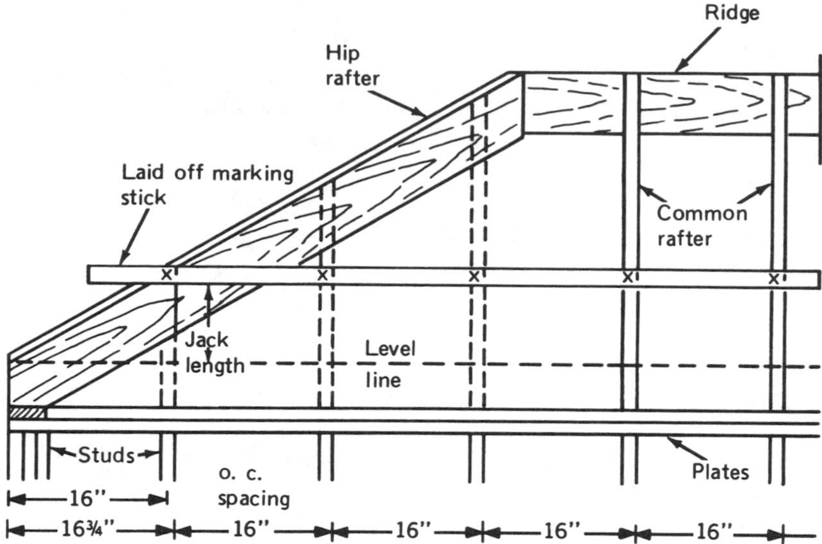

Figure 9–6 Practical method of defining the length of hip jack rafter, shortest rafter first.

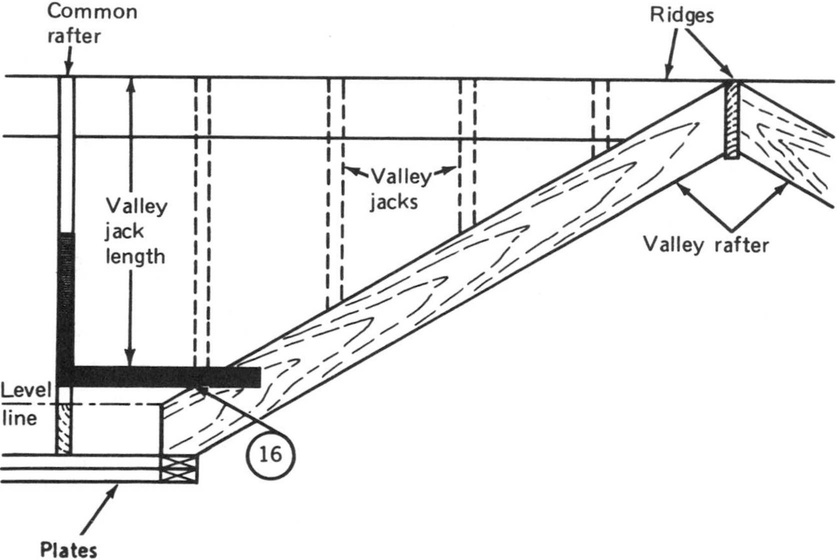

Figure 9–7 Practical method of laying out on-center spacing and defining length of longest valley jack rafter.

the valley rafter. The measure of the valley jack is from this point to the ridge. Notice that the point is the tip of the bevel; however, this point is not the longest part of the rafter layout since the pitch cut must be made from this point as well. This idea is brought out only to preclude the reader from assuming incorrectly that a rafter of roughly this length is sufficiently long. Allowance must be made for the plumb cut.

The laid-off marking stick may also be used as an aid in determining the valley jack rafter length. If used, it must be aligned parallel with the ridge rather than the plate. One final word about this method: It is equally well suited to determining hip or valley jack rafter lengths for on-center spacings of 24 in. or other on-center spacings.

Framing Square Table Method

Many of the more expensive framing squares provide tabular data on the length and difference in length of jack rafters. They are frequently titled "Difference in length of jack rafters" because the length of the first shortest or first longest hip or valley jack is variable. However, once the length of either is known, the table data are appropriate to use.

Table 9–1, which corresponds to Figure 9–8, lists most of the common roof pitches as well as one or two extremes. Notice that this table is for the difference in length of jack rafters spaced 16–in. o.c., as might be found on a framing square. Also notice the 12 right triangles below the table. These are used to help explain the table and orient the carpenter.

Assume that a roof has a *3-in-12* pitch. The difference in length of the jack rafter in Table 9–1 is shown as 16.49 in. The triangle with the 3 included shows how this value compares with (1) the 16–in.-o.c. spacing and (2) the hip or valley length per foot of run. A translation to meaning is that for each 16 in. of spacing along the plate or ridge line the jack rafter's length must be increased 16.49 in. or 16½ in.

TABLE 9–1 DIFFERENCE IN LENGTH OF JACK RAFTERS (16 IN O.C.)

Rise (in.)	3	4	5	6	7	8
Diff./16-in. o.c.	16.49	16.87	17.33	17.875	18.52	19.23
Rise (in.)	9	10	12	15	18	24
Diff./16-in. o.c.	20	20.83	22.625	25.61	28.84	30.72

TABLE 9–2 DIFFERENCE IN LENGTH OF JACK RAFTERS (24 IN. O.C.)

Rise (in.)	3	4	5	6	7	8
Diff./24-in. o.c.	24.74	25.30	26	26.83	27.78	28.84
Rise (in.)	9	10	12	15	18	24
Diff./24-in. o.c.	30	31.24	33.94	38.42	43.27	

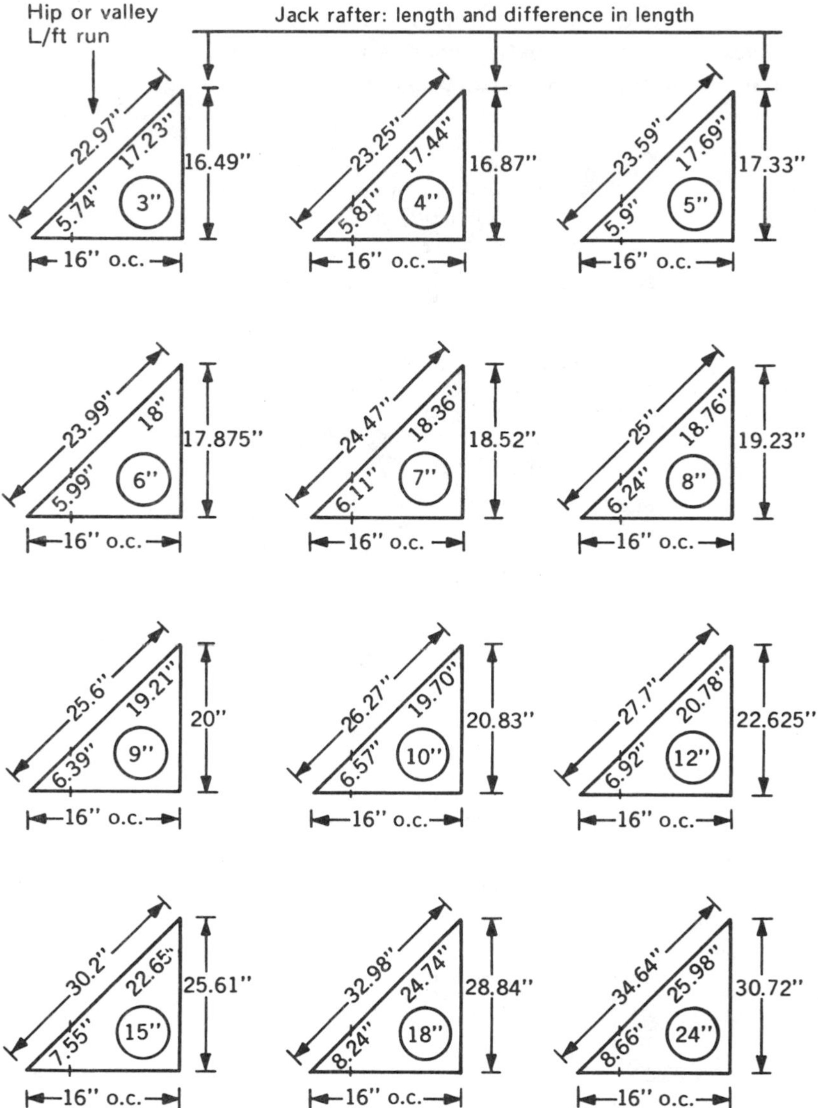

Figure 9–8 Mathematically determining the length and difference in length of jack rafters for numerous roof pitches.

The length of the jack rafter from the point of view of the hip or valley must be increased or decreased in length 22.97 in.

Table 9–2 provides quick reference of the same kind of data but for 24-in.-o.c. spacing of jack rafters. For a 3-in. pitch where jacks are spaced 24-in.-o.c., the difference in length is 24.74 in. or 24¾ in.

The table and right triangles shown in Figure 9–8 do not always lend themselves to translation that is meaningful. Therefore, four examples have been chosen that should make their meaning clear.

Example 1

Figure 9–9 illustrates two *hip roof jack rafter* situations that make use of the tabulated data. In Figure 9–9a a pitch of 3 in 12 is shown. An orientation of the figure is necessary before explanation can proceed. Locate the following points:

Hip rafter line: line that the hip rafter follows from plate to ridge.

Wall plate line: horizontal wall plate that the jack must seat upon.

16-in. o.c.: spacing of the jack rafters

3 hip jack rafters

16.49 in.: difference in length of the jack rafters taken from Table 9-1 (16.49 in. converted to 16½ in.)

44.1°: angle of one miter of the compound miter

The *dashed line* that is parallel with the *wall plate line* begins at the intersect point of the shortest jack and bisects both jacks to the right. From this line the difference in length of jacks is determined from Table 9-1. The length needed to be added to the second jack is 16.49 in. or 16½ in. Therefore, its total length is x + 16½ in. where x is the length of the first jack. The third jack has a length

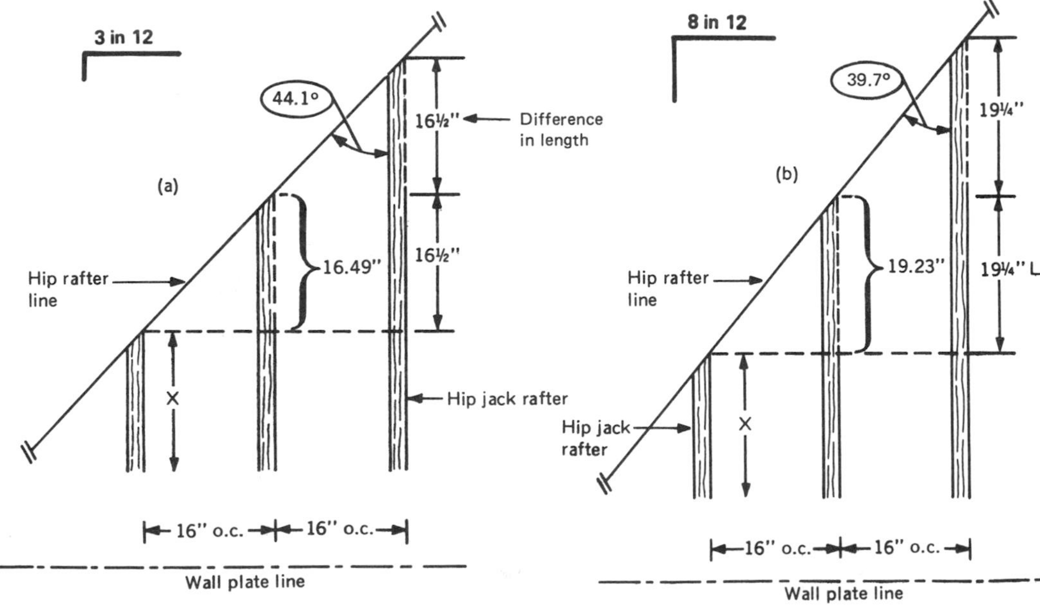

Figure 9–9 Laying out the hip jack rafter for difference in length and defining the number of side-cut degrees.

of x + 16½ in. + 16½ in. because a step of 16 in. to the right calls for an additional 16½ in. of jack rafter length.

If a ruler were used to measure the hip rafter line between any two similar points where jacks intersect the hip, its length would be 22.97 in. or almost 23 in. One other important factor in the drawing is the 44.1° miter needed at the tip end of the hip jack. Each pitch results in a different angle. Zero pitch, for example, would result in a 45° miter. (A table with these miters is given later in this section.)

Example 2

The significant difference in Figure 9–9b is that all data are for an *8-in-12* pitch. The identities of hip rafter line, wall plate line, three jack rafters, and 16-in.-o.c. spacings are the same as in Example 1. However, a much steeper roof is being constructed when the pitch is 8 in 12. Therefore, the difference in length of the jack is predictably longer. Table 9–1 calls out a difference in length of 19.23 in. The second jack shows that its total length is x + 19.23 in. or x + 19¼ in. L.

What is the L for? Since 0.23 in. is just short of 0.25 in., which is equal to ¼ in., the 19¼-in. mark must be termed *light.* That is, the carpenter makes a mark on the lumber at a length of x + 19¼ in. He or she then cuts the stock so that almost all or all of the mark is cut away, too. By this adjustment to the cutting, he or she accounts for the slight discrepancy that was introduced in converting hundredths to fractions. It also means that when laying out the third jack, there must be two light (L) adjustments. One he or she makes by using the very leading edge of the pencil mark, and the other by cutting away the line. Needless to say, these are sophisticated adjustments that are used by the experienced carpenter and must be learned by the apprentice.

Example 3

Figure 9–10 provides data on *valley jack rafters* for the *10-in-12* and *5-in-12* pitch roofs similar to that given in Figure 9–9. However, there are several differences.

Roof ridge line: must be the horizontal reference for valley jacks instead of the plate line

Valley rafter line: position of the valley rafter

Difference in length: may be considered either from shortest to longest, or vice versa

The *dashed lines* parallel with the *roof ridge line* (Figure 9-10) show the point of intersection of the valley rafter and jack rafter. The difference in length for this pitch is 20.83 in. So the length of the jack *at the shortest point* of the second jack is equal to x + 20.83 in. or x + 20¹³⁄₁₆ in. F (F stands for full). No problem exists using this method since additional rafter stock needed for the compound miter is already understood to be necessary. Note also that the angle of the miter is 37.5°, which is quite a bit different than the 44.1° used with the *3-in-12* pitch.

Example 4

Figure 9-10b shows a *5-in-12* pitch roof with valley and valley jacks. Data are similar to the other three examples except that the difference in length is 17.33 in. as Table

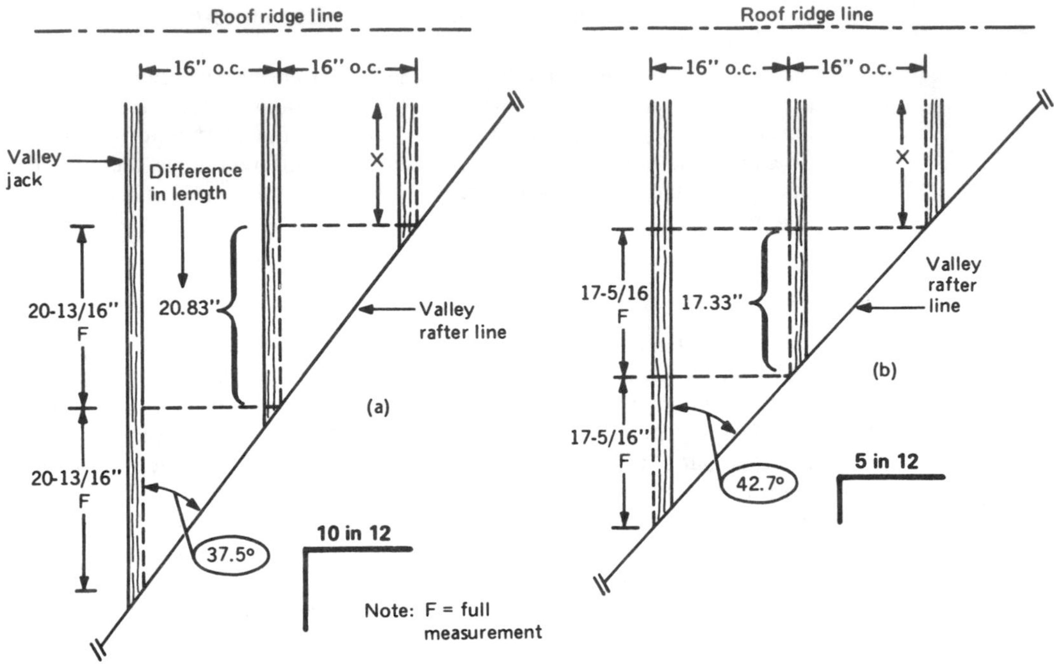

Figure 9–10 Laying out the valley jack rafter for difference in length and defining the number of side-cut degrees.

9–1 in Figure 9–8 lists. As before, the total length of the longest jack shown is x + 17⁵⁄₁₆ F + 17⁵⁄₁₆ F.

Full means that the thickness of each marking line must be used and it must still be on the stock after the cut is made. In the conversion of 0.33 to ⁵⁄₁₆ in. there is a discrepancy: ⁵⁄₁₆ in. = 0.315. The thickness of a pencil line will just about make up the 0.015 shyness.

Jack Rafter Compound Miter Cuts

With this orientation and determination of the jack rafter length the compound miter cut can be examined. Since the four examples just studied provide a variety of pitches as well as examples of hip and valley jacks, their characteristics are used in explaining the compound miter cut.

It should be understood that the compound miter cut of all jacks fit against either the hip rafter or valley rafter. It is also important to know that as the rise of a roof (pitch) increases the bevels of the compound miter become more acute. Figure 9–11 shows the four jack rafters illustrated in Figures 9–9 and 9–10. Notice the following data as support for the previous statement of pitch versus miters:

Pitch	Degrees
3 in 12	44.1
5 in 12	42.7
8 in 12	39.7
10 in 12	37.5

It is shown that the plumb-cut angle increases with a rise in pitch and also that the side-cut angle becomes a more acute angle. Obtaining the plumb-cut layout with the framing square is relatively easy, but laying out the side cut can be difficult if degrees are used. There is a purpose for listing the degrees as shall be shown.

Three methods are available for use in laying out and cutting the side cut of the jacks.

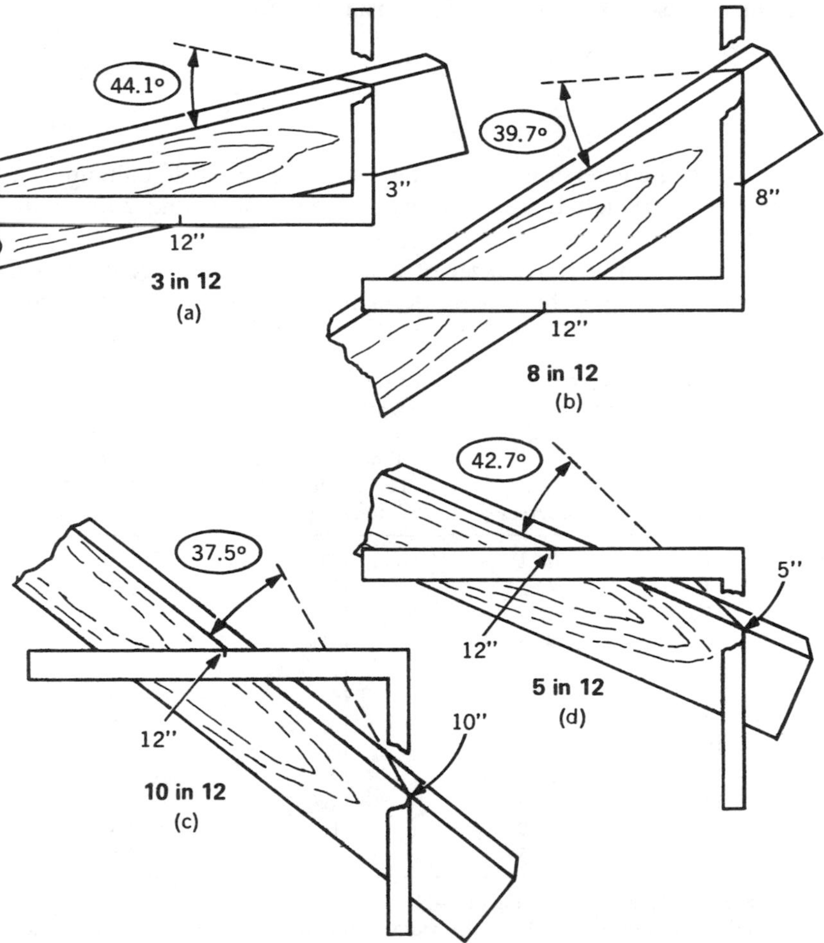

Figure 9–11 Laying out the compound miter cut for various pitch roof jack rafters.

Method 1. The framing square provides one solution. Either the table imprinted on the square for side cuts should be followed or the one in Table 9–3 should be used. If Table 9–3 is used, column 3 provides the combination of values needed for the proper degree miter required. For example, to lay out the side cut for a 3-in-12 pitch, place 12 in. (blade) and 12 7/16 in. (tongue) such that this last value intersects with the longest edge of the plumb cut; then mark alongside the tongue.

Method 2. The second method uses a bevel or miter square and framing square as Figure 9–12 shows. A bevel square is set to a predetermined value and locked tightly with its wing nut. This tool is then used to lay out all side cuts for jacks that have common rises. (Refer to Table 9–3 and Figure 9–12 throughout the explanation.)

The blade of a framing square is 2 in. wide; the number of degrees for side cut per rise is known; therefore, using the formula

$$x = \frac{y}{\tan}$$

provides a conversion that defines the values listed in column 4 of Table 9–3. Note in a right triangle these values would be the base, and 2 in. would be the opposite side.

Since the *outer corner* of a well-used square receives much abuse, its exact position may not be exactly defined. Therefore, the setting of the bevel square

TABLE 9–3 SIDE CUTS FOR JACKS AND CRIPPLE RAFTERS

(1)	(2)	(3)	(4)	(5)
		Framing square	Miter square on framing square (in.)[a]	
Rise	Degrees	12 in. on blade tongue (in.)	($x = y$)/tan[b]	1-in.[c]
3	44.1	12 7/16	2 1/16	1 9/16
4	43.5	12 5/8	2 1/8	1 5/8
5	42.7	13	2 3/16	1 11/16
6	41.8	13 7/16	2 1/4	1 3/4
7	40.8	13 15/16	2 5/16	1 13/16
8	39.7	14 7/16	2 7/16	1 15/16
9	38.6	15	2 1/2	2
10	37.5	15 5/8	2 5/8	2 1/8
12	35.3	17	2 13/16	2 5/16
15	31.9	19 1/4	3 3/16	2 11/16
18	29.0	21 5/8	3 5/8	3 1/8
24	20.5	32	5 3/8	4 7/8

[a]See technique shown on Figure 9–12.
[b]Formula assumes $y = 2$ in., the width of framing square blade; values rounded to the nearest 1/16 in.
[c]Readings on inner blade scale; miter square's steel edge set on 1 in.

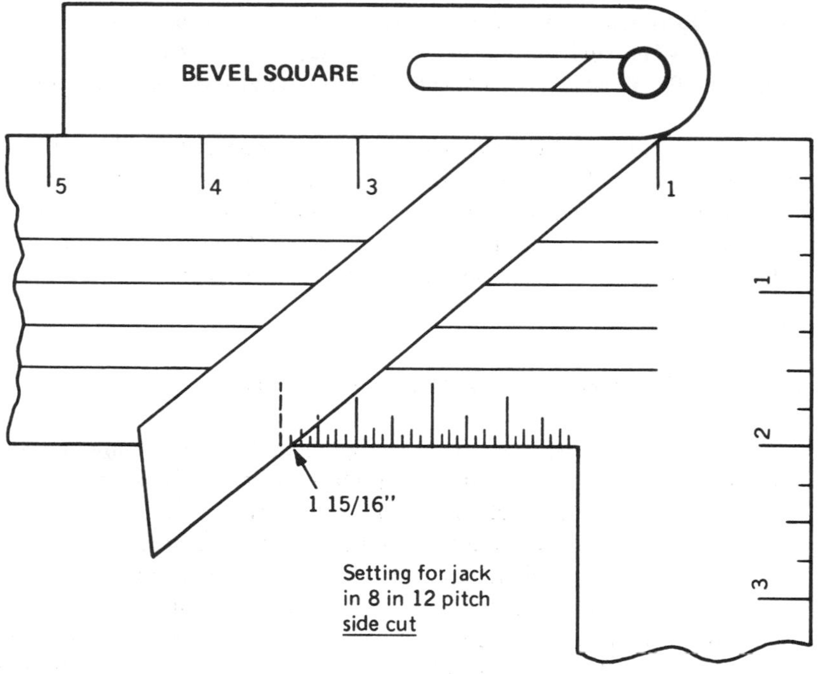

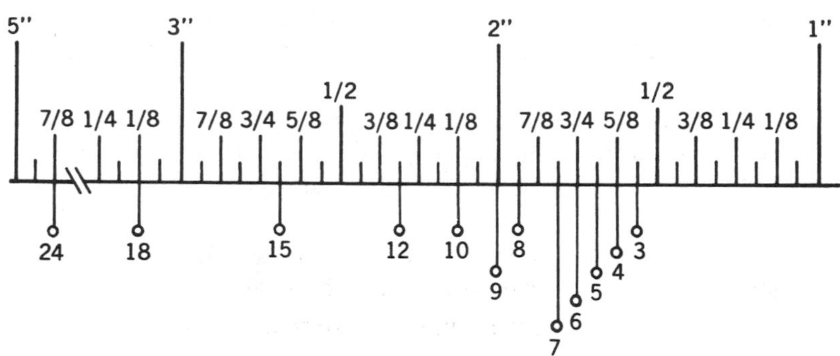

Figure 9–12 Bevel square setups for jack and cripple rafter side cuts.

should be made from the *1-in.* mark as shown in Figure 9–12. It is also standard that the framing square's tongue be 1½ in. wide, so this value has been subtracted from the value in column 4 of Table 9–3 and the remainder is shown in column 5.

The blade on the miter square should be set according to column 5 of Table 9–3 and Figure 9–12 for each side-cut bevel needed. For example, if an 8-in-12 pitch is used, the bevel square is positioned as shown in Figure 9–12a, where the edge of the blade is set against the 1 in. on the outer numbers and 1 15/16 in. on the inner numbers. The scale at the bottom of Figure 9–12 is expanded to show the points that should be used with various pitches if the bevel square technique is used.

Method 3. The third method uses the degrees listed in Table 9–3 and a radial arm power saw. This power saw's tilting head is set to the number of degrees shown in column 2 of Table 9–3 and its arm is set for the plumb cut. No marks for side cuts need to be made on jacks *after* a test cut has been verified as accurate. With the saw set properly, a single operation cuts the compound miter.

LENGTH AND BEVEL CUTS ON A CRIPPLE JACK

Once the proper orientation of a cripple jack rafter is understood, the practical determination of its length and compound miter cuts become relatively easy. Refer again to Figure 9–4, which shows the orientation of the cripple jack to common rafter. This figure shows that the rise per foot run of the cripple jack is the same as the common rafter. This also means that the hip and valley jack and cripple have common elements. Their compound miters are the same. One of the bevels of the compound miter is the plumb cut and the other is the side cut. These are found using the method for hip or valley jacks from Table 9–3 and Figure 9–12.

Practical Method of Determining the Cripple Jack Length

The simplest and most practical method of finding the length of the cripple jack is shown in Figure 9–13. It involves the use of a 1 × 2 piece of stock, a ruler, and mason line.

Assume that cripples are spaced 16-in. o.c. Assume further that the roof being framed resembles the one shown in Figure 9–13 and that a hip jack is already in place. The object is to find the length of the first and second cripple jacks.

1. Mark a point on the valley rafter which is 16-in. o.c. to the right of the hip jack, and partially drive a nail into the valley.
2. Install a 1 × 2 as shown, making sure that it is parallel to the plate. Mark on the 1 × 2 a point 16-in. o.c. as shown.
3. Install a mason line over the nail and center it over the mark on the 1 × 2, and where the line contacts the hip rafter (b), make a pencil mark.
4. Remove the mason line and measure the distance between points a and b. This gives the length (L) of the cripple.

Notice that the difference in length can be used; however, the length from point a to the beginning of the difference in length point c needs to be measured. So why complicate the practical method?

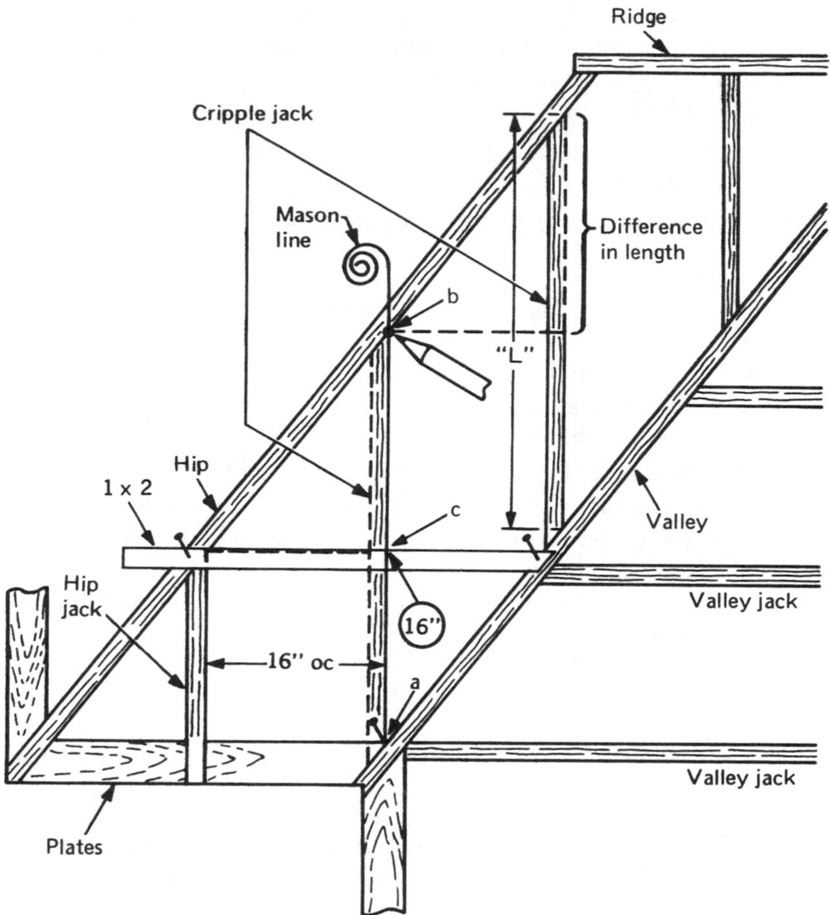

Figure 9–13 Practical method for defining the length of the cripple jack rafter.

Compound Miter Cuts on Cripple Jacks

With the length of the cripple defined, the compound miters need to be laid out and cut. The procedure is exactly the same as that used for hip and valley jack rafters.

Rather than elaborate the methods explained previously, suffice it to say that they may be used if the following suggestions are taken:

1. Use the same pitch as on the main common roof rafter to mark the plumb cuts.
2. Use the proper setting of bevel square or framing square.
3. Make sure that the lines drawn on the cripple jack stock are properly oriented according to their position of installation.

Since both cripples in Figure 9–13 and in most actual situations are exactly the same, cut two or as many as needed. Remember that when nailing these members in place they must align with their on-center spacing and be properly oriented for height to backed or dropped hip and valley rafter.

PROBLEMS AND QUESTIONS

1. What cut must be made to a common rafter to make it a jack rafter?
2. A hip jack rafter employs a
   a. ridge cut **b.** bird's-mouth cut
3. A valley jack rafter uses a
   a. ridge cut **b.** bird's-mouth cut
4. Why does a cripple jack rafter have a compound miter cut on each end?
5. Are the rise and run values for a jack rafter the same as for *common* or *hip* rafters?
6. Describe in a few paragraphs the practical method of determining the shortest jack rafter length.
7. Assume that a roof has a 6-in-12 pitch and rafters are spaced 24-in. o.c. The shortest jack is 23 in. long; what is the length of the next jack?
   a. 46 in. **b.** 49⅝ in. **c.** 50⅞ in.
8. Why is the jack rafter's side cut line an angle less than 45° when measured along the crown of the rafter?
9. Lay out and mark a 5-in-12 pitch for a common ridge cut on a scrap 2 × 8. Set your portable circular saw to 45° and cut along the line. Set your miter square so that its body is even with the top edge of the 2 × 8 and adjust and tighten the blade to the bevel cut made with the saw (not ridge cut). Next lay the miter square as shown in Figure 9-12 and compare the values. Did the cut produce a value equal to that in column 5 in Table 9-3? (See note c in the table.)

PROJECT

If you have been completing the project of building the porch, you know that in Chapter 8, we (on paper) designed and installed the valley rafter. In this project our concern is the design of the jack rafters needed to complete framing of the roof. Locate the work that we must perform by turning to Figure 9–1a. The rafters we will design are numbered 6, 7, and 8.

1. Recall that the approximate length of porch ridge from the point of the last common rafter to the end extending over the valley is about 16 ft (1 ft over the overhang and 15 ft over the valley area). Since the rafters are installed on 16–in. centers, there should be approximately ______________ jack rafters.
2. There are two methods for the design of the valley rafters. One is to continue from the

point where the last common rafter was installed and install one every 16 in. Two, we could match the position of each valley rafter to its counterpart on the main roof (see Figure 9–1a, rafters 1, 2, and 3). Since the design requires basically the same effort, we will opt for matching valley rafters on the porch roof to those on the main roof. Since the common rafter for the porch is approximately 18 ft long, we know that all valley rafters must be shorter. The first step is to find the length of the longest valley jack by using the practical method. Since we are using the common roof jacks as our reference, what should we do?

a. Should we use a combination square and mark a perpendicular line across the longest point where the longest roof jack is nailed to the valley? If no, what should we do?
b. Should we apply the concept shown in Figure 9–7 next, except that the distance may differ from 16 in.? If no, what should we do?
c. Suppose that the distance measured is 15 in.; what measurement would you make and mark on the porch ridge from the last common rafter?
d. On which side of the mark on the valley and ridge would you place an "X" signifying the location of the valley jack?

3. Now we would take a tape and measure the jack's length. Let's assume that this measurement is 15 ft 4 in. We know that the overall length of the rafter will be longer because of the compound miter that fits against the valley. But we can safely predict that a ____________ -ft 2 × 8 will be long enough.

4. Before designing the cuts needed on each jack, let's produce a list of jack lengths that will meet the needs of the valley. The list should start with the known information, the longest jack. *Specs:* Rise = 6 in 12, o.c. spacing = 16 in. Use Table 9–1 for making the calculations; also use Figure 9–10. To be sure that we agree, indicate the difference in length here ____________, and check your answers; then complete the fill-ins below.

a. Longest valley jack	= 15 ft 4 in.	Order a ____________-ft rafter
b. Number 10	= ____________	____________
c. Number 9	= ____________	____________
d. Number 8	= ____________	____________
e. Number 7	= ____________	____________
f. Number 6	= ____________	____________
g. Number 5	= ____________	____________
h. Number 4	= ____________	____________
i. Number 3	= ____________	____________
j. Number 2	= ____________	____________
k. Number 1	= ____________	____________

5. We could spend a lot of time marking the bevel and plumb cuts on each rafter as the data in the chapter tell us and show us in Figure 9–12. If we were cutting the rafters by radial arm saw or hand saw, these lines would be needed. But we are going to use a portable power saw, so once we set the saw blade for 45° we will only need the plumb line marks for the valley cut. So to make the lines on each valley jack, we should use a framing square and the numbers ____________ and ____________ to make the line.

6. The ridge jack layout should be made with the framing square. Since this cut is the same as was made on the common rafters, what framing square numbers will be used?

Note: You should always try the first jack for fit before proceeding with the remainder. This frequently saves materials and time and ensures accuracy.

Answers:

1. 12 − 1 = 11, the last one can safely be omitted
2. **a.** yes, this is the best way to locate the rafter position
b. yes, you could use a ruler versus a framing square
c. the same, 15 in.
d. As you face the valley part of the roof, the "X" must go to the *left* of the marks
3. 16 ft
4. 17.875 or ≈ 18 in.
a. 16
b. 13ft. 10⅛ in. 16
c. 12 ft. 4¼ in. 14
d. 10 ft 10⅜ 12
e. 9 ft 4½ in., 10
f. 7 ft 10⅝ in., 10
g. 6 ft 4¾ in., 8
h. 4 ft 10⅞ in., 7
i. 3 ft 5 in., 4
j. 1 ft 11⅛ in., 3
k. 0 ft 5¼ in., 1 (skip this one)
5. 6, 12
6. 6, 12

Chapter 10

The Roof—Part IV: Other Requirements and Sheathing Techniques

Many times a picture and a few carefully chosen words can develop more fully an understanding of a complex topic. If a serious study has been undertaken thus far, the items in this chapter will easily fall into place with the understanding gained previously.

To avoid tiresome repetition of adaptations of principles covered previously, only the aspects of principles and tasks related to other framing needs are given. Subjects included are dormer construction, collar ties, gable end studding of a gable roof, installation of a fly rafter, and framing for a chimney that passes through a roof. The remainder of the chapter identifies sheathing techniques used on roofs. Sheathing of proper thickness grade and installation are all important aspects of roofing that must be carefully attended to.

DORMER CONSTRUCTION

Dormer without Sidewalls

The first in the series of dormers examined is the *dormer without sidewalls.* Figure 10-1 illustrates the framing technique used. Inclusion of such a dormer is rather expensive. Note the complexity of its construction. Initially, a frame of headers and

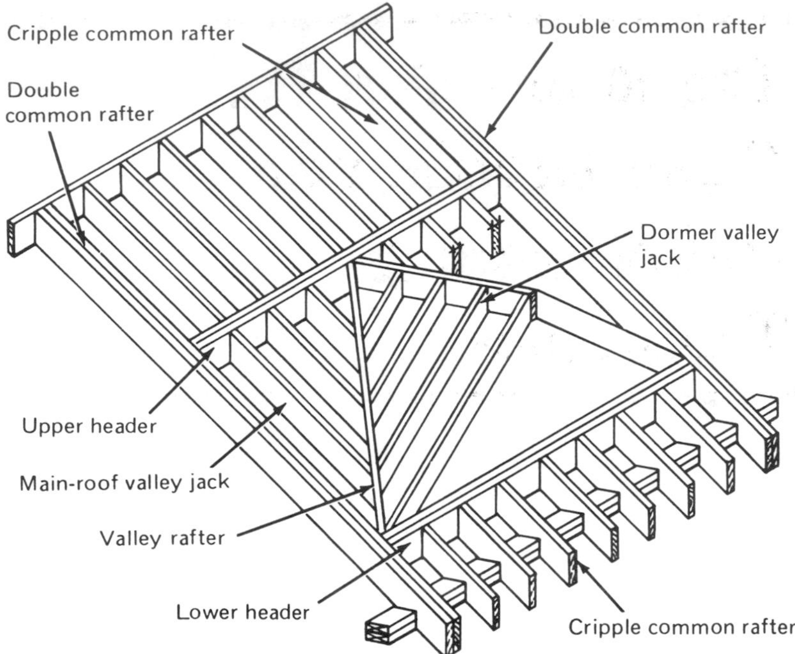

Figure 10–1 Dormer without sidewalls.

double rafters are constructed. Double headers are installed perpendicular to plate and ridge. The double common rafter, sometimes called a *trimmer rafter,* is needed to supply additional strength. *Cripple common rafters,* sometimes called *jack common rafters,* are installed with normal on-center spacing. Two valley rafters are installed next, followed by the installation of the main roof jacks and the dormer ridge and jacks.

From the details shown here, one new principle of roof construction is identified. It is the framing of the opening with headers and double common rafters. Note that the headers are cut from rafter stock. Their flat surfaces are not installed plumb as is the ridge; rather, they are nailed so that both top and bottom edges are flush with the rafter.

What significance does this header installation have on valley rafters and dormer ridge board layout, cutting, and installation? For one thing, each valley rafter has a single side miter cut rather than compound miter. The top cut is probably a 45° bevel; the lower end of the valley rafter uses two 45° cuts to fit into the corner. Neither end uses a plumb cut, which is different from normal valley rafters. Installation is the same as explained in Chapter 8; no special problems should be encountered.

Main roof jacks will only need single miter cuts since they, like the valley rafter, are on a flat plain. The opposite (upper) ends use square cuts. In contrast, however, standard procedures need to be used for laying out and cutting the dormer jack rafters since these elevate to a different ridge. Chapter 9's methods should be employed here.

Cripple common rafters need to be laid out from the pattern for common rafters and then measured for length, cut, and installed.

Dormers with Sidewalls

A dormer with sidewalls may be constructed where the walls rest on double rafters or where they are built up from studs resting on the ceiling joists. Both types are shown in Figure 10–2. In either case the studs and plates are usually constructed after the full roof joists are installed.

In the dormer shown in Figure 10–2a, the studs are cut and installed to rest on doubled rafters and headers. Two valley rafters are used and they connect to the header near the roof's peak and rest on the double rafters. Four jack rafters are needed as well as several short common rafters for the dormer roof.

Once again specialized layouts and cuts need to be made on the valley rafters where they are nailed to the header and dormer ridge. The lower end of each valley intersects with dormer plate line and common rafter, so a special layout is needed here as well. The cripple common rafters below the double header need to be *plumb cut* since it is important that the header be installed plumb.

In Figure 10–2b a standard wall-framing situation exists that simplifies the roof construction somewhat. A single lower header is all that is needed to support the rafters. Recall, though, that if the header length exceeds 4 ft, double headers should be used.

In framing the roof two short valleys are needed as well as four jack rafters. Notice that for the lower end of the dormer jacks, a compound miter must be cut similar to those studied in Chapter 9. Both situations shown in Figure 10–2 require individualized fitting of valley and jack rafters.

Shed Dormer with Sidewalls

Figure 10–3 shows a typical shed dormer. The dormer opening must be framed in first. Double headers are installed plumb, which means that common cripple rafters need plumb cuts where they contact the headers. Once the headers and common rafters are in place, the sidewalls are framed. Each shed roof rafter must be cut to fit against a previously installed common rafter. Obtaining the angle cut needed for a good fit is easily accomplished with a bevel square and uncut rafter or scrap 2 × 4 and rafter. After cutting a sample rafter and assuring that it fits well, all others should be cut from this pattern.

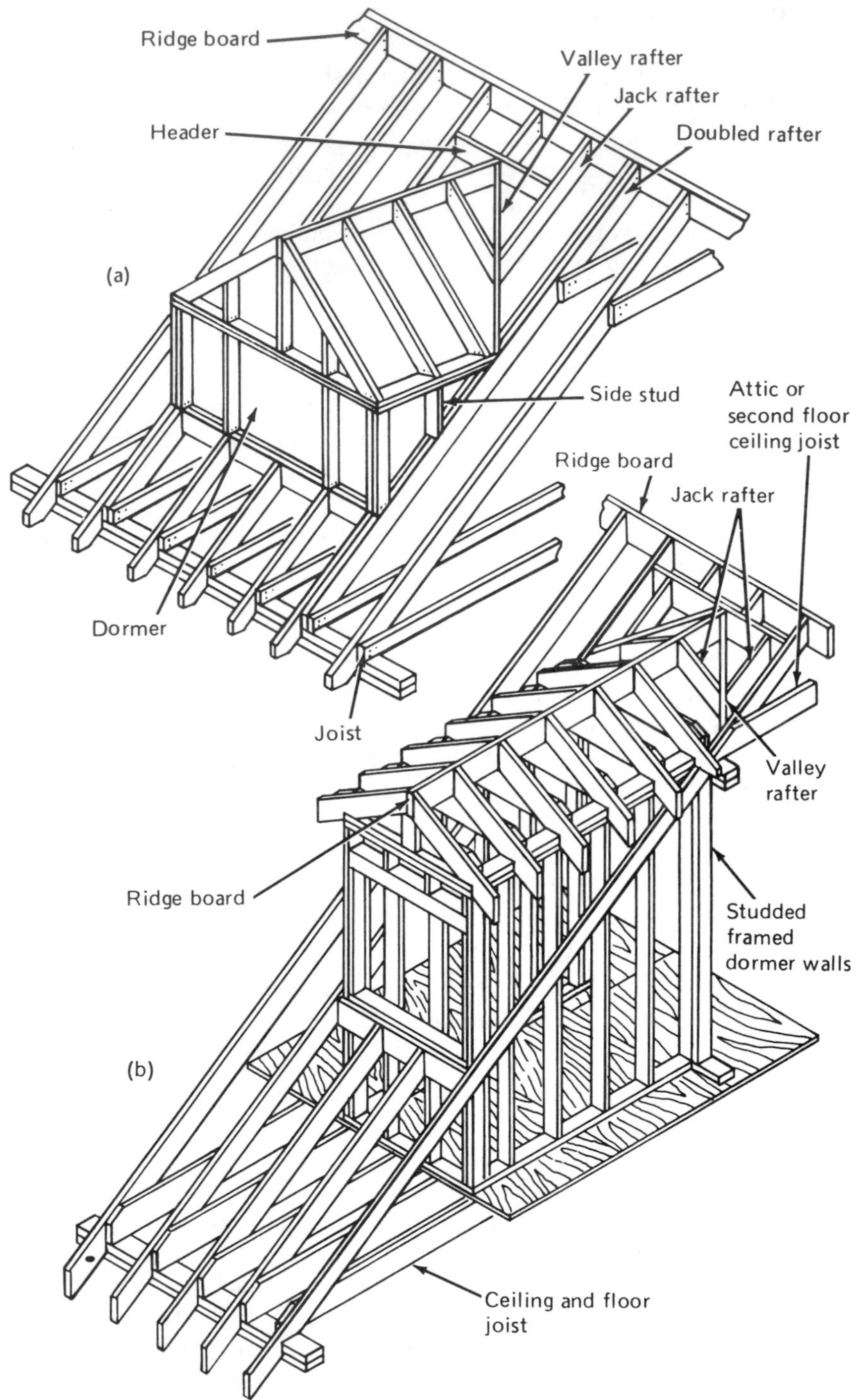

Figure 10–2 Dormers with sidewalls extending to the second floor or fastened to double rafters.

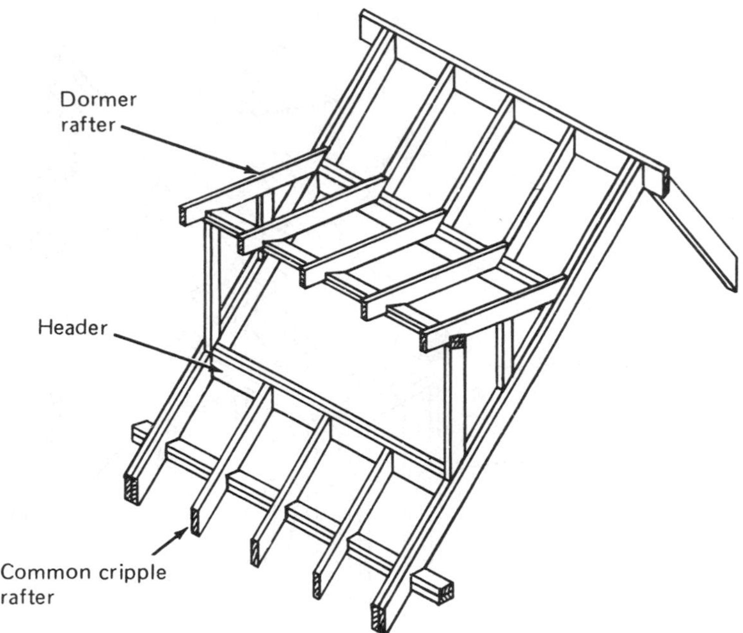

Figure 10–3 Dormer with a shed roof design.

GABLE END STUD INSTALLATION

Once the rafters have been installed, the gable ends on the gable roof need to be cut and installed. There is more than one way to perform this task. However, one principle should always be used regardless of the method selected. That is, the gable end studs should align over the wall studs, which are in the end walls.

Figures 10–4 and 10–5 illustrate the two techniques for installing gable end studs. In Figure 10–4 each gable end stud is measured and cut for length. Then a cutout is made along its face edge so that its outer edge aligns with rafter and wall plate. Toenailing must be used on the lower end of each stud, but 12d or 16d nails may be used on the upper end since these may be driven straight through.

In Figure 10–5b, a special 2 × 4 plate is installed so that its top surface is exactly positioned as a seat for the lookouts. Gable end studs are cut with a bevel and the assembly is made accordingly. Once again, notice that each gable end stud is in alignment with the wall studs below.

FLY RAFTER INSTALLATION

A *fly rafter* is used on a gable roof where the gable end needs an overhang. Two methods are employed in Figure 10–5. In Figure 10–5a, standard common rafters

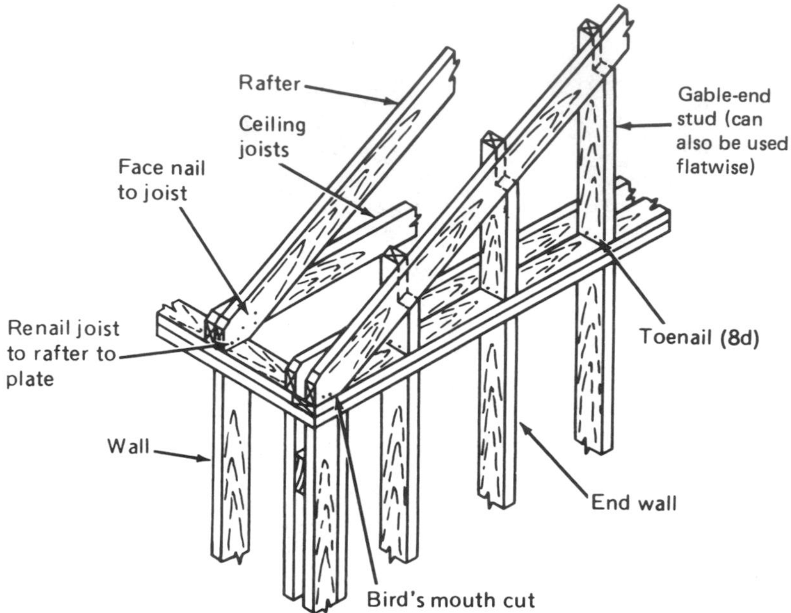

Figure 10–4 Installing gable end studs.

are installed, and gable end studs are attached as explained previously. A method of layout for *lookouts* is selected such as 16-in. or 24-in.-o.c. spacing from lower edge to ridge, or vice versa. Notch marks are laid out and cut as shown so that each lookout sits neatly in place.

When all lookouts are installed, the fly rafter, which is usually a 2 × 4 or 2 × 6, is cut and then nailed to each lookout. Cornice is then applied over this frame.

The second technique is shown in Figure 10–5b. In using this technique,the last rafter before reaching the end wall is doubled, metal hangers are installed at 16-in.- or 24-in.-o.c. spacings as shown. The gable ends are completed. The nailing block shown aids in stabilizing the lookouts.

The fly rafter is cut to fit, then nailed against the lookouts. The fly rafter is usually as wide as the lookouts but may on occasion be wider.

FRAMING FOR CHIMNEY AND CHIMNEY SADDLE

Whenever a chimney is built to pass through the roof a *chimney well* must be framed. This well is similar to a stairwell in a floor assembly, but in most cases much smaller. Figure 10–6 illustrates a completed chimney well. This example shows that one common rafter is cut to allow the construction of the chimney well. It also

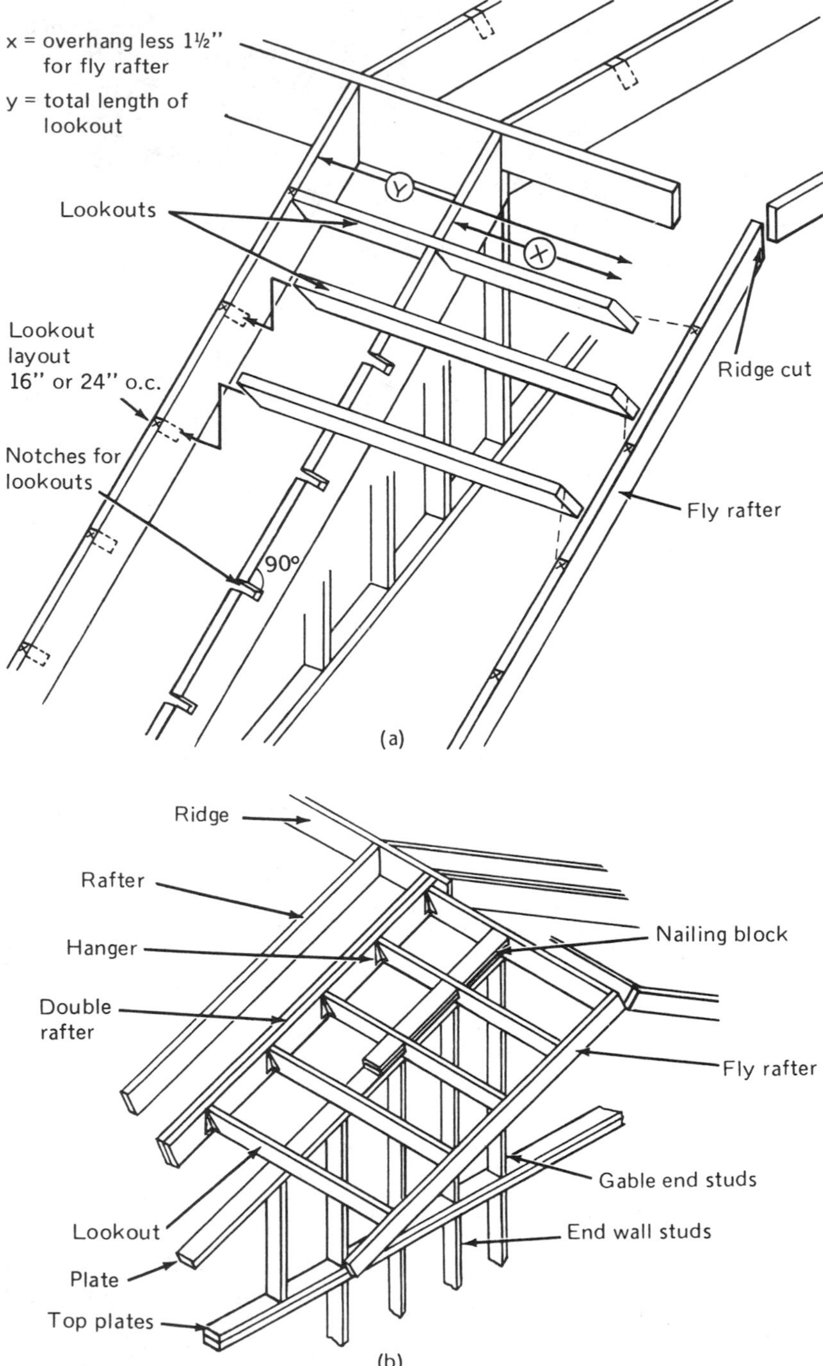

Figure 10–5 Fly rafter installation techniques.

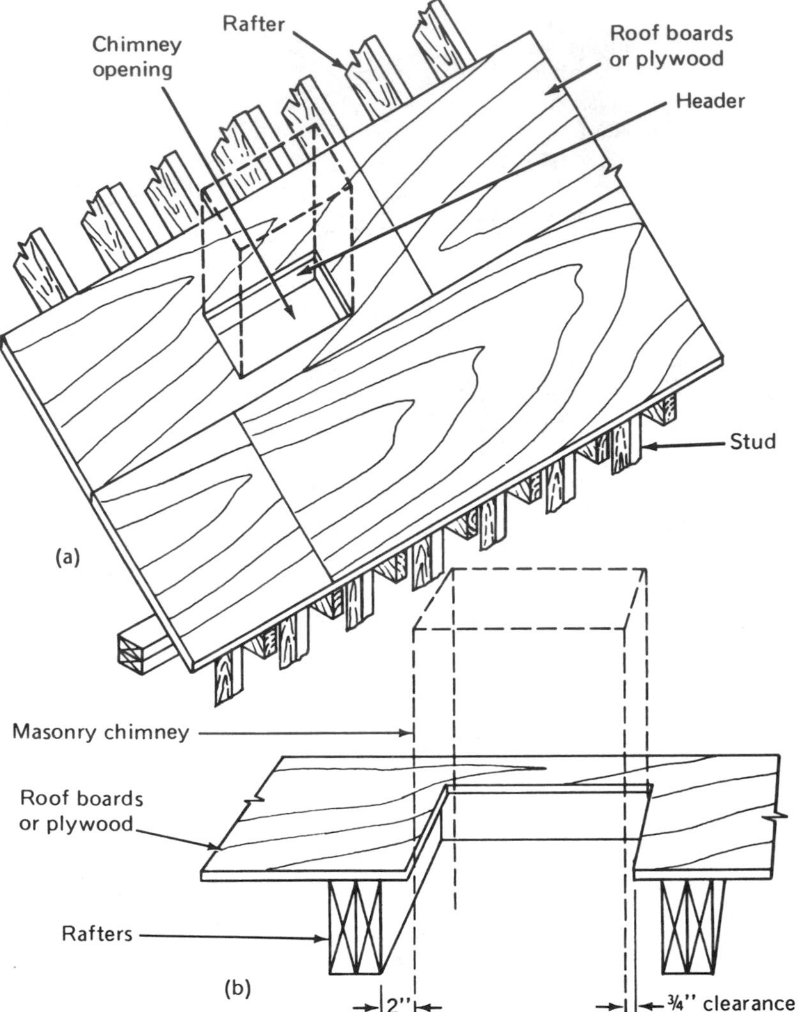

Figure 10–6 Framing for chimney.

indicates that doubled rafters are used on either side of the opening and headers complete the opening.

Figure 10–6b illustrates a cutaway view of the opening to dramatize the two basic standards that several building codes specify: they are the 2-in. clearance between framing and chimney and the ¾ in. clearance between sheathing and chimney. These two requirements should be followed faithfully since a fire hazard is greatly reduced. Another reason for following this rule is to allow for roof movement, expansion, and contraction as weather and wind exert their forces on the roof. The chimney is rather stiff by comparison and the spacing removes the possibility of rupture of the chimney.

The final part of roof framing around a chimney well is the construction of a *chimney saddle.* Figure 10–7 shows the saddle as a built-up area on the back side of the chimney. The use of the saddle causes the normal water runoff on the roof to shed to either side of the chimney rather than buffet it.

The peak of the saddle is level. The supports for it are made from nominal 2-in. stock tapered to form its shape. A 2 × 4 is usually used as the horizontal member and backing of its top edge is done on the best jobs. Sheathing is cut and nailed over the frame. Composition shingles similar to those used on the main roof should not be used on the saddle because of the chance that water will flow back under them and cause roof leaks. A special sheet metal cover should be made and installed. This cover should be made so that (1) it runs under several courses of shingles, (2) covers the saddle, and (3) bends up the side of the chimney approximately 4 to 6 in. Chimney flashing will be installed over the saddle metal cover and leaks will not occur.

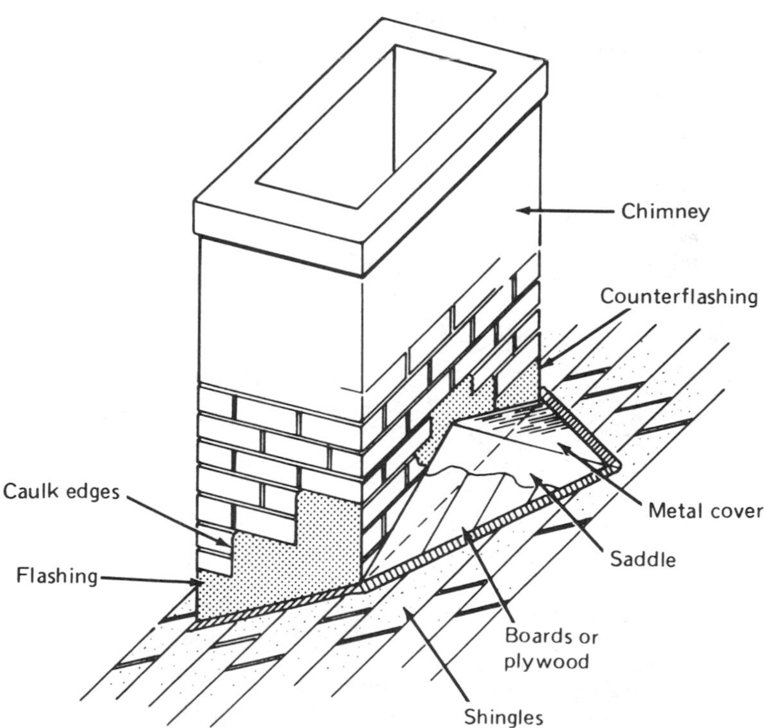

Figure 10–7 Chimney saddle framing and sheathing.

PLYWOOD SHEATHING

Plywood is customarily installed on the roof as sheathing. It is easy to install, covers the roof quickly, and various thicknesses may be used depending on loads and rafter on-center spacings. The thickness to use is easily determined from tables such as Table 10–1. The first column, titled "Panel Span Rating," lists the rafter on-center separations left of the slash beginning with 12 in. and ending with 48 in. Numbers to the right of the slash are floor or ceiling joist on-center separations. The 1⅛-in. plywood designations are especially thick plywood panels most appropriately used in plank and beam construction. In a roof designed as an open cathedral or exposed rafter-beam type, these last three types of plywood could be used as sheathing provided that the spans between rafter beams not exceed 48 in. (column 3).

If, for example, a roof is constructed with rafters spaced 16 in. o.c., either 5/16- or ⅜-in.-thick plywood panels may be used to sheath it. Under the live load column, notice that a 50-psf live load maximum is indicated. To assure that this load will be borne by the plywood attached to the rafters, only the grades listed below the chart should be used. If deflection is a concern, the plywood thicknesses and grades are applicable to $S/180$ for combined live and dead loads or $S/240$ for only live loads.

Figures 10–8 and 10–9 illustrate several parts of the techniques of installing plywood on the roof. In Figure 10–8 ply-clips are shown being used between rafters

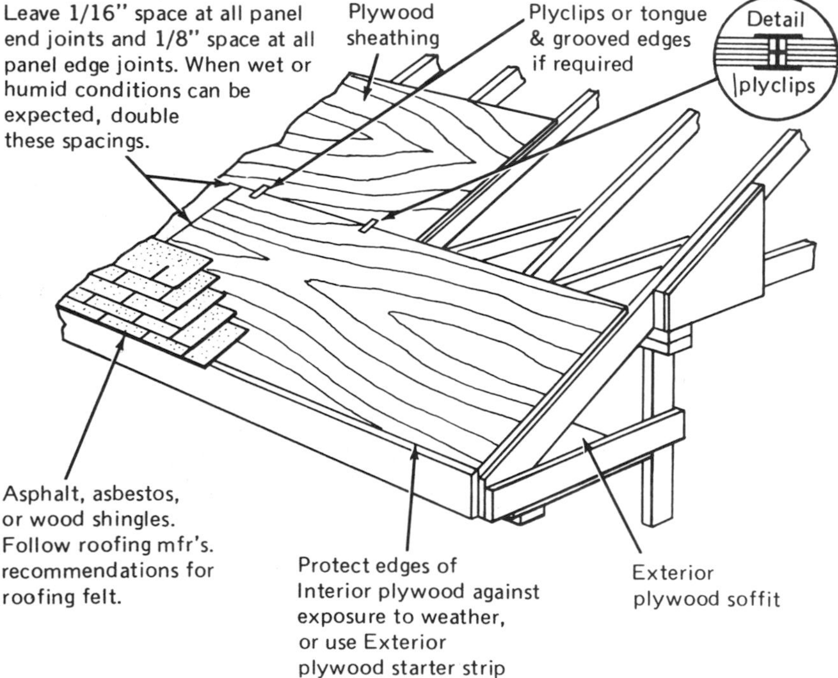

Figure 10–8 Details for sheathing a roof.

TABLE 10-1 RECOMMENDED UNIFORM ROOF LIVE LOADS FOR APA-RATED SHEATHING AND APA-RATED STURD-I-FLOOR WITH LONG DIMENSION PERPENDICULAR TO SUPPORTS

(1)	(2)	(3)	(4)	(5)	(6)	(7)	(8)	(9)	(10)	(11)	(12)
		Maximum span (in.)		Allowable live loads (psf)[b] Spacing of supports center-to-center (in.)							
Panel span rating	Panel thickness (in.)	With edge support[a]	Without edge support	12	16	20	24	32	40	48	60
				APA Rated Sheathing							
12/0	5/16	12	12	30							
16/0	5/16, 3/8	16	16	55	30						
20/0	5/16, 3/8	20	20	70	50	30					
24/0	3/8, 7/16, 1/2	24	20[c]	90	65	55	30				
24/16	7/16, 1/2	24	24	135	100	75	40				
32/16	15/32, 1/2, 5/8	32	28	135	100	75	55	30			
40/20	19/32, 5/8, 3/4, 7/8	40	32	165	120	100	75	55	30		
48/24	23/32, 3/4, 7/8	48	36	210	155	130	100	65	50	35	
				APA Rated Sturd-I-Floor							
16 o.c.	19/32, 5/8, 21/32	24	24	135	100	65	40				
20 o.c.	19/32, 5/8, 3/4	32	32	165	120	100	60	30			
24 o.c.	11/16, 23/32, 3/4	40	36	210	155	130	100	50	30		
48 o.c.	1 1/8	60	48				290	160	100	65	40

Source: American Plywood Association.

Note: Loads apply only to C-C EXT-APA, structural I C-D, Int-APA, and structural I C-C EXT-APA. Check availability before specifying.

[a]Tongue-and-groove edges, panel edge clips (one between each support, except two between supports 48 in. o.c.), lumber blocking, or other.

[b]10 psf dead load assumed.

[c]24 in. for 1/2-in. panels.

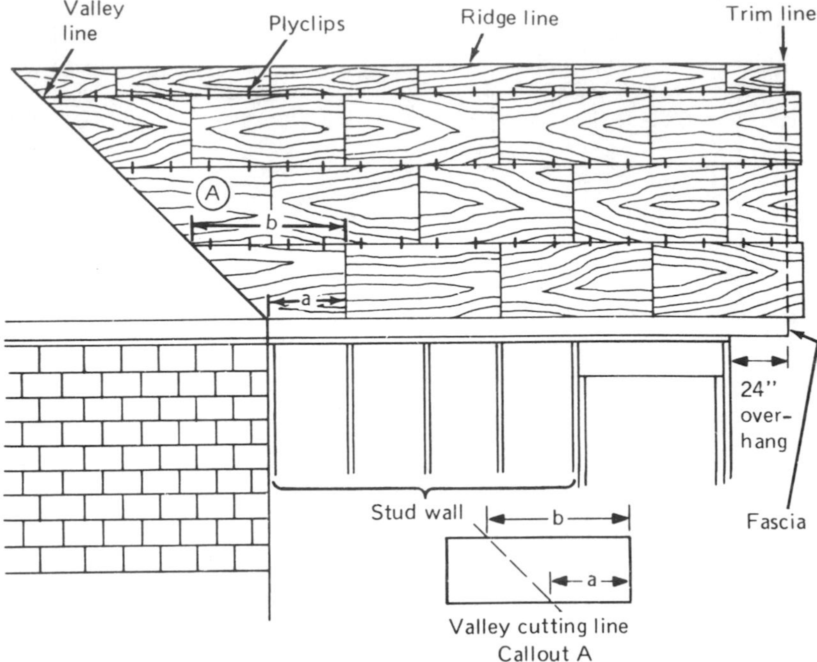

Figure 10–9 Sheathing a roof with plywood that includes a valley.

to aid in supporting the edges of the plywood. These clips are frequently used with 5⁄16-, 3⁄8-, and 1⁄2-in. plywoods, where rafters span 16 or 24 in. o.c. They are not as a rule used where 3⁄4-in. plywood sheathing is installed over rafters with these on-center spacings. However, should rafters be spaced 32, 42, or 48 in. o.c., ply-clips should be used with the thicker plywood panels.

Figure 10–9 shows two installation techniques. One is the sheathing of the valley. Pieces of plywood are easily laid out and cut to fit, as callout A shows. These panels are best cut on the ground for safety sake. The other technique is trimming excess plywood along the rake of the roof. Normally, a chalk line is snapped, then a power saw is used to trim the plywood.

Nailing should follow the rule that the portion of nail penetrating the rafter should be at least one-half the nail's length. From this rule Table 10–2 is established.

Nailing should follow the schedule of 6 in. along the ends of the sheet and not

TABLE 10–2 NAILS FOR SHEATHING

Nail size	Plywood thickness (in.)
6d	5⁄16, 3⁄8, 7⁄16, 15⁄32, 1⁄2
8d	19⁄32, 5⁄8, 23⁄32, 7⁄8, 3⁄4
8d or 10d	1 1⁄8

more than 8-in. separation on intermediate rafters. Closer spacing of nails than indicated adds strength to the roof and aids in its ability to withstand forces placed on it.

Gluing panels to rafters has not been customary in the past. However, there is considerable merit to the idea of using glue. A much more uniformly sound structure is created. The bonding accomplished with the glue does make a panelized structure of considerable strength.

PROBLEMS AND QUESTIONS

1. What is the basic difference in construction between a dormer without sidewalls and one with sidewalls?
2. Why is a set of headers and doubled rafters installed first when framing a dormer without sidewalls?
3. What name is given to the doubling (second) rafter?
   a. jack **b.** trimmer
4. Explain where cripple common rafters are used around the area of the dormer.
5. How many valley rafters does each dormer framework with sidewalls require?
   a. 2 **b.** 4 **c.** 6
6. In a dormer with sidewalls, do the studs always extend to the floor joists of the second floor?
7. Where should the upper end of each shed-type dormer rafter end?
8. What is the principle of gable end stud alignment?
9. Which type of roof may use a *fly rafter?*
   a. hip **b.** gable
10. What is the purpose of a lookout in fly rafter installation?
11. What is the standard acceptable clearance between chimney and framing for residential houses?
    a. 1 in. **b.** 2 in. **c.** 3 in.
12. What is the recommended clearance between sheathing and chimney?
    a. ½ in. **b.** ¾ in. **c.** 1 in.
13. Describe the purpose of a chimney saddle.
14. What would be the *minimum* thickness of plywood sheathing for use on a roof with rafters spaced 16 in. o.c., where a live load of 75 psf or less is predicted?
    a. ¼ in. **b.** 5⁄16 in. **c.** ⅜ in. **d.** ½ in.

PROJECTS

There are several projects to try in this chapter: gable end stud installation, fly rafter installation, and plywood sheathing of the roof.

A. Gable End Stud Installation

Refer to Figure 10–4, which shows gable end studs, and Figure P10–1 in this project. Our design project is to determine how many studs there will be and how much material to order for the job.

1. Since the run is approximately 15 ft and each stud will be spaced 16 in. o.c. there will be ____________ gable end studs on each run and a total ____________ studs.
2. We know that the total rise is approximately 6 in. for every foot of run; therefore, 6 in. × 15 ft = 90 in. We can now determine the material needed for each run. The formula is easy to follow and is: longest, longest + 6 in., and so on. The + 6 in. is necessary for nailing (see Figure 10–4). The next longest is equal to longest—(reduction for decrease in span of 16 in. + 6 in. for nailings).
   - a. Longest = 96 in.
   - b. Next longest = ____________
   - c. Next longest = ____________
   - d. Next longest = ____________
   - e. Next longest = ____________
   - f. Next longest = ____________
   - g. Next longest = ____________
   - h. Next longest = ____________
   - i. Next longest = ____________
   - j. Next longest = ____________
   - k. Last = ____________
   - Total ____________ in. = ____________ lin. ft
3. Determine the total material needed for both runs (2 × total for one run − longest).

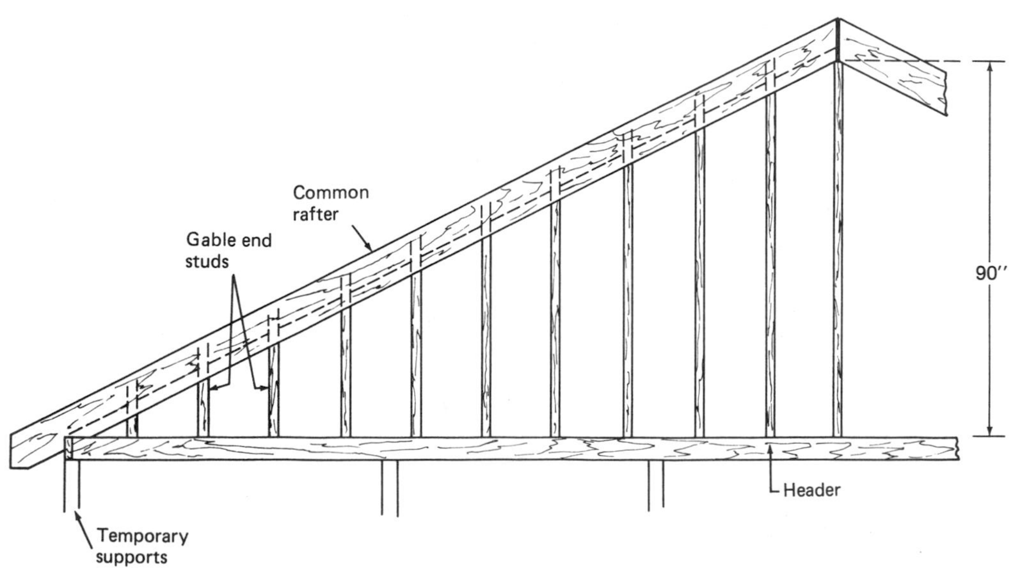

Figure P10–1

B. Fly Rafter Installation

Refer to Figure 10–5a for the design we shall use. The fly rafter must extend 12 in. beyond the wall line to conform to the overhangs on the remainder of the roof. After laying out the positions of each lookout and making these cuts, we needed to determine the material needed.

1. Where should the notches be made into the end rafter?
2. What size must each notch be?
3. What length must each lookout be?
4. Approximately how many lookouts will be needed for each half of the gable end? How many for both halves?
5. How many linear feet of 2 × 4 will be needed for the fly rafters?
6. Now let's summarize the material requirements.

    a. Lookout 2 × 4 material = ____________ ft
    b. Fly rafter 2 × 4 material = ____________ ft
    Total = ____________ ft

C. Sheathing the Roof

1. Use Table 10–1 to select the plywood thickness for the porch roof. The thickness needs to be ____________ inches.
2. Using Figure 10–9 as a model, let's determine the plywood needs for one side and then double it for the total roof.

    a. Calculate the amount needed for the common rafter section first using the formula length (ft) × width (ft).
    Common rafter section = ____________ ft × ____________ ft
    = ____________ ft^2
    b. Calculate the amount needed for the valley jack rafter section next using the formula length (ft) × width (ft) divided by 2.
    Valley jack rafter section = ____________ ft × ____________ ft/2
    = ____________ ft^2
    c. Calculate the total (must include a 5% allowance for waste and enough to cover both sides of the roof).
    Common rafter section = ____________
    Valley jack rafter section = ____________
    Allowance for waste = ____________
    Total square feet = ____________
    Total number of sheets at 32 ft^2 per sheet = ____________.

Answers:

Project A: **1.** 11, 22;

2. (A = reduction in height, B = nailing material, C = needed length)

	A		B		C
a.	90	−	0 + 6	=	96 in.
b.	90	−	8 + 6	=	88
c.	88	−	8	=	80
d.	80	−	8	=	72
e.	72	−	8	=	64
f.	64	−	8	=	56
g.	56	−	8	=	48

h. 48 − 8 = 40
i. 40 − 8 = 32
j. 32 − 8 = 24
k. 24 − 8 = 16
Total 610 in. = 51 ft

3. 51 + 51 − 7½ ft = 94½ ft

Project B: **1.** Starting from either bottom (rafter tail) or ridge, make a notch 16 in. o.c.
2. 1½ × 3½ to fit a 2×4
3. 16 + 10½ = 26½ in.
4. 18-ft-long rafter/ 16 in. = approximately 14, 28
5. 36 ft
6. **a.** 94½ lin. ft
b. 36
Total 130½ lin. ft

Project C: **1.** 5/16 or 3/8 in.
2. **a.** 20 × 18 = 360 ft^2/side × 2 for both sides = 720 ft^2
b. 15 × 18 = 270/2 = 135 ft^2/side × 2 for both sides = 270 ft^2
c. 720 $ft.^2$
270
50 (allowance for waste)
1040 ft^2 divided by 32 ft^2 per sheet = 33 sheets

Chapter 11

Waterproofing Techniques Used with Siding Installation

If anything can be said about the techniques of waterproofing joints between wood siding and other materials, it would probably be that today it is simpler than in years past. This does not mean that techniques used 50 years ago are not in use today; on the contrary, they are very much in use. But today, vinyl and other types of caulking are available, and these materials are so improved that they retain their elasticity or plasticity for many years. This, of course, means that they may be compressed and stretched as building components expand and contract.

Caulking is effective in sealing joints between wood and wood siding, but these two materials frequently expand and contract at the same rates. This could be severe; however, experience has shown that there is little problem encountered here.

Where caulking is used between brick or block and wood, the situation is different. These materials are considerably different and their expansion and contraction rates are quite dissimilar. Wood will expand much more than brick. The unequal changes in these materials place the greatest stress on caulking, so the best nonhardening, nondrying types should be used.

Felt paper, either 15 or 30 lb, is an old standby that is used as backing to prevent leaks. It is usually installed horizontally and overlapped in the direction of rainfall. However, felt is frequently omitted from the exterior sheathed wall, especially if bituminous-coated sheathing panels are used.

Even over this type of sheathing, felt is used as a waterproof backing in several places, at the corners of a building, where strips of it are installed vertically, and behind window and exterior door assemblies, where strips are doubled and nailed in place before these units are installed.

The third waterproofing material is also an old standby: metal flashing. It is usually made from three metals: copper, aluminum, and galvanized steel. In special cases zinc is used. Flashing metal is usually purchased in rolls or by the foot from a roll; it is cut on site as needed and shaped to any contours that are required for it to be effective. Two of the most common areas for its use are flashing between sidewall and intersecting roof and over the rain cap on wood-framed windows and doors.

WATERPROOFING THE JOINT BETWEEN CORNER BOARDS AND SIDING

Regardless of the type of sheathing used, but especially where insulation sheathing is used, strips of felt vertically installed will seal corners from water that may penetrate the joint of corner board and siding. Figure 11–1 illustrates this technique. A single strip of felt should be cut to reach from top plate to sole plate for each interior and exterior corner of the building. These pieces are then installed as shown in Figure 11–1. Notice that the felt is applied over the sheathing board. This method of using felt develops a fairly wide surface either side of the corner.

Corner boards are often prepared on the bench rather than constructing them on the wall. One piece is usually left at full width (a nominal 1 × 4). The other piece of the corner is usually ripped to 2¾ in. wide so that when nailed to the 1 × 4, two finished surfaces equal in width to the 1 × 4 are made.

Where ¾-in.-thick sheathing is used over the wall, a corner board of this size is very effective. It is attractive, not too skimpy, and not too wide. Further, siding can be nailed easily to studs behind the sheathing. Or, in installations where siding is installed directly over studs (not shown in Figure 11–1), adequate nailing surface is present. Notice the bottom drawing of Figure 11–1a; it shows the amount of direct contact or overlap of the corner board and corner studs. This top view also shows a good nailing surface for siding.

Where 2-in. rigid insulation board is used for sheathing to create a thicker, more thermally insulated wall, felt is still used in the same manner as Figure 11–1b shows. The corner board is installed over this covered corner; but notice in the lower illustration that only a small overlap exists between the corner board and the stud. Toenailing will be required to hold the corner board in place. It probably would be better to use a 1 × 5 corner board. This extra width would provide a bit more overlap and sounder nailing, and a 1 × 5 corner would not look too wide.

Lap siding is shown butted to the corner board; however, other types, including cedar shingles, may also be used. The type of siding used is not the significant issue in waterproof installation at the joint. It is the way the joint is fitted. For most installations a finely cut close fit is most appropriate. Several master carpenters

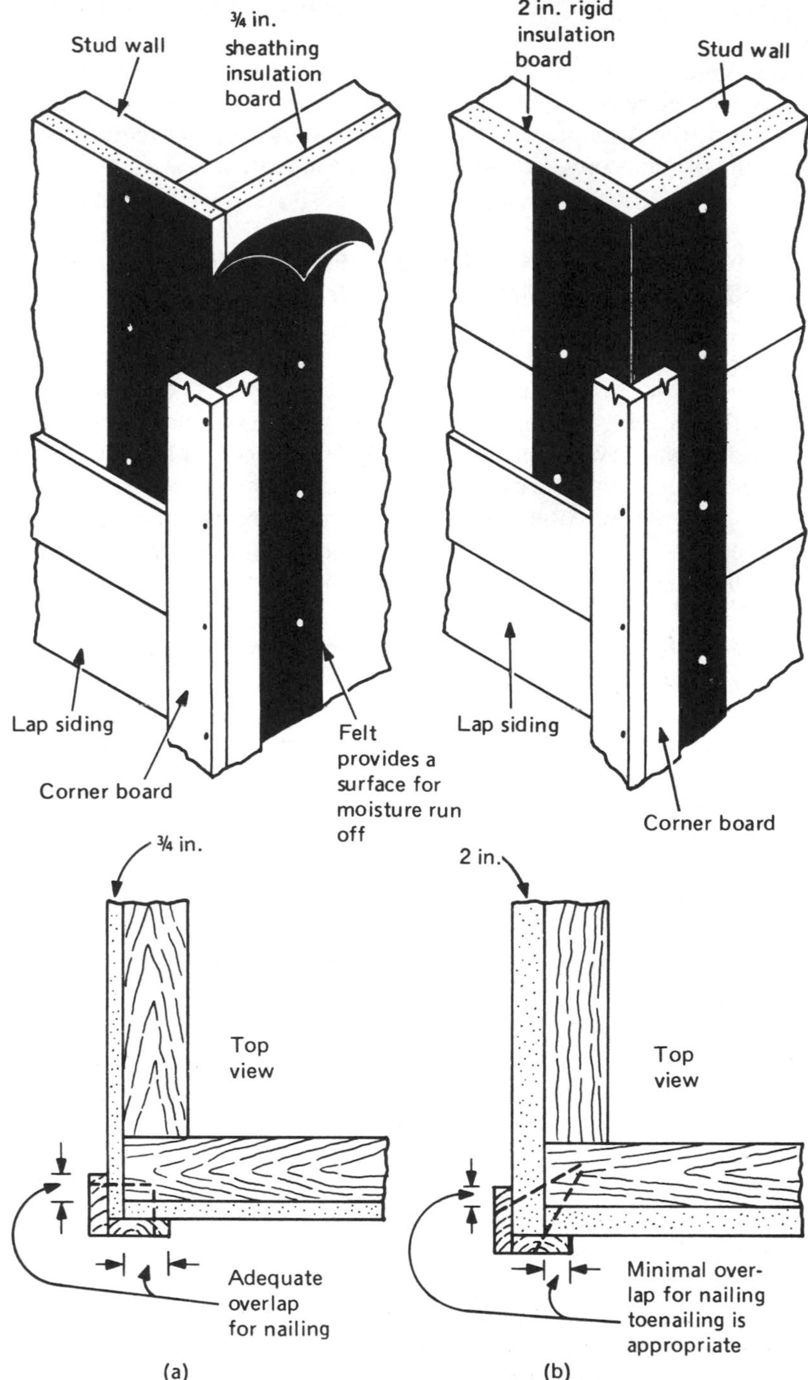

Figure 11-1 Waterproofing exterior corners where corner boards and lap siding join.

recommend a slight *undercut* to the siding's end. This ensures a snug fit. As long as each piece is snugly and neatly fitted *and does not* create pressure on the corner board, the fit is proper. No caulking should ever be required and normal applications of paint or sealer will complete the seal.

WATERPROOFING THE JOINT BETWEEN SIDING AND BRICK WALL OR BRICK VENEER

It is quite commonplace to use brick veneer on the front wall of a house (see Chapter 2). At other times brick walls intersect with walls constructed with wood. In each of these occurrences a condition exists where two dissimilar materials adjoin. This means that these materials expand and contract at different rates, and that they also react differently to wind and seismic pressures.

Several possibilities exist for waterproofing these joints. To make the joint effective and long lasting, some understanding of the problems and construction techniques of these walls needs to be developed.

Waterproofing between Siding or Corner Board and Brick Veneer

Figure 11-2 illustrates a typical technique used to make a watertight joint. Carefully examine the illustration to determine the construction of the wall. For example, notice that a 1-in. air space exists between the interior face of the brick and outer surface of the sheathing. This is an acceptable and preferred condition. Next note that the corner board (1 × 4) is nailed as closely as possible to the brick. This arrangement creates a small space where caulking can be applied.

Caulking is only a part of this waterproofing technique. Felt paper, as illustrated earlier, should be installed under the board and lap around the corner. Any moisture accumulated between wall and brick will fall to the foundation and escape through the weep holes between bricks.

Fitting the corner board to the brick is not usually a difficult task. However, fitting one to a fieldstone or sandstone veneered wall is quite difficult but no less necessary. Scribing is usually required, and once the cutting is complete, caulking is needed to seal the joint.

Waterproofing between Siding That Is Butting to a Brick Wall

The brick wall may be part of a room or the side of a chimney or fireplace, as Figure 11-3 indicates. The reason for joining the two types of wall is unimportant; what is important is that the joint must be waterproof.

It is frequently necessary to hold the framing away from the brick, but at other times, as when little chance of structural problems or fire exist, the studs are fas-

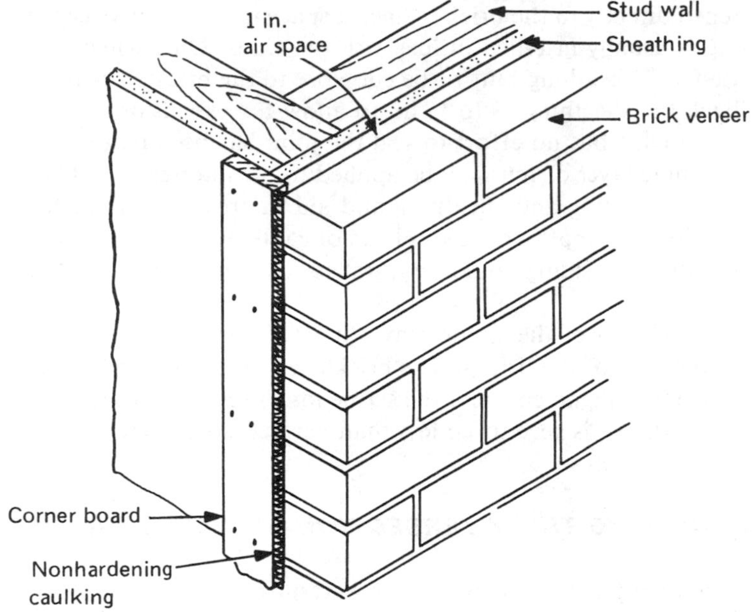

Figure 11-2 Waterproofing the joint between siding and brick veneer.

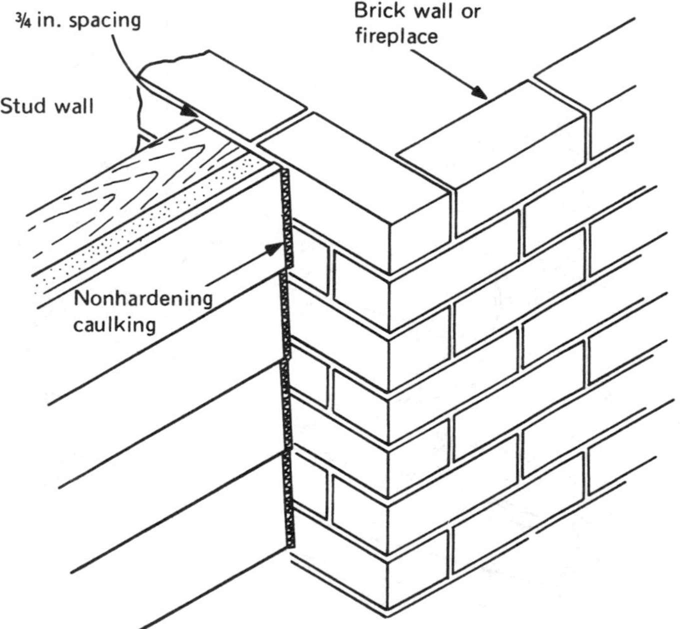

Figure 11-3 Waterproofing a joint between lap siding and brick wall.

tened directly to the bricks. Since a space may exist between stud and brick, sheathing is usually installed in line with the studs. Only when the brick wall is installed first will sheathing butt or be cut close to the brick. It is prudent, however, to hold back the sheathing ⅛ to ¼ in. to allow for expansion.

Thus far, no effort to seal the joint has been made. Prior to siding the walls, a double layer of felt may be applied to the intersection. This would be compressed between siding and sheathing and aid in creating a tight joint. A second method would be to apply a generous bead of caulking between sheathing and brick. Following this, the siding could be installed and caulking applied as shown in Figure 11–3.

The fit of the siding may require the use of a scribe because of irregularities in the face of the bricks. The bricklayer's technique of installation may also create imperfect alignment, and the scribe may need to be used again for fitting. In either event the fit is sure to be less than perfect, so caulking is a necessity.

WATERPROOFING THE INTERSECTION OF WALL TO ROOF

One of the oldest techniques of waterproofing uses flashing to create the tightness necessary; but from outward appearances it is not very attractive. Figure 11–4 illus-

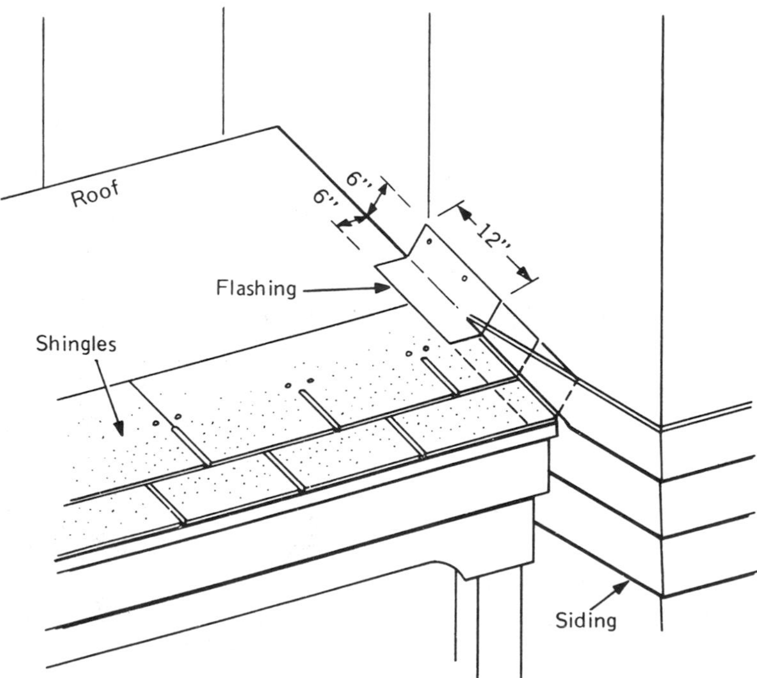

Figure 11–4 Applying flashing at a wall and roof intersection.

trates the technique used. The roof, in this case a patio, breezeway, or porch roof, intersects with the second-story wall of a two-story house. Although this illustration depicts this situation, remember that all other applications require the exact same treatment as explained here.

Before the shingles are applied to the roof and siding is applied over the wall, pieces of flashing need to be cut and folded. It is customary to cut these pieces as long as a shingle is wide (12 in.) and have the piece at least 12 in. wide. Folding through the center creates a piece as shown.

The first piece of flashing is installed overhanging the roof's lower edge by 1 in. *before* the first course or starter course of shingles is laid. The second piece of flashing is laid over the first course of shingles but held back 1½ in. from the lower edge of the first course. This metal is all but concealed when the second row of shingles is installed. Each piece of flashing is laid as described previously until the ridge is reached; there an overlap of flashing needs to be made.

Siding is fitted along the roofline, but *must not* make contact with the shingles. To do so would cause a moisture trap and decay of siding and flashing might result. Rather, a ½- to ¾-in. clearance evenly maintained throughout the length of the intersection between siding and shingles is the appropriate way to treat the joint. Nails driven into the siding should be kept as high as possible from the crease in the flashing.

WATERPROOFING AROUND THE CHIMNEY WITH FLASHING

Techniques similar to those illustrated in Figure 11–4 are used when flashing is used around a chimney that protrudes through a roof. Pieces of flashing are cut and installed up the side of the chimney and a single piece of molded flashing is used on the horizontal sections of the roof and chimney line. Figure 11–5 illustrates these pieces.

What makes this installation different from the flashing along the sidewall of the house is that two layers of flashing are formed; one is laid onto the roof and bent up alongside the bricks or blocks. The other (counterflashing) is fitted into the mortar joints between bricks or blocks. Once the flashing and counterflashing are in place, mortar is forced into the joint and this weatherseals the metal to brick. The overlap of the shingles covers the flashing on the sloping sides, but the flashing usually shows along the top and bottom surfaces.

A close look at Figure 11–5 shows that flashing is installed as the roof is being shingled. As soon as the row of shingles that requires a cutout for the chimney is laid, the lower flashing piece is set in place and nailed to the roof. Next the side pieces are installed row by row as the shingles are laid. Finally, the top piece, sometimes called the "saddle cover," is installed. This procedure provides the basis for a watershed off the roof and around and past the chimney. However, counterflashing must be installed next. It laps over the roof flashing and fits into joints that are later tuck-pointed with mortar.

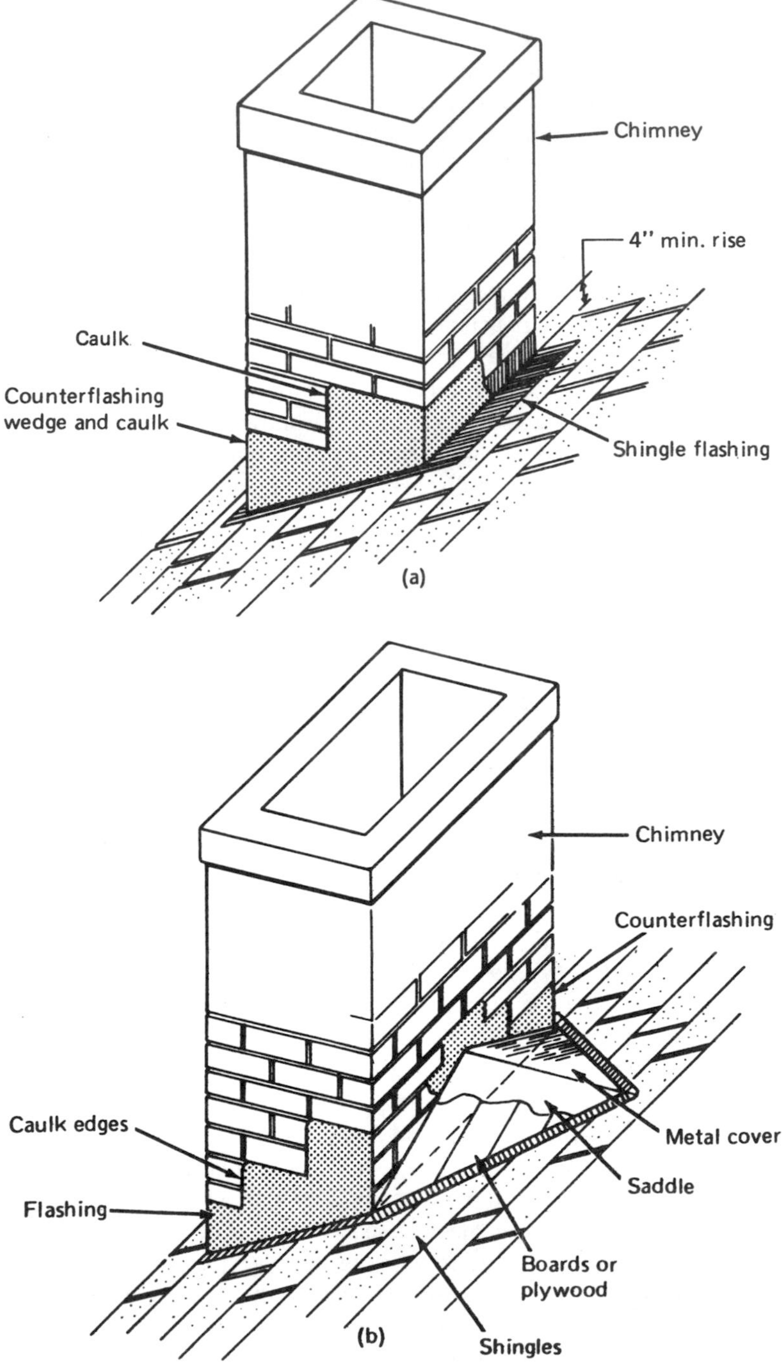

Figure 11–5 Types of waterproofing techniques for flashing around chimneys.

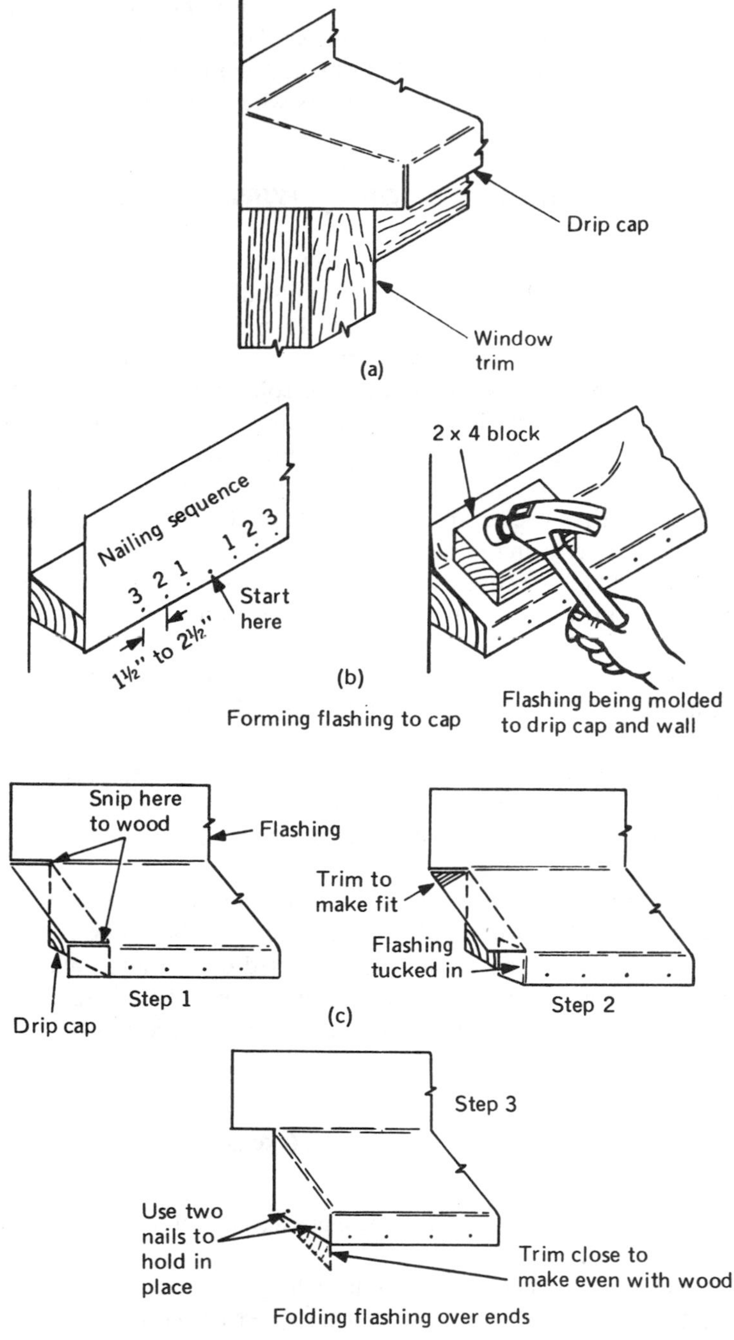

Figure 11-6 Flashing the head of a door or window frame.

Installing flashing around a chimney is not a simple task. It must be done properly and soundly. Therefore, the skill of a sheet-metal worker is frequently used to fabricate the pieces and install them. The mason that laid up the chimney usually is required to tuck-point all the joints when the metal is in place.

FLASHING THE HEAD OF A DOOR OR WINDOW FRAME

Metal flashing must be installed over each wooden window and door. Most units are purchased with wooden drip caps installed. If one is not on the unit, it is recommended that one be installed. Figure 11–6 shows the flashing installed on a drip cap over either a window or door. Metal, probably aluminum, must be purchased from a local lumberyard to do the job. It is sold by the running foot with widths starting at 12 in. Unless there is a specific reason to do otherwise, a 12-in.-wide strip should be split in two because this will give enough metal for two windows or doors. Measure the drip-cap length and *add* 1¾ in. when cutting the metal for length. The added length is needed to seal the ends of the drip cap and the wall area to prevent the entry of blowing rain.

The machine edge of the flashing is held even with the bottom of the drip cap. A corrosion-resistant nail (aluminum for aluminum flashing) is driven about midway from the ends through the flashing into the drip cap (see Figure 11–6b). From then on nails are spaced evenly at 1½ to 2½ in. apart; work first on one side of the center, then on the other side until the ends are reached.

Following the nailing, the metal is formed by using a block of 2 × 4 12-in. long and hammer where it is shaped to the slope of the cap and the wall (Figure 11–6b). When the flashing is molded, again working from the center out, press the molding to the cap and wall and nail it near the top into the wall. Four or five nails will hold it.

Figure 11–6c shows the three steps that are used to cut and form the ends of the flashing. The areas indicated for snipping should be cut, then the front piece is folded back. The top piece is then pressed over it and nailed. With all the window and door caps flashed and the building paper installed, the preparation tasks are complete. The layout of siding must be considered next.

PROBLEMS AND QUESTIONS

1. What characteristics of caulking make it so useful for a building made of different materials?
2. List two common uses for metal flashing.
3. Explain why it is desirable to use a strip of felt paper behind corner boards.
4. What type of fit should a lap siding board to corner board be?

5. What would be a reason why studding, sheathing, and siding is kept away from a brick wall or brick chimney?
6. Why are short pieces of flashing most desirable when flashing a shingled roof to a side-wall?
7. When siding is cut to the slope of the adjacent roof, as in Figure 11-4, rot is minimized if a ______ clearance is maintained between shingles and siding.
   a. ¼ in. **b.** ½ to ¾ in. **c.** over 1 in.
8. What is counterflashing?
9. Is the saddle cover behind a chimney a single piece of flashing or more than one piece?
10. What condition would have to exist for flashing to be used above a door or window frame?

Chapter 12

Column Construction for One- and Two-Story Houses

Columns used to support roof sections and their attached ceilings and for decoration are varied in design and construction. They may be nothing more than a 4 × 4 post to circular columns crowned in doric, corinthian, or ionic orders (Figure 12–1). The modern trend is away from the ionic and corinthian orders; however, the doric order, because of its simplicity, is found in use today. Actually, other titles are given to columns as a whole and these are very limited and somewhat localized. For example, the circular column is generally called a "colonial column," and turned columns milled from 4 × 4, 5 × 5, or larger stock are titled according to style (Figure 12–2).

Thus far the only requirement a carpenter has when using these columns is to trim them for length and install them. There are, however, a variety of columns that are frequently built on site, and these are the ones this chapter develops an understanding of. They are the one-story and two-story built-up types.

ONE-STORY COLUMNS

Cased-Post Column

The *cased-post* type of built-up column is one that uses two 2 × 4s as a core. These 2 × 4s are spiked together. This, of course, makes a 3½ × 3 in. post around which the casing is constructed. Figure 12–3 illustrates this type of column, and is used throughout the following discussion.

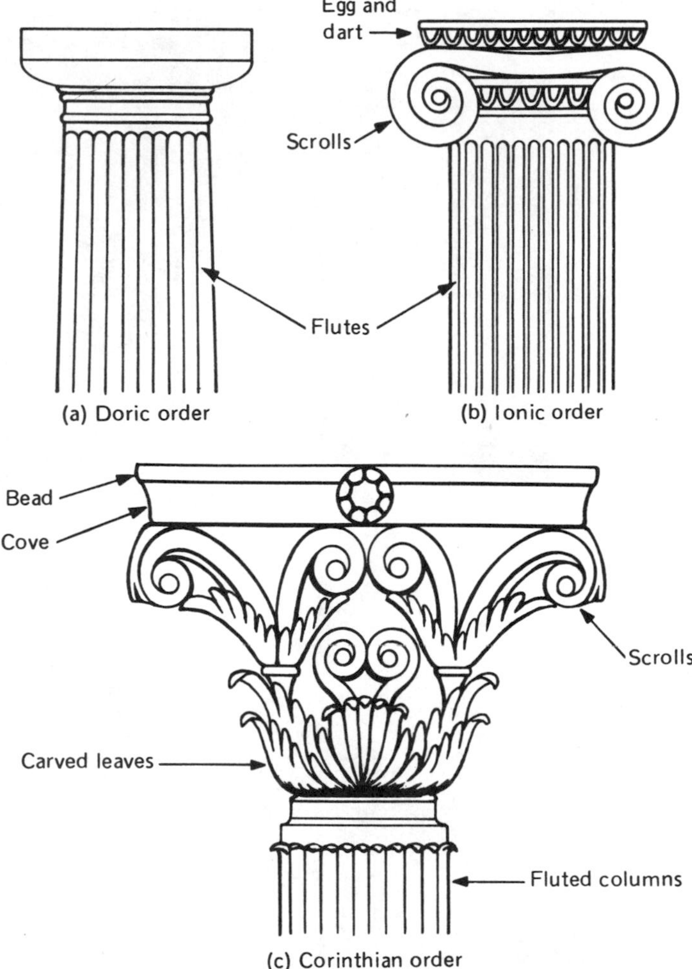

Figure 12–1 Crowns for columns: Doric, Corinthian, and Ionic.

The reason for using double 2 × 4s instead of one 4 × 4 post for the core is that 4 × 4s are generally sized differently from 2 × 4s. Also, a 1 × 4 which is used as part of the casing is usually sized to be as wide as a 2 × 4. A close look at Figure 12–3a shows why this is significant. Notice that the casing begins with the fitting and fastening of 1 × 4s to each side of the post. Since 2 × 4s and 1 × 4s are equal in width, little problem should be encountered fastening these casing pieces. Next, 1 × 6s are trimmed in width to be equal to the distance from outer surface of a 1 × 4 to the outer surface of the opposite 1 × 4.

Columns of this type may easily be built-up on the bench rather than in place. To do so one must determine several critical factors: (1) What is the overall height

Figure 12-2 Turned, milled columns. (Photo courtesy of Nord Co./Division of JELD-WEN.)

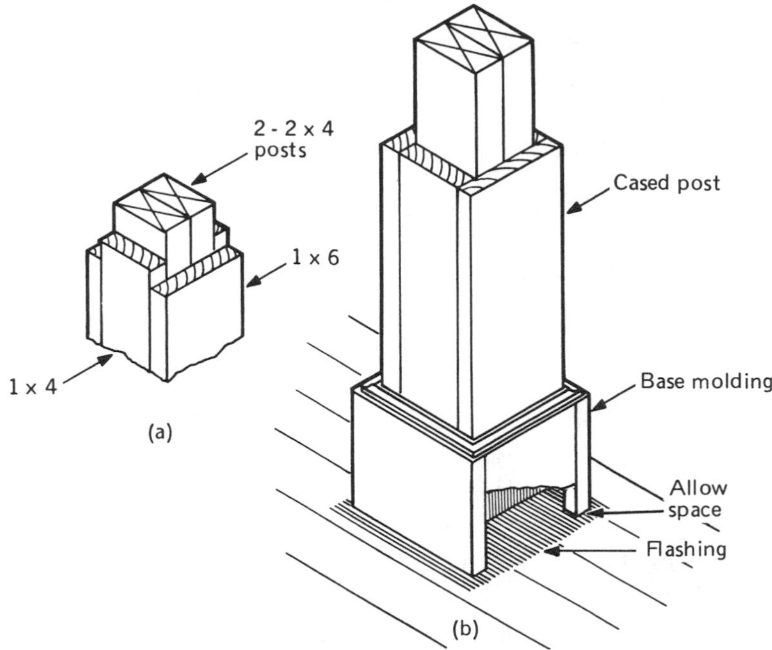

Figure 12–3 Cased column: cutaway view.

of the column? (2) How will the column be attached to the headers or plates of the wall or roof assembly? (3) What adjustments must be included if cornice and soffits are to be fitted around the column? (4) Will the column need a crown? Finally, (5) what base molding is to be used?

The height of the column can easily be found by measuring with a ruler at the place where it will be installed. However, it is much wiser to use two pieces of scrap 1 × 2s as measuring sticks. These are inserted in the place where the column is to be installed, then nailed together as Figure 12–4 shows. No height error can occur if this method of measuring is followed.

It may be that the column's core 2 × 4s will extend up into the framing to a plate area, or possibly one of the two 2 × 4s will extend higher into the ceiling or framing. If either situation exists, additional length must be added to the core 2 × 4s before they are cut. This in no way causes the length of the 1 × 2s to change, since they are set for the overall height of the column, which should be its casing height. In fact, once the core studs are cut to their required lengths and nailed together, a set of pencil lines should be made around the top of the studs using the 1 × 2 measuring stick as a guide.

The next important step is to determine if all four sides of the column will require casings of the same length. Here several possibilities exist, as Figure 12–5 shows.

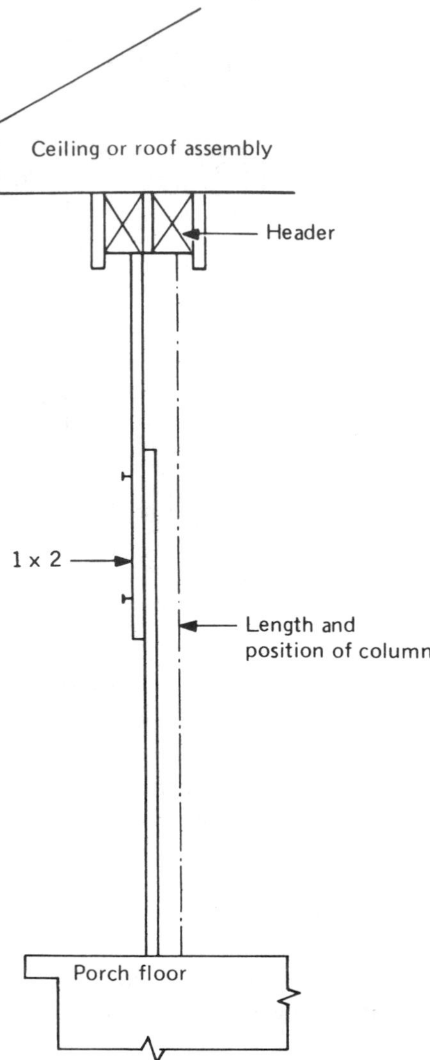

Figure 12–4 Determining column height using measuring sticks.

In Figure 12–5a the cornice boards, *fascias,* or *friezes* extend an equal distance below the headers. The soffit board is not as a rule extended to sandwich between the column and headers since it is made of materials that are subject to crushing. This situation indicates that the 1 × 4-cased sides of the column will extend the full length of the column. However, the 1 × 6-cased sides will be trimmed shorter than the full length of the column. The amount to be trimmed can be either measured with a ruler or marked on the 1 × 2 measuring stick.

The major difference between the cornices in Figure 12–5a and b is the width of the outer fascia board. Notice that it is made much wider so that it may be

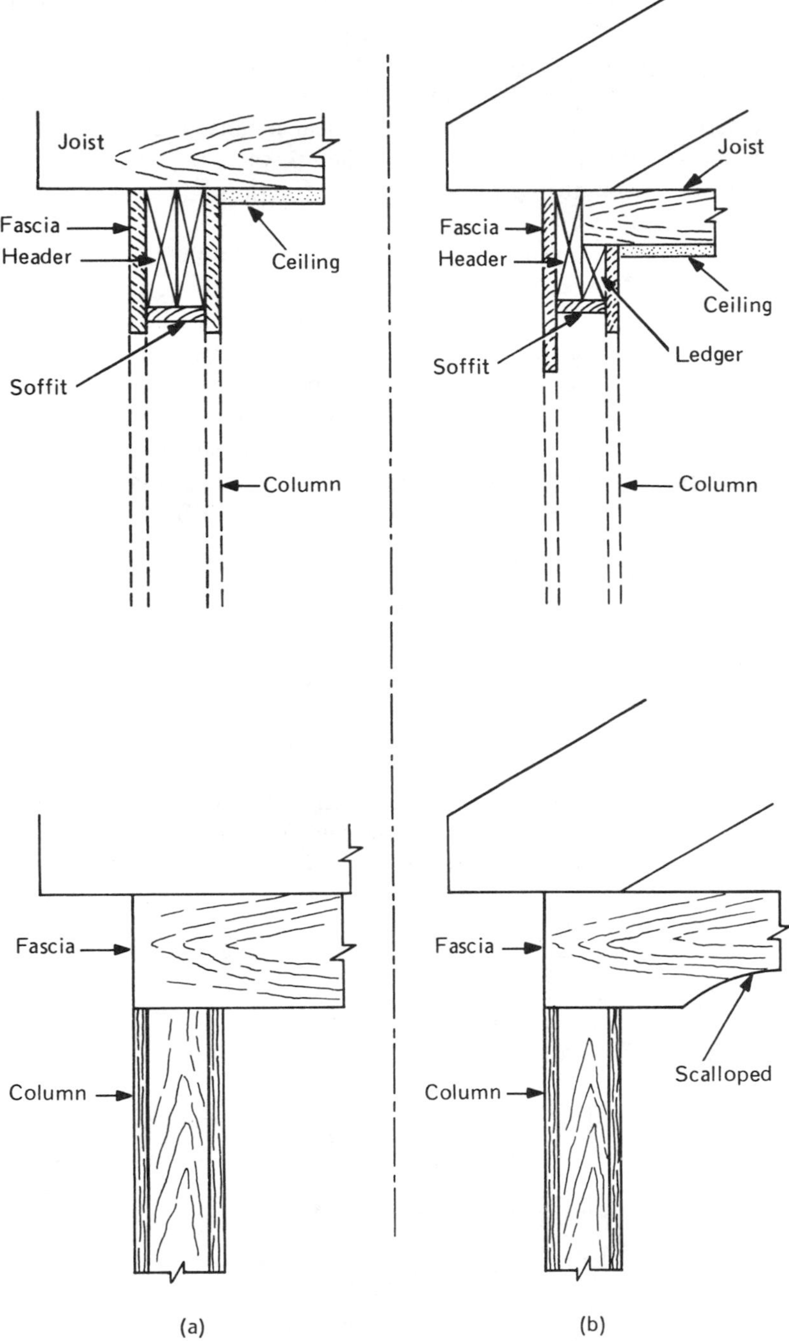

Figure 12-5 Fitting details that affect the casing.

scalloped. This means that each 1 × 6 of casing on the column must be cut to different lengths. This is not difficult to do, but individual measurements must be made or a pair of marks must be made on the 1 × 2 measuring stick.

A final factor that may need to be resolved concerns the soffit board. If it is cut to allow only the core studs to seat against the headers or plates, the 1 × 4 casings would need to be trimmed the thickness of the soffit. This is a decision that must be made during the cornice construction and carried through to the column construction.

Suppose, on the other hand, that the column is to be built with a crown; what steps must be taken? Figure 12-6 shows a simply designed crown. First the overall length of the column must be trimmed equal to the thickness of the crown material. The crown needs to be cut and installed, and cove or crown molding is used in the joint below the crown. The cornice stock must be cut away to allow the crown to make contact with the framing material. One very important characteristic that is easily added to the crown is beveling of its top surface on the sides exposed to rain. If a slight bevel of ⅛ in. is made on these sides, rain, dew, snow, or water from washing down the house will readily fall away from the surface and joint. This simple alteration eliminates the risk of rotting that so frequently results with crowned columns.

If the column is cased with a 1 × 4 and 1 × 6 as described previously, it lacks a certain formality and has an unfinished appearance. It needs base molding. A simple addition of a 1 × 6 base molding as shown in Figure 12-2 and here in Figure 12-7 provides the final touch.

Although many different widths of molding could be used, standard 1 × 6 stock lumber seems most fitting for the single-story column. As shown in Figure 12-7, the molding is cut to fit around the base of the cased post. Obviously, a square-edge cut could be used; however, four raw edges would show. Painters would need to take special steps to seal them. The results could be good or bad; time would

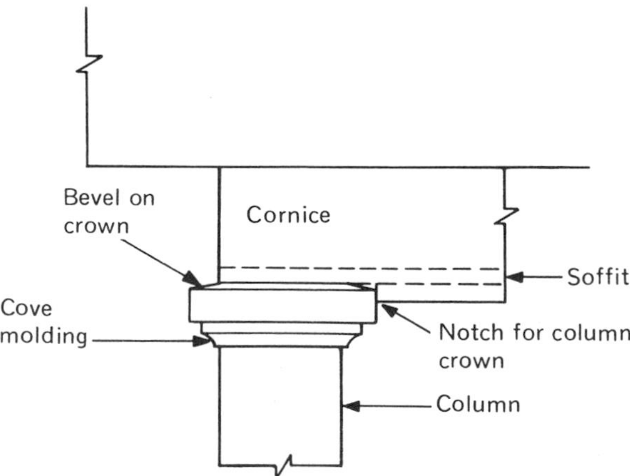

Figure 12-6 Details for adding a crown to a column.

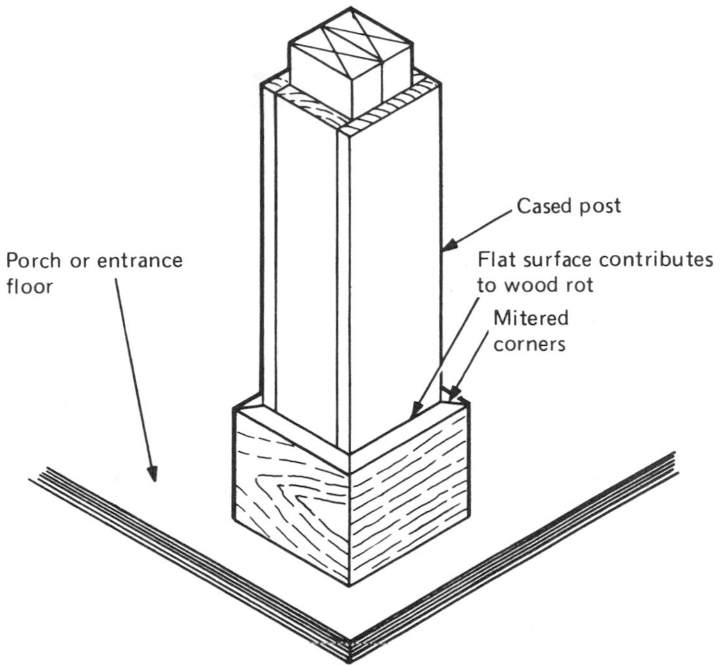

Figure 12–7 Adding a base molding to a cased column using mitered corners.

surely tell. Therefore, miter cuts are customarily used. The mitered corner is neat, fits well, and eliminates the raw-edge problem.

The addition of the base molding is not the perfect answer if installed as shown in Figure 12–7, for several reasons. One, it is very square, and second, it contributes to rotting. To be more appealing and to be a deterrent to rot it should have the final appearance of Figure 12–8. The top edge of the base molding should be beveled in the direction of rain (Figure 12–8a). This slight bevel would act as a rain cap and readily shed water from the casing and molding joint. Or a rain-drip molding (Figure 12–8b) could be installed on top of the base molding; it also adds a bit of refinement to an otherwise simple, but nice, base.

All of the preceding refinements described have been added to a column that is straight. That is, its thickness is the same throughout its length. In most single-story applications this *nontapered* design is the appropriate one to use. However, suppose that a tapered column is needed?

Cased Tapered Column

The two-2 × 4 stud core is still the basis for this column; however, the casing of the post would consist of four tapered pieces of 1-in. stock. In addition, nailing blocks would need to be installed approximately 24 in. o.c. from bottom to top,

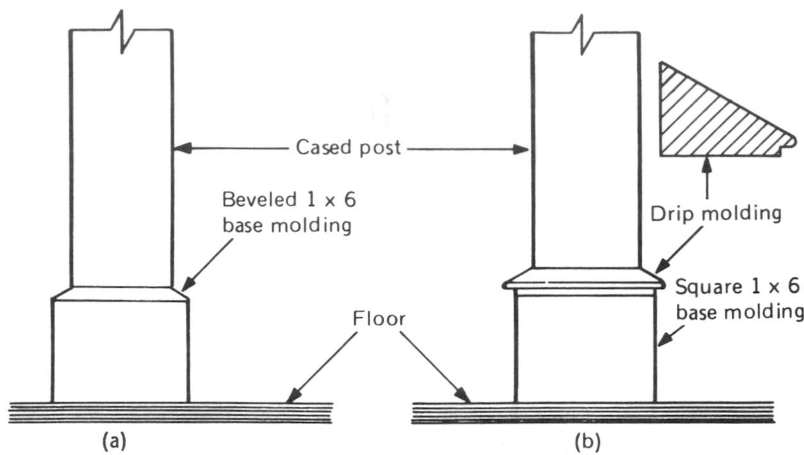

Figure 12–8 Beveling the top of the base molding or adding a drip molding to shed water and reduce the chances of decay.

much like those shown in Figure 12–9a. If a 2-in. taper were to be built into the column, 1-in. blocks would be installed at the bottom of the column and nailed to the stud core. None would be used at the top. A straightedge laid as Figure 12–9b suggests could quickly define the thickness of blocks at each 24-in. point along the core. Once cut and nailed in place, the core and blocks would be ready for casing.

At this point two options are available. One, you can cut the casings for length and mark the centers at each end for width and then measure from the center points to each outward edge a distance equal to *one-half* the casing width, as Figure 12–10 suggests. Using a chalk line, snap a line connecting the points previously marked. These lines serve as guides for sawing and planing.

The second option is to use one machined edge of lumber, measure the width of casing from this edge, snap a chalk line, and then cut and smooth the piece as in Figure 12–10b. However, you are now faced with a situation of making a cut at each end of the casing that will properly fit to floor and cornice. This is not easily done for several reasons: A cut from the bottom end improperly made causes the entire piece to be either too short or too narrow; a cut from the top causes the piece to be too short and hence too wide as it is raised up.

The method shown in Figure 12–10b can be used, but several things must be done. The length of the casing must be found, along with its width at top and bottom. Then a centerline can and should be snapped with a chalk line. From this line a *right angle* can easily be made with a framing or combination square. Once these are drawn, the single cut for taper can be laid out and all cuts can be made.

A third solution is also possible. Tack nailing the slightly longer than necessary casing onto the modified stud core positions one edge correctly. The opposite edge can be scribed along measurements transferred to its surface from below (Figure 12–11). The top and bottom lengths can also be transferred from the core to the

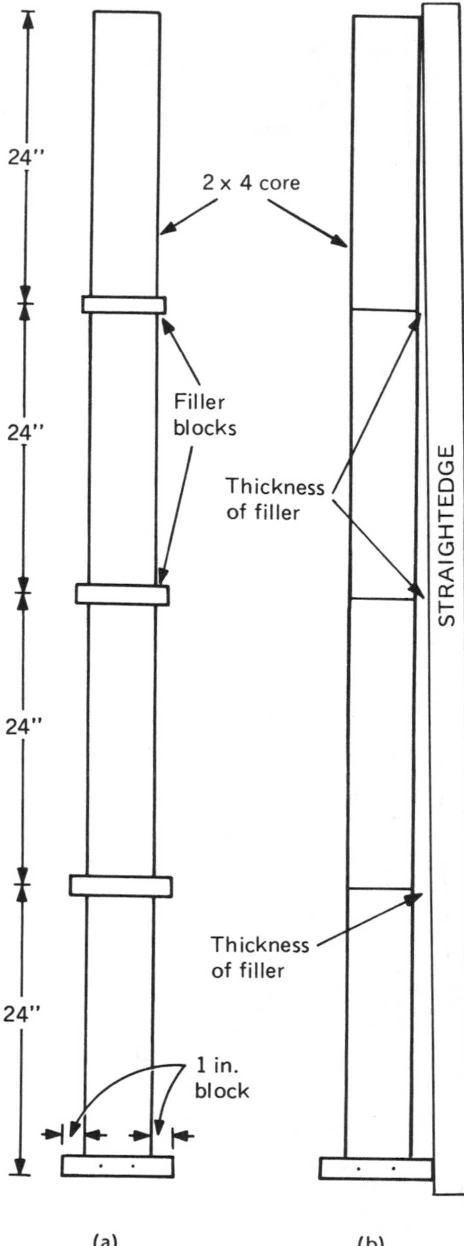

Figure 12-9 Fitting the core with filler blocks for a tapered cased column.

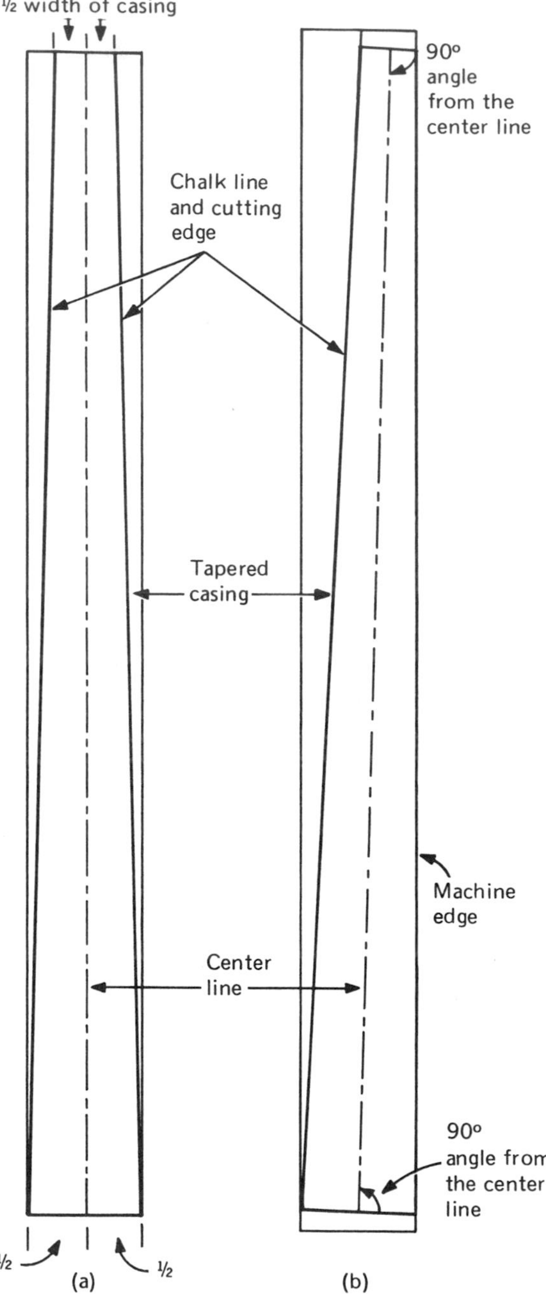

Figure 12-10 Layout of tapered casing pieces: measurement method.

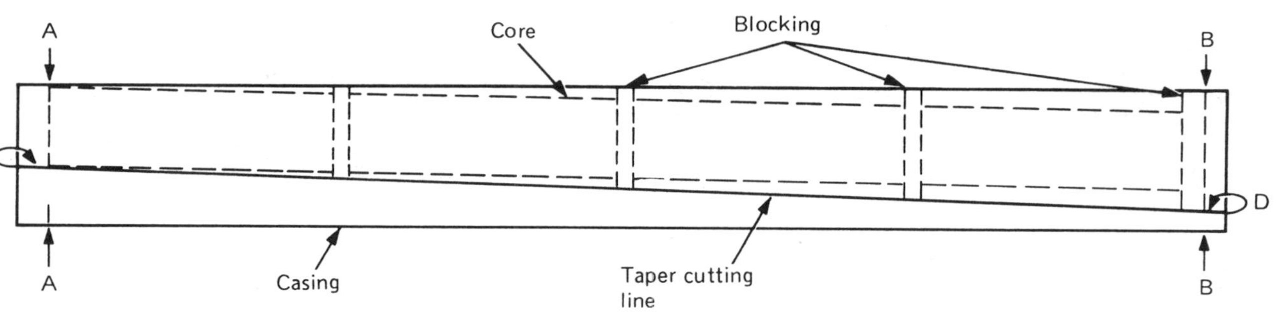

A-A transfer points for top end cut

B-B transfer points for lower end cut

C-D transfer points for taper cutting line

Figure 12–11 Layout of tapered casing pieces: overlay method.

casing. All four casing pieces can be laid out in a similar manner and cut accordingly.

There is one sure way of making each joint professional looking. While dressing (planing) the long-side edge of the casing, slightly undercut each throughout its full length. The nails will draw the boards together beautifully. In fact, if this technique is used, each casing piece should be cut about a full 1/16 in. wider than necessary, so, when the bevel is added, the inner or narrower width exactly equals the thickness needed.

Moldings added to the top and base over the casing will not fit accurately if cut square and mitered. They need to be tapered and mitered. This is done quite easily by using a *bevel square*. Once it is set, all pieces can be laid out with it making the crosscutting lines; then a combination square should be used for the 45 degree miter. A word of caution is needed here: Be sure to use the bevel square from the same side of the molding stock each time—all top or all bottom. Switching results in improper fits, open gaps, and frustration, not to mention the waste of lumber.

Anchoring a Column

It is not sufficient that a column be set in position and left unfastened. The weight of the porch or roof section would, of course, bear upon it to some degree. This is not the solution. It must be anchored properly at top and bottom. The top is frequently held in position by toenailing, although sockets and straps or bolts may be used.

The bottom of the column may be positioned over a previously embedded J bolt, steel pin, or plate and pin (Figure 12–12). In these situations a hole must be center bored into the core, so that the bolt may snugly slide into it. This is adequate for most installations, but it does not afford any restriction to keep the column from twisting.

A galvanized metal or aluminum L plate installed under and alongside the core or casing is more rigid and provides sounder anchoring (Figure 12–12b). In this

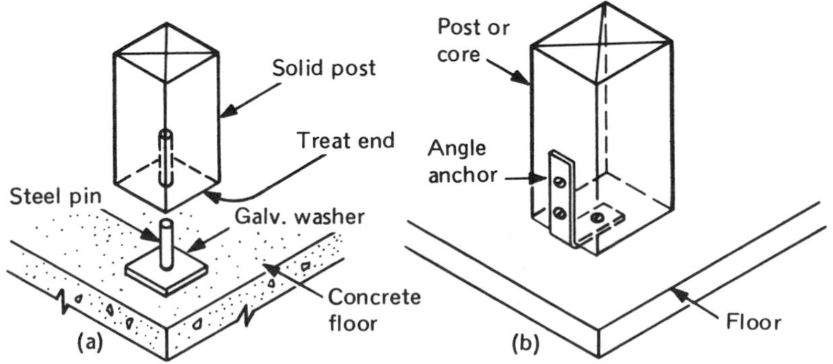

Figure 12–12 Anchoring the bottom of a column using a pin or angle bracket.

method the wood core, casing, or base molding is chiseled out to a depth equaling the thickness of the flange. Bolts or screws are used to secure the column to the plate. As a rule the plate's vertical flange does not show because the molding or casing and molding are installed over it.

A final touch to complete the bottom of a column is to keep the molding off the floor to allow for air circulation and free passage of water. Most of the time a carpenter will use one segment of a folding ruler as a thickness gauge to keep the molding off the floor. By simply laying it flat on the floor and positioning the cut molded piece on top, the reference is established. Another technique that accomplishes the same thing is using a small strip of felt shingle instead of a ruler. After each piece of molding is nailed fast, the ruler or piece of shingle is removed.

TWO-STORY COLUMNS

There is a distinct correlation of principles of construction between the one- and two-story column. The 4 × 4 is not used at all, but the cased-post arrangement is used extensively. The major correlation exists in the techniques used in casing the core, and fastening it to the roof or ceiling assembly and floor. The variations in the core structure may vary widely.

The extremes of core design range from no core at all to an extensive built-up core. The conditions usually prevalent where no core is used is one where the columns are purely decorative and do not support any appreciable load. For example, a small projection of joists built with cantilever construction would be self-supporting. The columns inserted between the lower edge or ceiling of this extension and floor would add beauty and style to the design of the house, but bear little weight.

There is sometimes a set of blocks that are included within the casing as seen in Figure 12–13. These blocks are included to provide nailing surface, but more important, to maintain the squareness of the column throughout its length. If the column is uniform in width, the blocks would all be equal in size; however, if the column is tapered, each block would need to be tailored to size at the point of installation.

If the column needs a core, many different arrangements can be made. However, the overall cross-sectional dimensions would restrict the variety of core assemblies. Two types are shown in Figure 12–14.

The ones in Figure 12–14a and b are made from two 2 × 6s and one 2 × 4 nailed so that an H or I structure is formed. Two different-size columns may be made if (1) the 2 × 4 is nailed on edge or (2) flat. If used on edge, an overall 7 × 8 column is possible; with shims an 8-in. square column can be made. If the 2 × 4 is used flat a 6 × 7 column is possible or, with shims, a 7-in. square column.

The third alternative shown in Figure 12–14c uses either two 2 × 6s or a 2 × 4 and a 2 × 6 in an L configuration. In this arrangement blocks similar to the

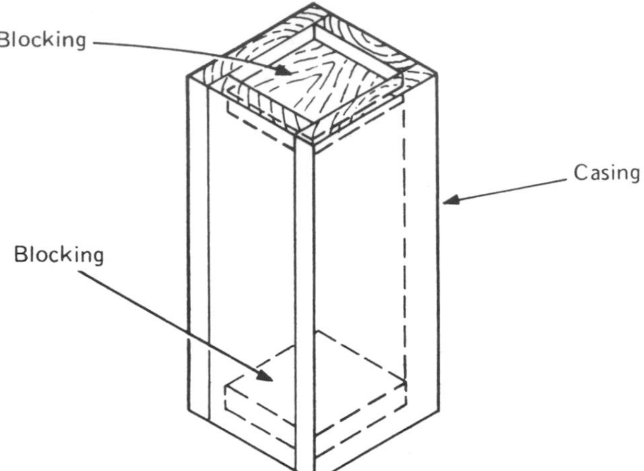

Figure 12–13 Building a hollow column that uses blocking to maintain column squareness.

ones shown in Figure 12–13 are used to complete the core. They establish the *square reference* needed to keep the column true throughout its length.

Although more materials and labor are used during the construction of a two-story column with crown and base, there is usually no great difference in principle or task when applying the casing crown and base molding. There is little, if any, difference in anchoring techniques either. Even the process of constructing a tapered column is the same.

What must be different is the cross-sectional dimension. A 4- or 5-in. square column would look very spindly and weak. It would have the appearance of a toothpick holding up a block of wood. Substance and mass are needed. If a group of columns are installed several feet apart, 7-in. columns are adequate. The human eye tends to blend things together and therefore the group of relatively slender columns

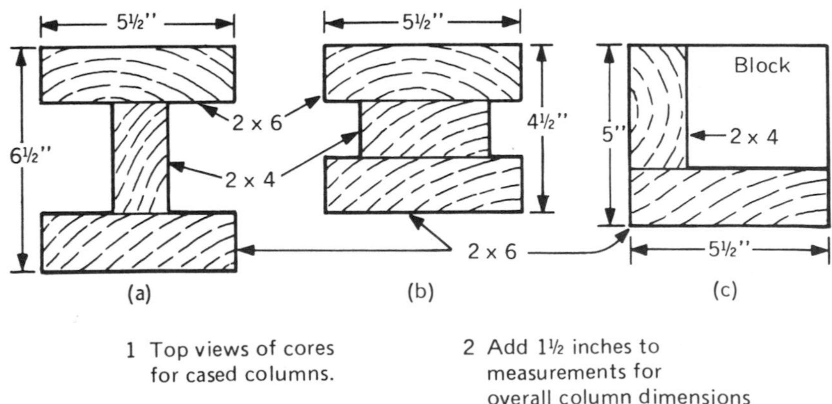

Figure 12–14 Alternative core arrangements for larger columns.

would take on mass. In contrast, if only four columns are used across a 20-ft front, each spaced 5 ft o.c., then columns with larger cross-sectional areas ranging from 8 to 12 inches are more appropriate.

These values should be kept in mind whenever decisions about column dimensions must be made. Along with the decision of column size, the dimensions of its crown and molding base should be selected to add a balanced appearance. Thicker crowns with wider bed molding may add needed mass. Also, the base molding may be more pleasing if 1¼ or 2-in. stock are used.

SUMMARY

On-site construction of columns is an exacting task. It requires sound applications of the principles of column construction, proper fitting, the use of cores in certain cases, and no cores in others. Straightness and squareness are two absolutely essential requirements. All efforts must be centered on these two characteristics because they define the appearance. The carpenter must use all the proper techniques to accomplish these needs and add whatever support materials are required.

PROBLEMS AND QUESTIONS

1. Of the three orders of columns, which one is most commonly used today?
2. If a cased column were to be constructed and its dimensions were 8 in. square and 9 ft long, what core materials would you use? Explain why.
   a. 2 × 4s **b.** 2 × 6s
3. Why should 1 × 4s be nailed to a core of 2 × 4s before the 1 × 6s are fitted?
4. What are the five critical factors that must be considered when on-site building of columns is done?
5. If you install a crown on a column and do not bevel its top surface, to what are you contributing?
   a. a poor fit **b.** a moisture trap **c.** excellent craftsmanship
6. Give one good reason for using a miter joint when fitting and installing column base moldings.
7. State the technique you would use on base molding to minimize the chances of rot.
8. Is the difficulty in casing a tapered column the
   a. fitting of the casing pieces along the length of the column?
   b. cutting the top and bottom of the casing pieces perpendicular with the centerline of the column?
9. What problem can occur with a column if a pin is used as an anchor on the base?
10. What technique is used in hollow-column construction that keeps the column square?

PROJECT

If you have been following the projects on building the porch, you know that the porch will need columns to support the roof. You can easily buy ready-made columns from many builder supply stores and lumberyards; but if you want to make your own, this project will provide design understanding by using materials and ideas from the chapter. The project has several phases: designing the column and base, anchoring on the porch floor, and joining to the headers and trim.

A. Designing the Column and Base

We shall make some assumptions. Our columns will appear like the one shown in Figure 12–7. The column will connect to the header as shown in Figure 12–5a. The soffit will fit between columns once the columns are in place. The fascia will extend down ½ in. below the soffit. The height from the porch floor to the underside of the 2 × 6 headers is 7 ft 8 in. There is a need for six columns, one about 10 ft away from the house wall, and four across the front of the porch.

1. How many 2 × 4 8-ft pieces will we need?
2. How many 1 × 4 8-ft clear pine or fir pieces will we need for casing the double 2 × 4s?
3. How many 1 × 6 8-ft clear pine or fir pieces will we need to complete the casing of the double 2 × 4s?
4. Looking at Figure 12–7, how long is the 1 × 6 needed for one side of the base molding? (Allow for cutting waste.) This means that our needs for encasing the entire column with a base requires ____________ ft. How many linear feet of 1 × 6 clear pine or fir is needed to make the base molding for all six columns?
5. What sequence of the following steps is appropriate for the construction of the column?
   a. Precut the 24 base molding pieces with a 45° miter. ________
   b. Cut the 2 × 4s for length and nail the two together, making six assemblies. ________
   c. Cut the 1 × 4 casing full length (same as the 2 × 4) and nail to the flat side of each 2 × 4 on each column. ________
   d. Trim the 1 × 6s casings for width and full length of the column and nail to the complete casing. ________
   e. Bevel the top edge of the base moldings. ________
   f. Mark the 1 × 6s at the top of the column for trimming to fit under the fascia. ________
   g. Custom fit each base molding and nail each in place. ________

B. Installation Preparation

The columns must be installed. We need to design a sequence that covers all aspects.

1. Since the porch floor slopes away from the house at a rate of ¼ in. per 10 ft, the total slope is equal to ____________ in. The headers must be level. The temporary supports should have accomplished this. However, you must check for level and correct any errors with shims.
2. The overall length of the four columns across the front of the porch must be ____________ in. longer than the two side columns.

3. Where will the six columns be located? We must mark their positions. What is the proper sequences for the following steps?
   a. Mark the position of the two side columns along the underside of the headers. ________
   b. Mark the location of the two corners on the underside of the headers. ________
   c. Divide the length between corner posts, marking the header so that there are three equal distances among the four columns. This positions the two intermediate front columns. ________
   d. Use a plumb bob to mark the porch floor where the column is expected to be positioned. ________
4. Once the columns are in position, plumb, and nailed, the cornice can be finished. (*Note:* Some carpenters will build the cornice in its entirety before the columns are installed. When they do this, the 1 × 4 soffit is placed under severe crushing stress. Therefore, it is best to have the strong header materials bear onto the columns.) Which pieces of the cornice (frieze and soffit) should be installed first and second?
5. Since the porch floor is tongue-and-groove wood, how do we fasten the columns to the floor?

Answers:

Part A:
1. 6 columns × 2/column = 12
2. 6 columns × 2/column = 12
3. 12
4. About 9 in., 3 ft, 18 lin. ft
5. Order: b,c,d,a,e,g,f

Part B:
1. ½ in.
2. ¼ in.
3. Order: b,a,c,d
4. soffit first, frieze second
5. toenailing with 8d casing nails

Chapter 13

Screening a Porch

As we explain in this chapter, screening a porch is clearly in the advanced skills area of the best carpenter. It requires a fair degree of design capability as well as working knowledge of cabinet joinery.

The method explained in this chapter is not the application of screen wire to 2 × 4 studs spaced 24 in. o.c. It involves the fabrication of panelized frames with joinery and sculptures in such a way that beauty of line and dimension are employed. If done properly, even economy of materials can be incorporated into the design. If done poorly, a considerable waste of materials will occur. If done properly, channels for rain and moisture runoff are included. If done poorly, rot and mildew form immediately.

What, then, does the overall appearance of this enclosure look like? One example is shown in Figure 13–1. Each panel extends from column to column or wall, and floor to soffit. In this porch each panel is approximately 8 ft high by 8 ft long. Actually, each width is slightly different, but as the photo shows, there is an apparent unity of dimension. All variations are distributed to the smaller side screen sections. The wider center sections are uniform in width to simplify layout, cutting, fitting, and assembly.

DESIGN

A key element in the planning stage of an undertaking such as screening a porch is the design. Two designs are shown in Figure 13–2. Given a specified panel width, each panel of screen wire can be made equal. This would create a uniform repetitive

Figure 13–1 Enclosed porch.

pattern. On most homes this pattern is conservative, adding a dimension of style and harmony. The horizontal member approximately 32 in. from the bottom is included for two reasons: it adds a measure of stability and beauty, and it reduces the chance for damage to the screen wire. Should furniture be bumped against the panel or a person walk into the screen by accident only a small panel of wire would be damaged.

The second design shown in the figure is one employing a large 36-in. center panel and smaller side panels. Here the harmony is experienced when adjacent panels are viewed from a short distance. The repetitive pattern is there but not as dynamic as in the first example.

Obviously, many other possibilities exist, but care must be used because panels that use wide widths of screen wire tend to look bold and are disproportionate on small houses. However, wide expanses of screen may be appropriate on very large houses. On the other hand, the wooden stiles and rails also add mass to the appearance. Too much wood can be as overpowering as too much screen wire. Therefore, for best results restrict the minimum size of screen width to 18 in. and the maximum width to 48 in.

Another design feature that must be considered is the thickness of the panel frames. Nominal 1-in. stock is not satisfactory. It tends to be too flexible and skimpy looking. Stock of 5/4-in. thickness is the most appropriate. It measures a full 1 to 1⅛ in. thick. This is stout enough to withstand the joinery used during construction as well as minor abuse, should any be forthcoming.

Still another design feature is the need to build panels ½ to ¾ in. shorter than

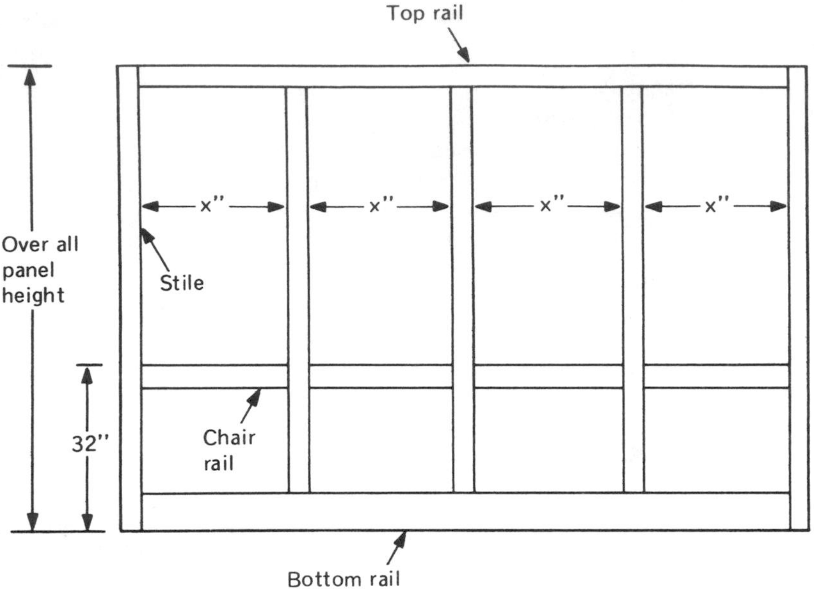

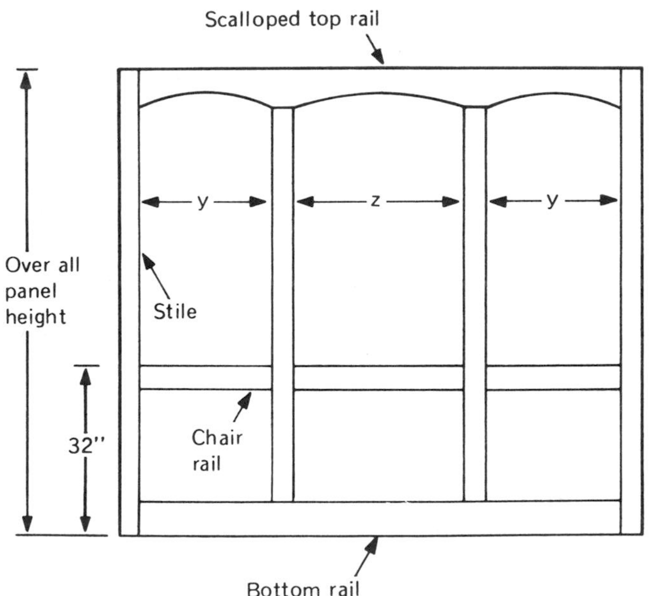

Figure 13–2 Screen panel showing stiles, rails, and overall designs.

the actual height of the opening. The space must be included along the bottom edge to allow rain and other water to escape from the floor of the porch. Lattice strips with inverted Vs cut each 24 in. and at corners are installed over the gap inside and outside to complete the job.

Still another optional feature is the small curve cut into the head rail of each screen section. This gentle arch breaks up the severe straight line. However, not all houses screened with this type of feature in the panel may look good—use caution.

ESTIMATING

Tools Needed for the Job

14-oz claw hammer	1 set hand chisels
Framing square	6-in. jack plane
Combination square	Block plane
Ruler	4-ft bar clamp
No. 8 hand saw	Electric (hand) jig saw (if curves are planned for)
⅜-in. electric drill	Radial arm saw (optional)
1 set dowel center attachments	Miter box with saw
5/16-in. drill bit	Stapler, with ⅜- or ½-in.-long staples
Portable power saw (optional)	

A Timetable

A timetable is developed in the following manner. The number of labor hours based on woodworking experience should be the judgment of how long it will take to (1) cut all pieces for length, (2) rip them if necessary, and dress them with a plane, (3) drill holes for dowels, (4) assemble each panel, (5) install screen wire and put on screen molding, (6) paint panels, (7) install panels, (8) hang screen door, and (9) apply final coat of paint.

Preparing an Estimate

No. ______ 5/4 in. × 8 in. × ______ ft for horizontal bottom rails
No. ______ 5/4 in. × 4 in. × ______ ft for chair rails
No. ______ 5/4 in. × 4 in. × ______ ft for heads of screen panels
No. ______ 5/4 in. × 4 in. × ______ ft for uprights
No. ______ 5/16 in. × 3-ft dowel sticks for doweling joints
No. ______ linear feet of ______-in.-wide screen wire
No. ______ linear feet of ______-in.-wide screen wire

No. ________ linear feet of screen molding
No. ________ linear feet of ¾-in. quarter-round molding
No. ________ linear feet of ¼-in. lattice molding
No. ________ dowels 5/16-in. diameter at two per joint
1 box staples
1 pair screen door butt hinges
2 lb 6d finish nails
1 lock set
2 lb 3d finish nails
1 screen door spring
1 screen door 3 ft 0 in. × 6 ft 8 in.
18 linear feet 1 × 2 (stops)
1 lb ¾-in. brads

All quantities and lengths needed for the panels must be supplied in the spaces provided.

CONSTRUCTION

The construction phase of the job includes three parts: (1) layout of both porch enclosures and each panel, (2) cutting and assembling, and (3) screening.

Layout

Figure 13–3 shows a typical porch floor plan. This plan is used as an example to aid in familiarity with layout methods. First the position of the screen panels is inserted

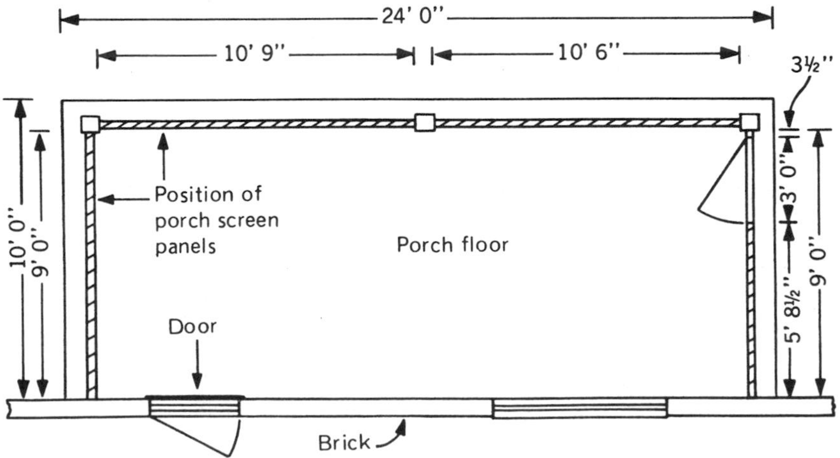

Figure 13–3 Floor plan of porch, defining length of panels.

into the standard porch floor plan. The screen door position is designated against one post for sound anchorage of the frame. The length of each panel is measured with a ruler and written into the plan. The results in our example show one panel 9 ft 0 in. wide, one 10 ft 9 in., one 10 ft 6 in., and one 5 ft 8½ in. wide. The door is a standard 3 ft 0 in. × 6 ft 8 in. screen door.

Now lay out the panels. For this example the 9 ft 0 in. panel and Figure 13–4 are used. (The others must be made using the same techniques.)

The panel frame is made of 5/4-in. × 4-in. stock for two full-length stiles, two intermediate stiles, one top rail, and a three-piece chair rail. The bottom is made

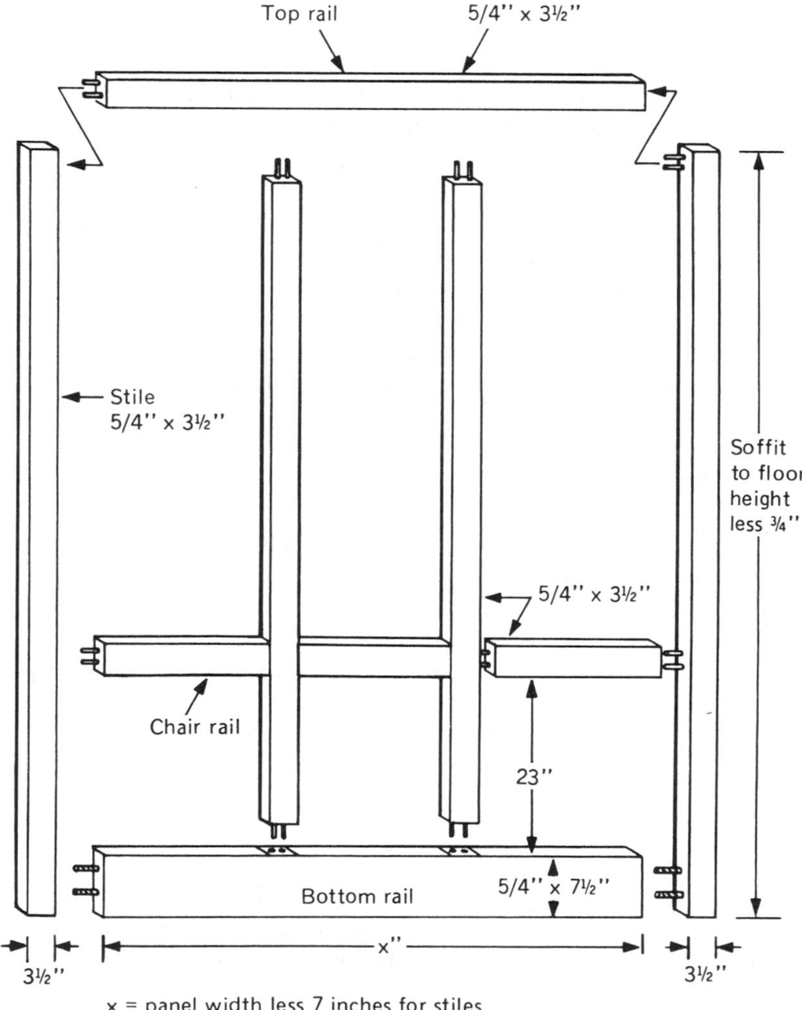

Figure 13–4 Panel illustrating doweling, approximate sizes of stiles and rails, and overall height.

from 5/4-in. × 8-in. stock. The finished panel frame must measure 8 ft 11½ in. wide (½ in. less than the opening and ¾ in. less in height than the total opening height). These variances are needed to allow for swelling, out of square, and moisture runoff.

The intermediate stiles are positioned to make three equal screen openings. The chair rail pieces are marked for position.

Cutting and Assembly

Mass cutting of all the pieces is preferred to individual layout and cutting. All outside stiles and intermediate stiles can be cut at one time. The top and bottom rail for each panel should be cut in pairs. Finally, the chair rail pieces should be cut and numbered. It is a good idea to number all pieces.

Lay all pieces for one panel on the floor and make sure that they fit together properly. Then bore holes for dowels at least 1 in. deep on each part of the joint. Pour or squirt some glue into a dowel hole before inserting the dowel and spread glue to cover the joint.

Start the assembly with the center chair rail and intermediate stiles. Next install the bottom rail, then the top rail. Follow this by installing one more chair rail piece and then the outside stile. Finally, install the last chair rail and the outside stile.

The frame must be square, so a framing square and a 1 × 2 brace nailed diagonally must be used while the glue is drying and while screen wire is being installed. After the glue has dried, sand both sides with a belt sander if you have one, or just plane any irregular joints flush.

Screening

The three upper panels must be screened first, and of these, the center panel should be done first. With the wire stapled, the screen molding is installed by cutting and mitering all corners and then nailing with ¾-in. brads. All excess screen wire is trimmed with a sharp utility knife after molding is in place.

The lower three panels are screened last. If children are at home, a ½-in. wire mesh installed over the screen wire and a heavier molding is desirable. *Note:* It is best to prime-paint the panels and molding before installing the screen wire.

PANEL INSTALLATION

Figures 13–5, 13–6, and 13–7 are provided for understanding of the installation of each panel. A ¾-in. quarter-round molding is used alongside each of the stiles and top rail. Lattice, ¼ in. × 1½ in., is used along both sides of the bottom rail.

1. Cut and nail quarter-round molding to the post and soffit.

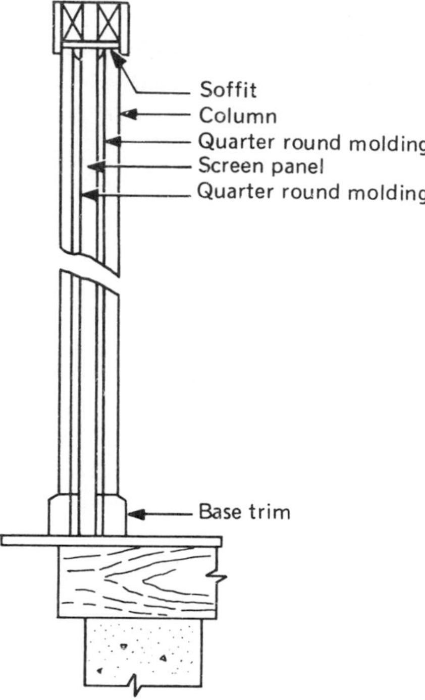

Figure 13–5 Full side view of panel in place and molding used to keep it there.

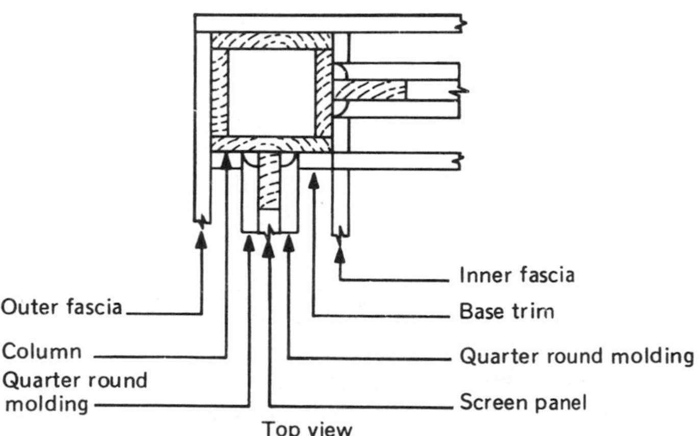

Figure 13–6 Top view of panels in place, showing quarter-round on each side of panel.

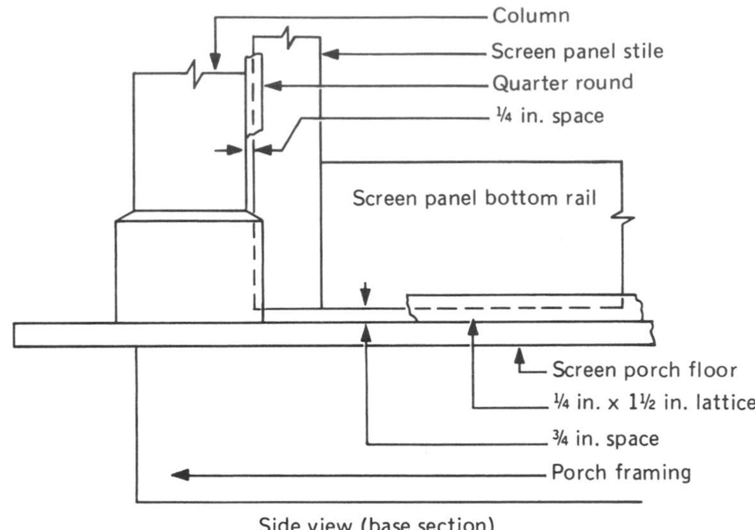

Figure 13–7 Side view of base section. Shows clearance between panel rail and floor and lattice covering space.

2. Place a panel against the quarter-round and raise to the soffit. Toenail into the posts.
3. Cut and nail quarter-round molding to fit against the post and stile and the soffit and rail.
4. Cut and nail the lattice to the bottom rail. It must touch the floor (see Figure 13–7). Before nailing, cut inverted Vs (∧) to allow water to escape.

SCREEN DOOR INSTALLATION

After the panels are installed, the screen door must be fitted and hung. Figure 13–8 shows a top view of this installation. First, measure the actual width of the door and then add ¼ in. for clearance. The remaining space between the door and post must be filled in with a piece of 5/4 × 3½-in. stock, ripped narrower if necessary. The 5/4-in. stock should reach from floor to soffit and be toenailed in place very firmly.

Butt hinges look more attractive than flush hinges. Above all, the door should open inward wherever there are steps outside. A passage lock needs to be installed to lock the door and keep it closed, and a screen door spring helps pull the door closed.

Normally, the height of the screen door is lower than the soffit; therefore, a filler piece of 5/4-in. stock is cut and fitted in this space. An allowance of a ¼-in.

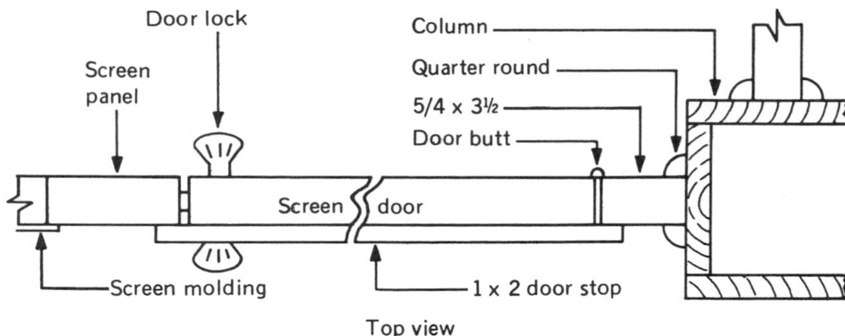

Figure 13–8 Top view of screen door installation. Note filler board between column and screen door.

space above the door is essential for proper door operation. Next, the 1 × 2 door stop needs to be cut and nailed in place on the outside, to the panel, the 5/4-in. stock, and the filler piece above the door.

SPECIAL CONSIDERATIONS

Frequently, a panel must fit against brick. If so, caulking can be used to fill the space between the panel and brick. If the space is large and quarter-round molding is difficult to install, a piece of lattice may be fitted in place of the quarter-round.

Where a panel must be anchored to a concrete porch floor, angle braces made from aluminum are used. Driving a stud through the metal into the concrete, or setting a bolt into the concrete (bolting the angle in place) are two methods of anchoring the braces to the floor. Screws through the other flange of the angle hold the panel in place. This should complete the project.

PROBLEMS AND QUESTIONS

1. If screen panels must vary a few inches in width, is it better to vary the width of the outside panels or the center panels? Explain.
2. When designing screen porch panels, list one or more factors that would influence you to choose three screen sections over two sections.
3. Which thickness of lumber is most appropriate for the construction of screen panels?
   a. ¾ in. **b.** 5/4 in. **c.** 2 in.
4. What width of stock would yield the smallest waste if your screen panel stiles were to be 2¾ in. wide?
   a. 5/4 × 4 in. **b.** 5/4 × 6 in. **c.** 5/4 × 12 in.

PROJECT

If you have done most of the projects so far, you know that this chapter on screen enclosure is the next logical step. Since the chapter provides all the tools necessary for the design, what we shall do is simply apply those tools for one section of the porch that has been designed thus far.

1. Let's prepare a design model for one of the front sections. If you recall (or go to Chapter 12), the overall porch length is 30 ft. Each column is approximately 5½ in. across and there are four columns. That means that the total space for panels is 30 ft − (4 × 5½ in.), for a distance of ____________.
2. Having four columns in line creates three spaces. Since there are three equal spaces, each measures ____________.
3. Let's assume that the height from the porch floor to the underside of the soffit is 7 ft 6 in. Refer to Figure 13–2, which shows the design of the finished screen panel. Notice that the two side panels are narrow and the center is wide. We will need to define the width of wire; therefore, we better agree on the opening widths. Wire is available in increments of 2 in. beginning with 24 in. to 36 in., 6 in. from 36 to 60 in. wide (i.e., 24, 26, 28, 30, ____, 36, 42, 60 in.) Given that we have approximately 100 in. of open area measured horizontally, which combination should we select to optimize the screen wire and look good?
   a. 30, 40, 30 = 100 in.: OK? Why or why not?
   b. 32, 36, 32 = 100 in.: OK? Why or why not?
   c. 28, 42, 28 = 100 in.: OK? Why or why not?
   d. 26, 48, 26 = 100 in.: OK? Why or why not?
4. Now turn to page 271, where "Preparing an Estimate" is shown, and complete an estimate for the panel.
   a. Using a pencil and paper, fill in each blank and compare your listing with mine, which appear in the answer section.
   b. What other support materials are necessary to build and install the panel?
   If your answers match mine, you have a true understanding of the project.

Answers:

1. 28 ft 2 in.
2. 9 ft 4⅝ in.
3. **a.** yes, pleasant proportions, 34- and 48-in. wire
   b. no, center panel is too close to the widths of the end panels, 36- and 42-in. wire
   c. yes, but the center panel may be a bit too wide, 32-ft and 48-in. wire
   d. no, differences between center and end panel widths are too great, 30- and 54-in. wire
4. **a.** Bottom rail: 1, 10 ft
   Chair rail: 1, 10 ft
   Head rail: 1, 10 ft (substitute a 5/4 × 6 piece here)
   Stiles: 4, 8 ft
   Dowels: 24
   Screen wire (option 3a): 16 ft 34 in. for outside panels
   Screen wire (option 3a): 8 ft 48 in. for center panel

Screen molding:	Horizontal	4 at 10 ft	= 40 ft
	Vertical	6 at 8	= 48 ft
	Total		88 ft

¾-in. between quarter and round molding: 4 at 8 ft. = 32 ft
2 at 10 ft = 20 ft
52 ft

Lattice 2 at 10 ft = 20 ft

b. 1 box staples, 1 pint waterproof glue, 1 lb of 6d galvanized finish nails, 1 lb 3d galvanized finish nails, ½ lb of ¾-in. brads.

Chapter 14

Staircases

A great deal of understanding, knowledge, and skill are needed by any carpenter to design and construct a set of stairs. Therefore, this chapter is restricted to defining and examining the principles of stair design for residential use. Residential use means that these stairs will be used by 10 or fewer people.

What must be learned are the terms used to describe stairs and their physical location. Providing a written definition of a *stringer* or *carriage* is not sufficient; a clear picture of the stair and its use provides the added dimension that completes the meaning. The principles of stair construction can be divided into two broad categories: (1) those parameters usually used to design residential stairs, and (2) the parameters used to prepare proper space for and anchoring of the stairs.

No construction skills and techniques are included in this chapter; they follow in Chapters 15 and 16, where the practical aspects of stair construction are developed.

TERMS

Most of the terms used to describe the parts of stairs are very old, some dating back several centuries. Consider the *carriage* portion of a stair shown in Figure 14–1. This is the principal support member, and it indeed acts as a carriage or, more appropri-

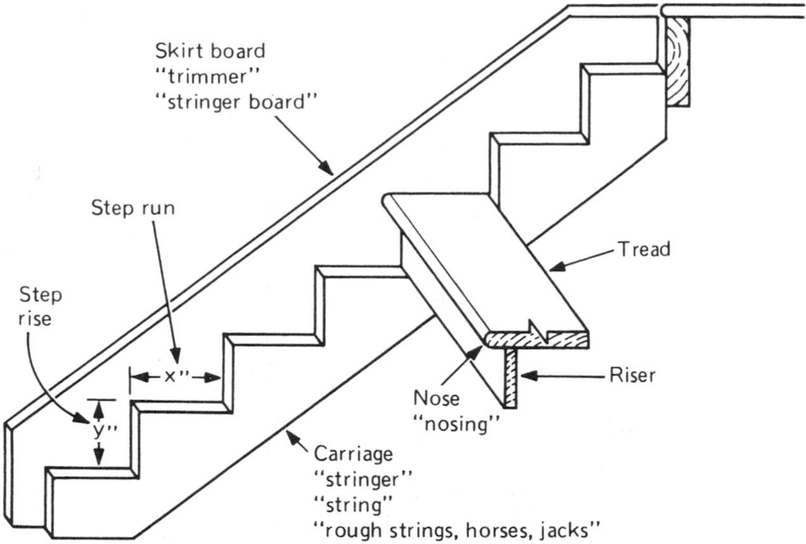

Figure 14–1 Terms used with stairs and their association.

ately, as *an undercarriage.* However, the term "carriage" is not used universally. This same support is also called a *stringer,* or *string* for short. Certain segments of the country approve of calling these supports *rough stringers, horses,* and even *jacks.*

Whenever a board of finish quality and grade is applied to the outer surface of the carriage or stringer, it may be called one of several names: *skirt board, trimmer,* or *stringer board.* Most of the time these are clear-grade stock that are used to cover the ends of the *treads* and *risers* as well as stringers. The *tread* is the board that is normally stepped on, as Figure 14–1 suggests. The *riser* is a board installed vertically and under the tread. This means that the outer edge of the tread protrudes about 1 to 1½ in. from the riser. Rather than call this a protrusion, the name *nose* or *nosing* is used. Nosings are irregular curves or form a perfect arc of 180°. On occasion nosing may be added to the tread and may include molding such as cove.

Associated with the tread is the *run of the tread* or *step run.* This is either the horizontal dimension cut into the stringer or is the width of the tread. The *rise* is associated with the riser since the rise is a measure of height from one step to the next.

Most building codes require stairs to be equipped with *handrails;* however, if the handrail is part of an assembly, it is frequently called the *balustrade.* Actually, the handrail and the *balusters,* which are the vertical supports of the handrail, form the balustrade. All these are shown in Figure 14–2. Where the handrail ends in an open area, as near the bottom step, a *newel* post is used. As a rule, this turned post terminates the balustrade.

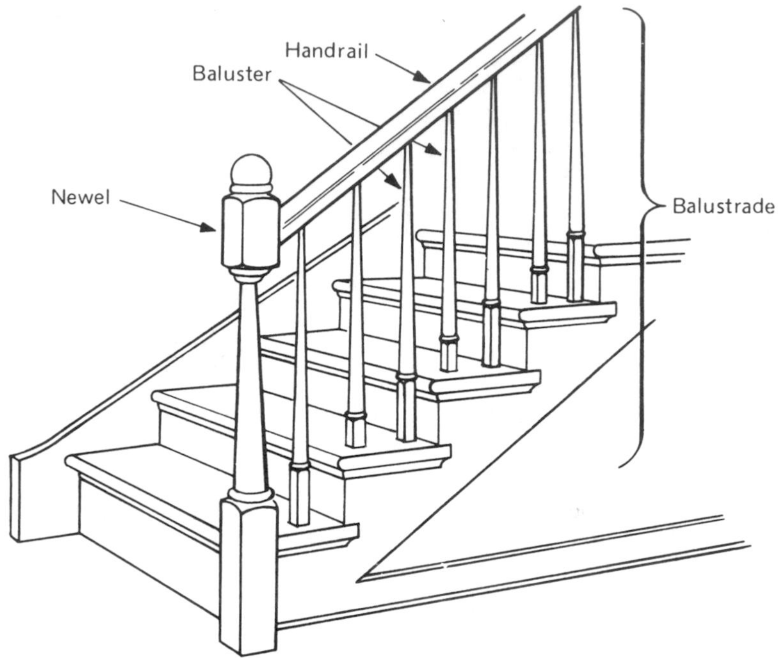

Figure 14–2 Terms used with balustrade, baluster, handrail, and newel.

If a set of stairs is built so that no trimmer is used, the stringers are dadoed for the risers and treads; these are called *closed* or *housed stringers* (Figure 14–3).

Another of the terms associated with stairs is one that is used with curved stairs called the *winder.* This is the tread that is cut in *pie* shape. There is also the *landing,* which, of course, is the flat area built whenever the stairs need to change direction.

PARAMETERS FOR RESIDENTIAL STAIRS

Width of Stairs

The usual range of stair width for residential use is from 30 to 48 in. Stairs made narrower than 30 in. are seldom used today. The trend has been for some time to use a 36- to 39-in. stair width. In larger houses, and frequently in split-level houses, wider stairs are used, some as wide as 48 in.

The 36- to 39-in. stair provides sufficient width to make climbing or descending comfortable and for the transport of furniture. This width requires a handrail

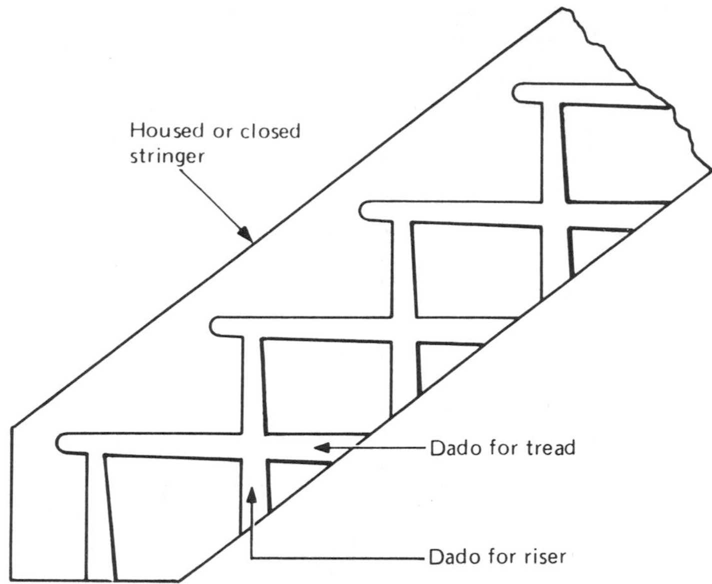

Figure 14–3 Housed or closed stringer.

on one side only. Stairs wider than these require two handrails according to several building codes.

The selection of stair width is usually made by the design architect. He or she selects it based on available space, purpose, and structural considerations which could, for example, include a landing.

Width and Depth of Landing

Landings are usually required where (1) more than 13 steps are needed to reach from one floor to the next; (2) a stair must make a turn, either a 90° or 180° turn, (3) a door is installed at either end of the stair.

If a landing is needed, its width is usually equal to the stairs; a 36-in.-wide landing for a 36-in. stair, for example. Its depth (Figure 14–4) is also equal to its width. As the figure shows, a depth of 36 in. is used for a 36-in.-wide stair.

If a landing is used where a door opens onto it, its depth must be adequate to allow opening of the door and to allow the person opening the door to stand safely on it. This suggests that a depth slightly in excess of the width of the step is necessary (Figure 14–5). On the other hand, a landing is not required if the door opens away from the direction of the step, provided that the rise from the last step to the floor does not exceed the normal step (Figure 14–6).

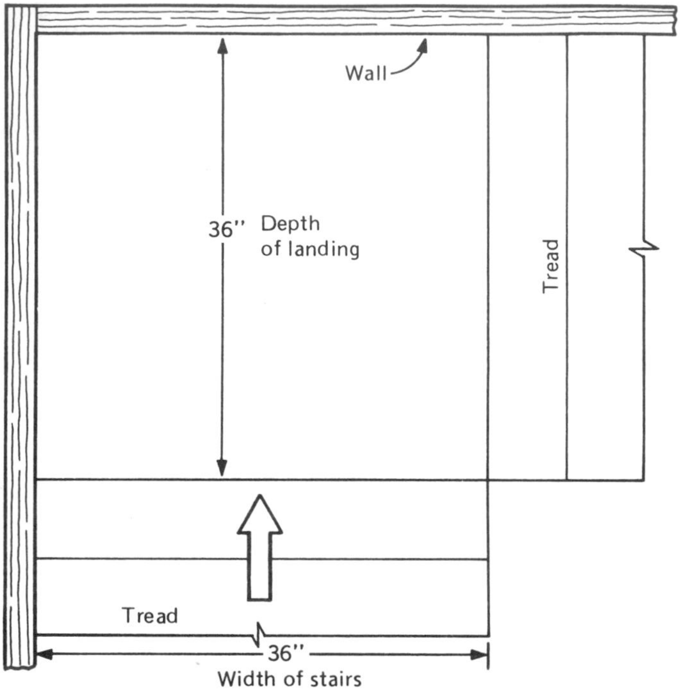

Figure 14–4 Landing dimensions.

Where a door opens onto a stair area at the lower level, a floor area must be provided; normally, this will be framed and floored or treated as a landing (Figure 14–7); however, a landing or floor area should be provided between the last step and door if the door opens away from the step (Figure 14–8).

Rise and Run of Individual Steps

Over a long period of time, carpenters have passed on by word of mouth several dimensions for the rise and run of steps. One of these is the "combination of 17 in." (Figure 14–9). This combination has been translated to many different proportions, the most common of which is an 8-in. rise and 9-in. run. In actuality the combination of 17 in. meant that a maximum of 17 inches was to be used. Still, this combination allows for a variety of rises and runs to be selected.

Another combination is the "combination of 25 inches." When using this combination, two rises and one run must not exceed a total of 25 inches. This could result, for example, in using an 8-in. rise and 9-in. run for a total of 25 inches (Figure 14–9).

Several building codes have set down the values of rise at 7½ in. and run at 10 in. But, for residential houses, an 8-in. rise and 9-in. run combination is acceptable.

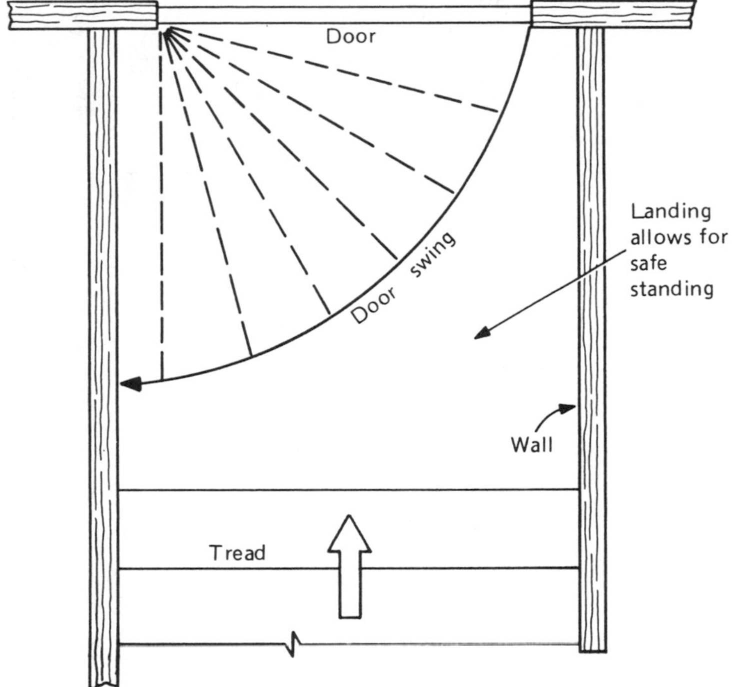

Figure 14–5 Landing where door opens toward the stairs.

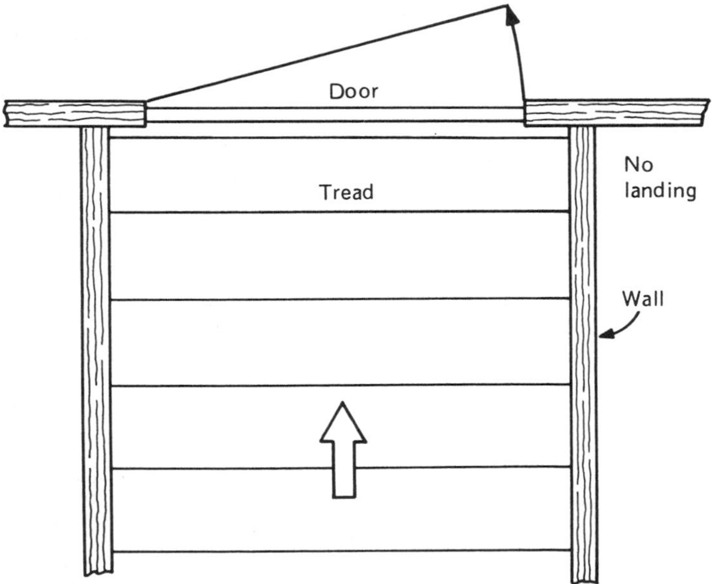

Figure 14–6 Stairs where door opens away from stairs—no landing.

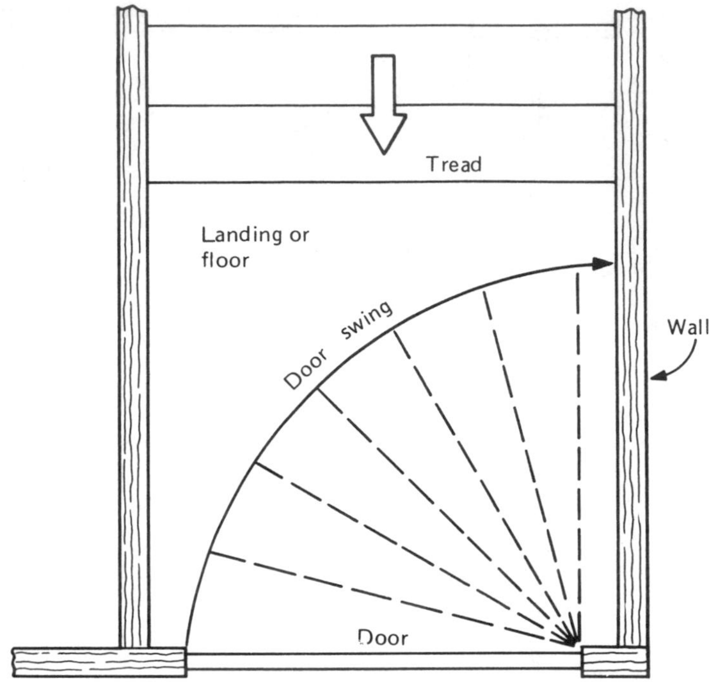

Figure 14–7 Lower landing where door opens toward stairs.

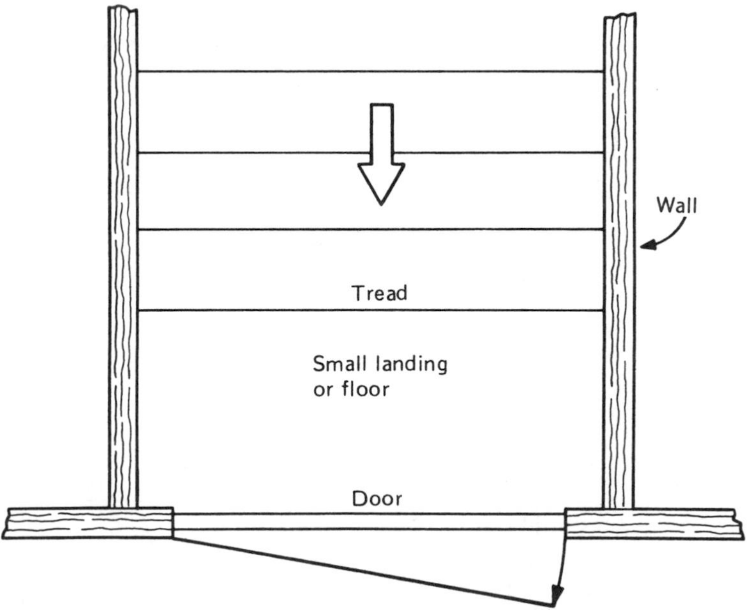

Figure 14–8 Small lower landing where door opens away from stairs.

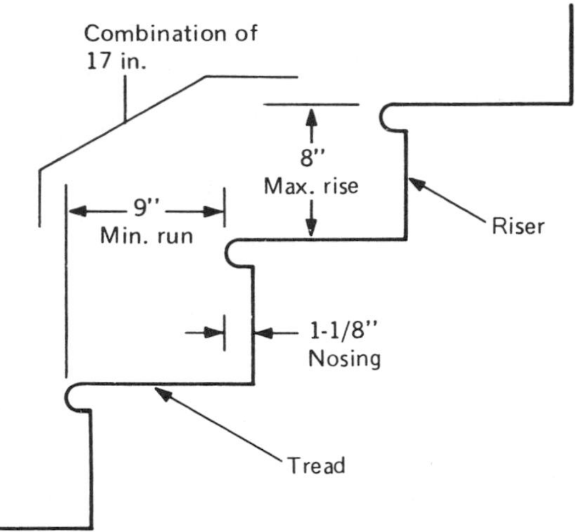

Figure 14–9 Parameters of step rise and run and nosing.

The selection of rise and run is not only determined to fall within these combinations but is also determined by the available space for the stairs as well as the total rise from one floor level to the next.

Rise and Run of the Stairs

Where, for example, a standard 8-ft ceiling is used and ceiling joists are 2 × 8 sheathed and floored, a total rise of 96 in. + 7½ in. + 1½ in. = 105 in. is measured. The aim of all good stair design is to have an equal rise for each step. Therefore, the total rise (105 in.) would be divided until an equal number is obtained with the proviso that the 8-in. rise is not exceeded.

In this case if 8 in. is tried first, the results show:

An 8-in. step rise with 105 in. of total rise produces the number 13.125 rises.

Therefore, the total number of rises must be increased to 14.

Dividing 105 in. by 14 rises = 7½ in. per rise.

This method may be used for any total rise measured.

If, for example, a more gentle climb is desired, 15 risers could be planned for, with the result that each rise would be exactly 7 in.

However, the decision of whether to use 14 or 15 risers alters the total run. If 14 risers are used, there will be 13 runs, as Figure 14–10 shows. These runs should be no narrower than 9 in. and may extend in width to 9½ in., in which case the total run is 9 × 13 or 117 in. *or* 9½ × 13 for 123 in.

If a 7-in. rise is used in lieu of a 7½-in. rise, 14 step runs are required. Again, not exceeding the 17 combination, a step run could range from 9 to 10 in. This translates to a total run of 9 × 14 or 126 in. *or* 10 × 14 or 140 in.

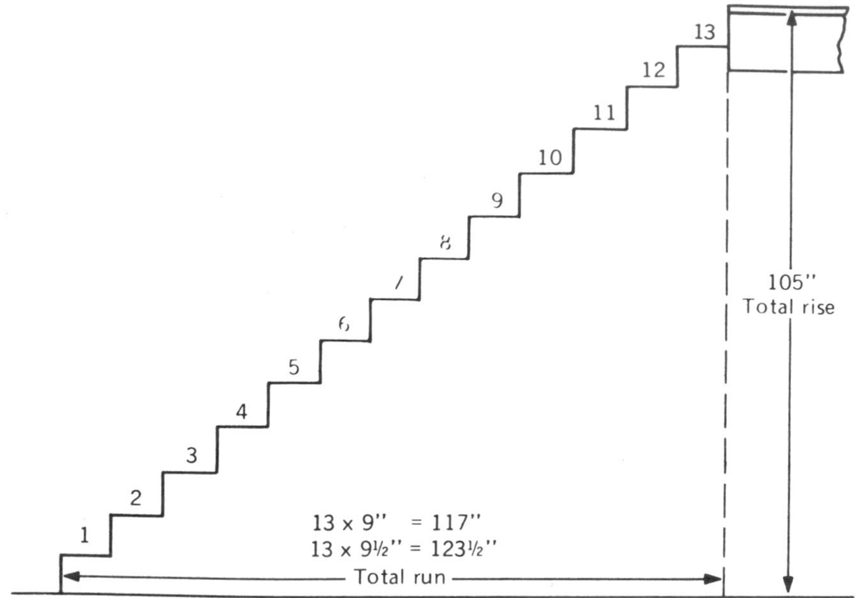

Figure 14–10 Determining total rise and total run of stairs.

The various combinations shown here allow considerable flexibility of design by the carpenter. If there was adequate room available, a stair with a 6⅝-in. rise and 10⅜-in. run could also be constructed. However, there is another factor that may be considered in stair design and it is the angle of the incline.

Angle of Stairs

All stairs used in residential houses should have an angle of slope between 30° and 40°. Figure 14–11 shows both extremes. Here are three cases:

1. The 30° angle for a typical total rise of 105 in. results in a wide step of about 11½ in. with a rise of 7 in.
2. The 35° angle for a typical total rise of 105 in. results in a step about 10 in. wide with a rise of 7½ in.
3. The 40° angle for a typical total rise of 105 in. results in a rise of 8 in. and a step of about 9½ in.

Number of Risers

There is no hard and fast rule for the minimum number of risers per staircase, but general guidelines and some building codes limit the number from 13 to 15. This means that if a situation arises where more than 15 treads are used, a landing should be built somewhere along the travel of the stairs to break up the long climb. Fortu-

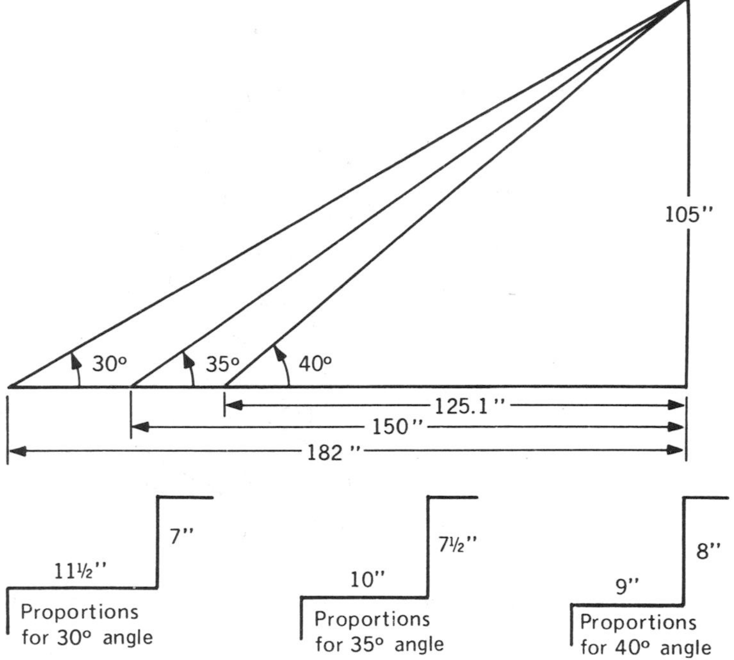

Figure 14–11 The steeper the angle, the steeper the stair and the less run distance required.

nately, most modern residential construction uses an 8-ft. ceiling, so use of a landing for the purpose stated above is not needed. If, however, a 10-ft ceiling were used, a check with local building codes would be required to see if a limitation were included. Then, too, the architect may include a landing in the plans for the stairs and thus eliminate the problem.

Handrails

In principle, a handrail should be included whenever a stair is built. Where the stair is installed between walls, a single rail installed 30 to 34 in. above the nose of the treads (Figure 14–12) is required and should extend slightly below the bottom step and past the top step. If either side of a stair is open, it should have a handrail as part of a balustrade assembly. Stairs that have both sides open must have a balustrade assembly on each side of the stairs.*

Many split-level homes have situations where short segments of stairs using four or fewer risers are made. If this is the case, no handrails need to be installed for safety; they may be installed for esthetic value. But where these stairs use more than four risers, handrails are usually required.†

*"Stairways," *Uniform Building Code* (1973), Sec. 3305, p. 480.
†Ibid., p. 481.

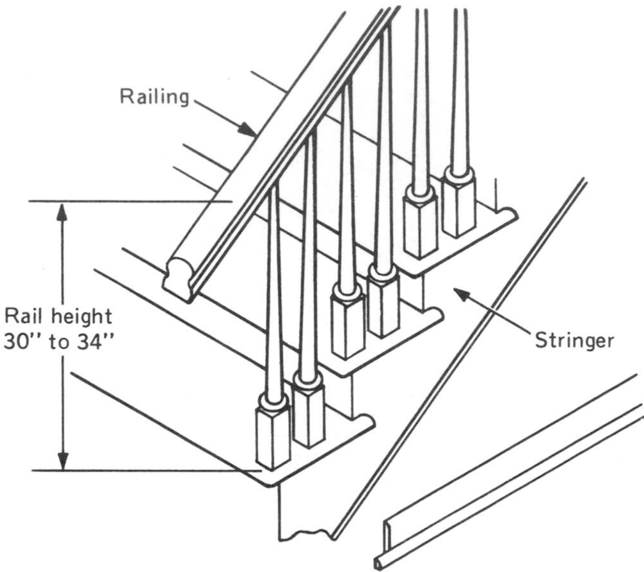

Figure 14–12 Handrails should be 30 to 34 in. up from the nose of the step.

PARAMETERS FOR PROPER HEAD CLEARANCE

There must be a proper amount of head room for people using the stairs, and there must be a *positive feeling* by all persons using the stairs that there is adequate head room. The first is clearly defined in plans and design of the stairwell, and if constructed properly, adequate head room is provided. The second, the feeling that there is adequate head room, can also be satisfied if proper design and construction are followed.

Defining the Stairwell Size

Most residential stairs require a stairwell, or opening, approximately 10 ft long by the width of the stairs (Figure 14–13). Several factors of vital importance can confirm or refine the accuracy of the 10-ft general opening size.

First, the vertical head room that satisfies both the *requirement and feeling* is 6 ft 8 in. taken from the nosing of the second step, as shown in Figure 14–14.This head clearance must be maintained for the remainder of the rise of the stairs. The stairwell length, therefore, should be equal to the total run of the stairs minus the width of the first tread. This dimension is a variable before a design is selected since the rise and run of the step determine the total run, as explained earlier. However, once a design is selected, a fixed length can easily be found. The following examples illustrate the point:

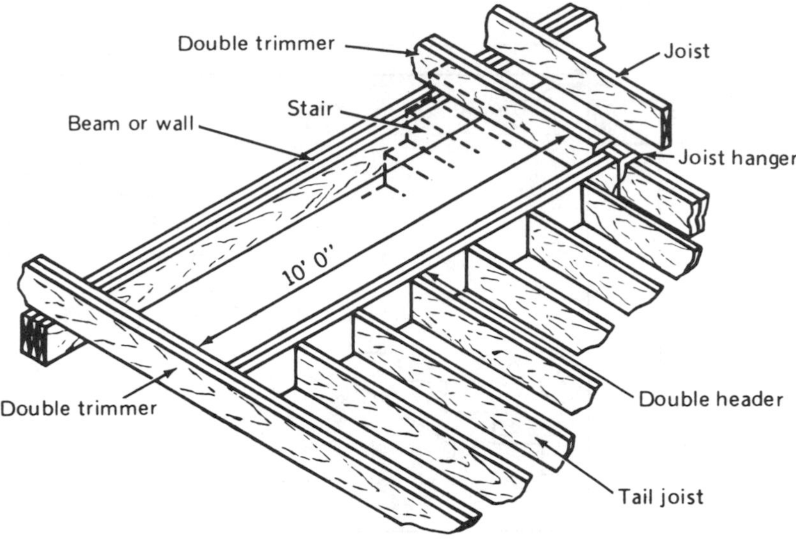

Figure 14–13 Framing the stairwell.

Example 1

The stairwell opening for a stair rising a total of 105 in. using a 7½-in. rise and 13 step runs is illustrated in Figure 14–14. Determine the length of the stairwell.

Solution

1. Measure down from the ceiling a distance 6 ft 8 in. and draw a horizontal line.
2. Since the line crosses the surface of the second tread, draw a vertical line up from its nosing through the ceiling joist.
3. The finished distance for the stairwell is 9 ft 6 in. The rough opening must be made larger to account for trim, riser, and a portion of stringer where the nosing or flooring could be nailed.

Example 2

In this problem the stairwell opening should be longer than the one in Example 1 since the total rise is the same (105 in.), but the rise per step is reduced to 7 in. and the treads (run) are 11 in. (Figure 14–15). Determine the length of stairwell.

Solution

1. Measure down from the ceiling a distance of 6 ft 8 in. and draw a horizontal line.
2. Since the line crosses between the second and third treads, the finished stairwell opening should be taken as perpendicular from the nose of the second step, even though a slightly greater distance for head room is made.
3. The rough opening for the stairwell should be made approximately 12 ft 6 in. to allow for trim of the stairwell and a 6-in. portion of the carriage or stringer.

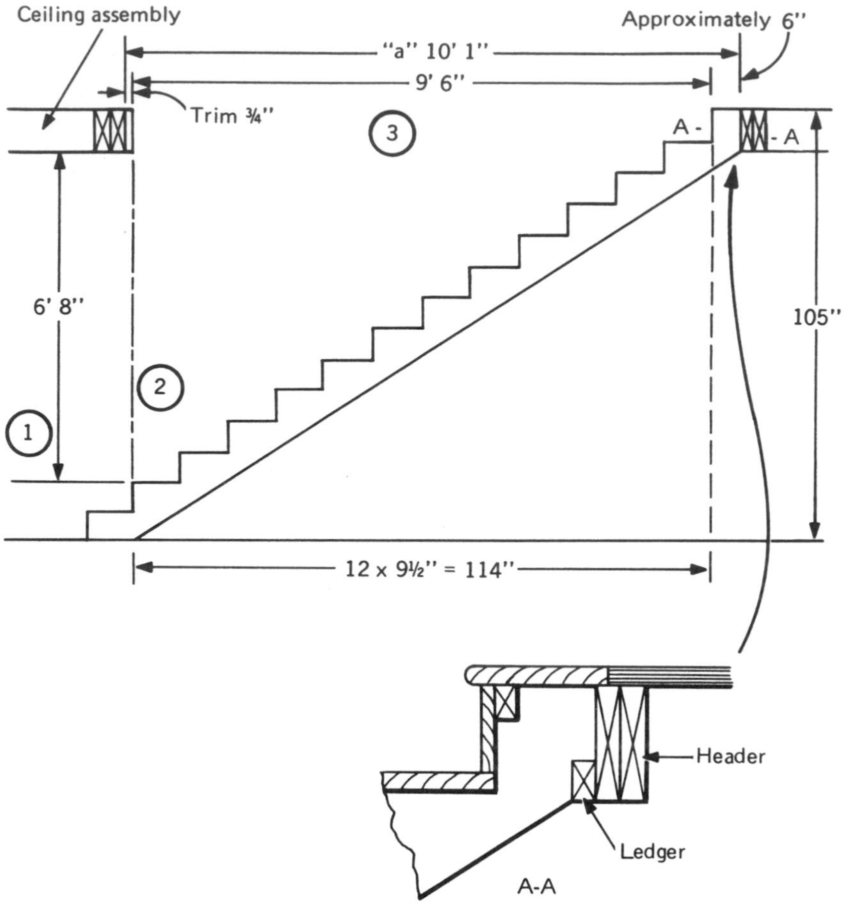

Figure 14–14 Defining the finished and rough opening of a stairwell (Example 1).

Conclusions

The general statement that a stairwell opening should be 10 ft long holds true for stairs whose rises are 7½ to 8 in. and have treads of 9 to 9½ in. However, greater head room clearance is needed when a stair is designed with risers less than 7½ in. and/or treads of 11 in. In this case the guidelines described in Example 1 or 2 should be followed *before* framing the stairwell.

If the ceiling over the stairs slopes at the same angle as the stairs as in Figure 14–16, head room clearance must be maintained from the second step to the top step. As long as 6 ft 8 in. is established and is then maintained, both proper design and a positive feeling by users of the stairs is accomplished.

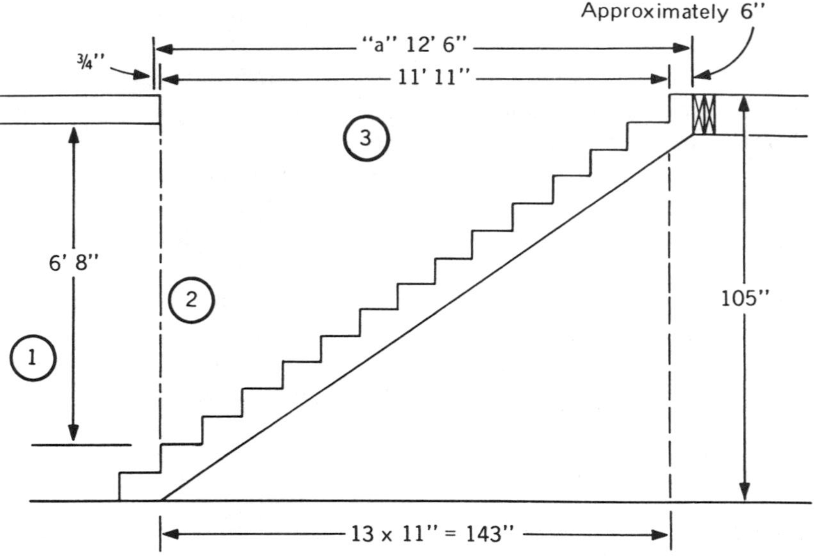

Figure 14–15 Defining the finished and rough opening of a stairwell (Example 2).

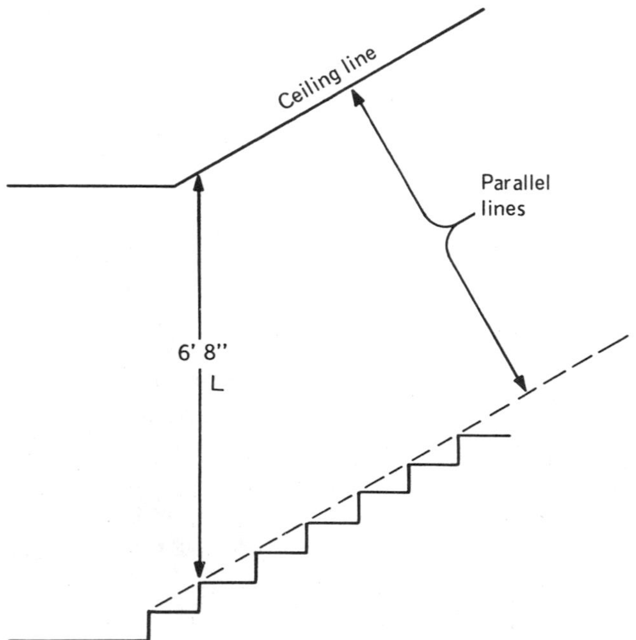

Figure 14–16 Minimum head clearance for main-floor to second-floor stairs is 6 ft 8 in.

Variance for Head Clearance for Basement Stairs

The previous examples were developed for stairs in the living quarters, as between the first and second floors. The basement stairs are generally constructed with the same principle of rise and run, but the allowance for head clearance is reduced to 6 ft 4 in. There does not seem to be any apparent reasoning for this reduction in allowance except tradition. There may be local building codes, however, that restrict this reduction in head clearance, so a check with the local building inspector should be made before construction starts.

As a general rule, basement stairs employ a full 8-in. rise and 9-in. run per step. The parameter, of course, reduces the length of the stairwell. However, even when using a 6 ft 4 in. head clearance, the edge of the stairwell falls (using a perpendicular line) between the noses of the second and third steps.

ANCHORING TECHNIQUES FOR CARRIAGES OR STRINGERS

A variety of techniques are available to choose from when anchoring a stair carriage or stringer is needed. It could be:

1. Toenailing the upper end of the stringers to the headers
2. Installation of a ledger nailed to the header and notched stringer
3. The use of steel angle braces lag bolted to both header and stringer
4. Anchoring the closed carriage to the side wall by nailing it to studs
5. Nailing a rough stringer to a previously installed trimmer board
6. Inclusion of a kicker plate in back of the first riser board
7. Any combination of these techniques (some are illustrated in Figure 14–17)

The selection of anchoring technique or combination of techniques must be made on site. It should be one that provides several important features to the stairs. First, the technique used must provide safety, so that no possibility of stair collapse can ever occur. Next, it must provide rigidity so that absolutely no feeling of sway, give, or motion is ever felt by persons using the stairs. In addition, the technique must be planned into the construction and used in that manner during the stair construction. The anchoring device used should not show in the finished product. However, if it must show, it must appear as a fundamental part of the stairs and not look like an add-on. In every set of stairs built with open stringers or jacks, the kicker plated will probably show unless, of course, a trimmer is cut and laminated over the rough stringer.

Many of the techniques listed before are easily used on the closed carriage type

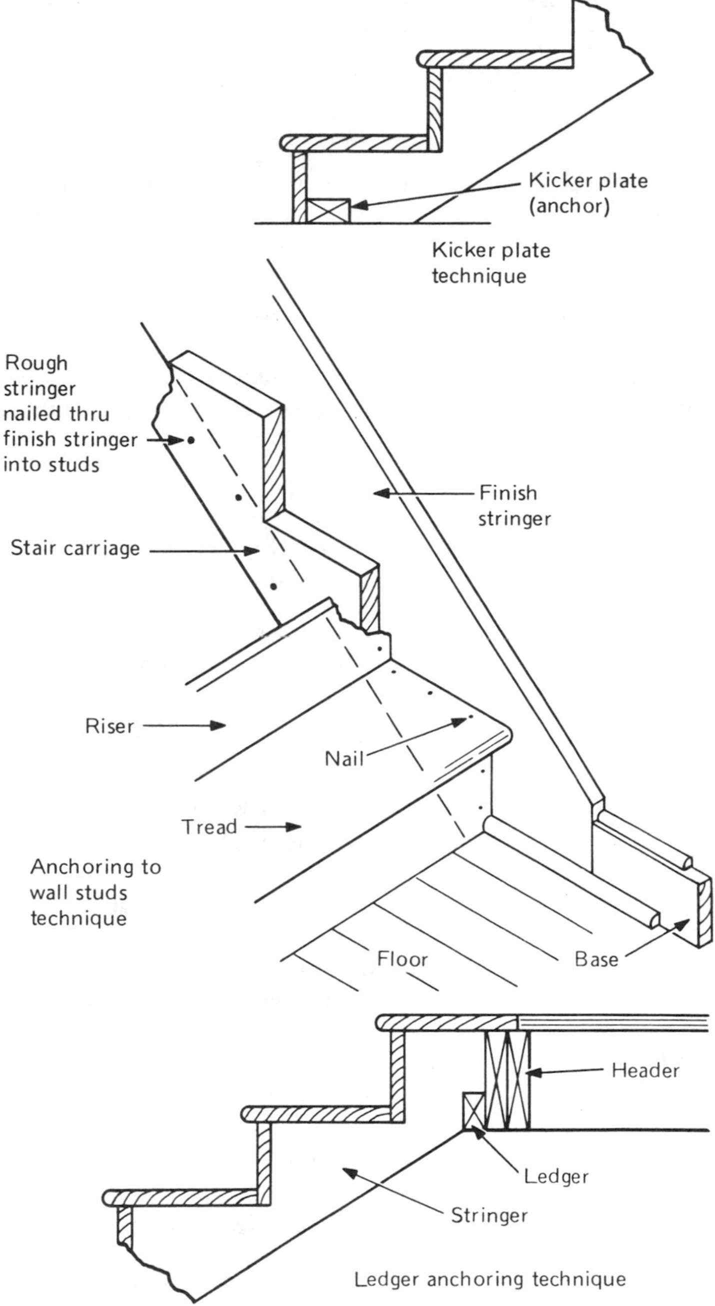

Figure 14–17 Anchoring techniques for stairs.

of stairs since ledgers, kickers, and even rough stringers are completely hidden. The finished product incorporates safety and displays expert craftsmanship.

SUMMARY

The principles of stair design and construction are varied and many. Those described and explained in this chapter are the most often used in the construction of straight stairs. Even a set of stairs with a landing is really two sections of straight stairs. Risers ranging from 6½ to 8 in. must be used along with a step run of 9 in. or more if good design is to be used. Further, good-quality materials must be used for rough stringers, carriages, trimmer boards, risers, and treads. Also, each step tread should have a nosing that extends beyond the face of the riser board by 1½ in. Finally, sound anchoring techniques must be designed and built into the stairs for safety and lasting quality.

PROBLEMS AND QUESTIONS

1. List three other names for a stair's carriage.
2. What is a trimmer board? What other names may it be called?
3. What is the difference between a *riser* and *rise*?
4. What is nosing? How much does it usually overhang the riser?
5. What parts make up a balustrade?
6. What is the range of stair widths for a residential building?
7. If a set of stairs is 39 in. wide how long should a landing be?
   a. 39 in. **b.** 45 in. **c.** 60 in.
8. Explain what is meant by a "combination of 17 in." and "combination of 25 in." in stair layout.
9. What is the maximum rise allowable for residential stairs? Minimum stair tread?
10. If the total rise from floor to floor is 108½ in., what would be the average rise if 8 in. is the maximum allowable rise per step?
    a. 7½ in. **b.** 7¾ in. **c.** 8 in.
11. Which angle of stair incline produces more stress on the person using it?
    a. 30° **b.** 40°
12. True or false: A handrail must be installed between 30 and 34 in. above the nosing of the tread.
13. Two different allowances for stair head clearance are established:
    a. For main floor stairs:________
    b. For cellar stairs:________
14. If a set of stairs has 13 step runs and each is 10 in., the rise is 7 in. per step, and the head clearance must be 6 ft 8 in., what is the length of the stairwell opening?
15. List four different anchoring techniques for stairs.

PROJECT

If you are into construction, the possibility of installing stairs is very good. Even if you buy premanufactured stairs, you will still have to frame the stairwell. Figure 14–11 shows three angles of stairs: 30, 35, and 40°. Figure 14–13 shows the framing associated with a stairwell. Figures 14–14, 14–15, and 14–16 show how to determine head clearance, which translates to the length of the stairwell opening. For this project we will create a table that includes all the opening lengths for stairwells with rises and step runs beginning with 6½ and 12 in. through 8 and 9 in. in increments of ½ in.

Floor-to-floor height is 105 in. (8 ft 9 in.).

Rise and run (in.)	Stair length	Stairwell opening (length)
6½ and 12	______	______
7 and 10	______	______
7 and 11½	______	______
7½ and 10	______	______
7½ and 11	______	______
8 and 9	______	______
8 and 9½	______	______
8 and 10	______	______

Use the following formulas to complete the table.

Stair length = [total rise/rise per step (rounded up to the next whole number)] × step run in inches. For example,

$$\text{stair length} = \frac{105 \text{ in.}}{7 \text{ in.}}$$

$$= 15 \times 10 \text{ in.}$$

$$= 150 \text{ in. (12 ft 6 in.)}$$

Stairwell length = stair length − 1 step run (in inches)

Answers:

Rise and run (in.)	Stair length (in.)	Stairwell opening (length) (in.)
6½ and 12	17 × 12 = 204 (17 ft)	204 − 12 = 192 (16 ft)
7 and 10	15 × 10 = 150 (12 ft 6 in.)	150 − 10 = 140 (11 ft 2 in.)
7 and 11½	15 × 11½ = 172 ½ (14 ft 4½ in.)	172.5 − 11½ = 161 (13 ft 5 in.)
7½ and 10	14 × 10 = 140 (11 ft 7¹⁵⁄₁₆ in.)	140 − 10 = 130 (10 ft 10 in.)
7½ and 11	14 × 11 = 154 (12 ft 10 in.)	154 − 11 = 143 (11 ft 11 in.)
8 and 9	14 × 9½ = 126 (10 ft 6 in.)	126 − 9 = 117 (9 ft 9 in.)
8 and 9½	14 × 9½ = 133 (11 ft 11 in.)	133 − 9½ = 123½ (10 ft 5 in.)
8 and 10	14 × 10 = 140 (11 ft 7¹⁵⁄₁₆ in.)	140 − 10 = 130 (10 ft 10 in.)

Chapter 15

Interior Stairs

This chapter details the planning and layout of basement and formal stairs. These two stairs may differ not only in appearance but in construction techniques.

BASEMENT STAIRS

Layout of Stringers

Basement stairs should be constructed with 2 × 12–in. stringers or carriages and 2 × 10-in. treads. Frequently, there are no riser boards used on these stairs. The example chosen for this section is a set of stairs using roughly an 8-in. rise and 9-in. run. The total height (rise) is equal to:

Ceiling clearance	7 ft 11 in. = 95 in.
Joist width	7½ in.
Subfloor and flooring	1½ in.
Total	104 in.

Fourteen risers are most commonly used for interior stairs, but in basements where headroom is sometimes lower, as in this example, 13 risers at 8 in. exactly equals the total height.

13 risers × 8 in. per rise = 104 in.

Next is the layout of the carriages. For this a framing square is used.

1. The 2 × 12 should be *sighted* and marked for its crown.
2. Since a 2 × 10 actually measures 9½ in. the step run should be set at 9½ in. (stairwell opening permitting this latitude of variance).
3. Allowance of nosing (overhang of tread) is set at 1½ in. (Figure 15–1).
4. This means that the 8-in. mark on the square's blade is used for the step run. (8-in. seat for the tread and 1½-in nosing (overhang) = 9½-in. tread width.
5. The 8-in. mark is used on the square's tongue for each rise (6½-in. carriage use + 1½-in. tread thickness = 8-in. rise).
6. Thirteen risers are laid out in order, starting away from the end of the 2 × 12.
7. The bottom cut is laid out next (Figure 15–1). Notice that the thickness of the tread (1½ in.) is subtracted from the rise so that when it is installed the first step is 8 in. high. The bottom cut must be parallel to the first tread run.
8. The top step run is measured for depth to allow for 1½-in. nosing, tread overlap on header if any, and rim of carriage. The sum total cannot exceed 9½ in.

Therefore, assume a 1-in. lap onto the header for the tread. This causes a reduction of the top step carriage run of 1 in. or a total of 7 in. (Figure 15–1b).

The stringer should be carefully cut along all the lines drawn and then another one can be laid out by using the first as a pattern. It too should be cut out.

Installation of Stringers

The first stringer to be set in place is the one against the wall, assuming that one or both sides of the stairs is against a wall. It should be placed with its top against the header and positioned properly. An 8d nail should hold it in place temporarily.

Before any final fastening is done, a spirit level must be laid across the step run to verify the levelness of the run. If the level indicates an out-of-level condition, the bottom must be marked and trimmed if too high or shimmed if too low (Figure 15–2).There really is no surprise to experienced carpenters that alterations are made. The concrete floor of most basements is not uniformly flat; also, the stringer was cut for an ideal condition.

The next operation is to determine the anchoring technique and apply it. Recall that a kicker board could be used at the bottom and nails could be used through the stringer into wall studs. Toenailing or angle brackets and lag screws could attach the top of the stringer to the headers.

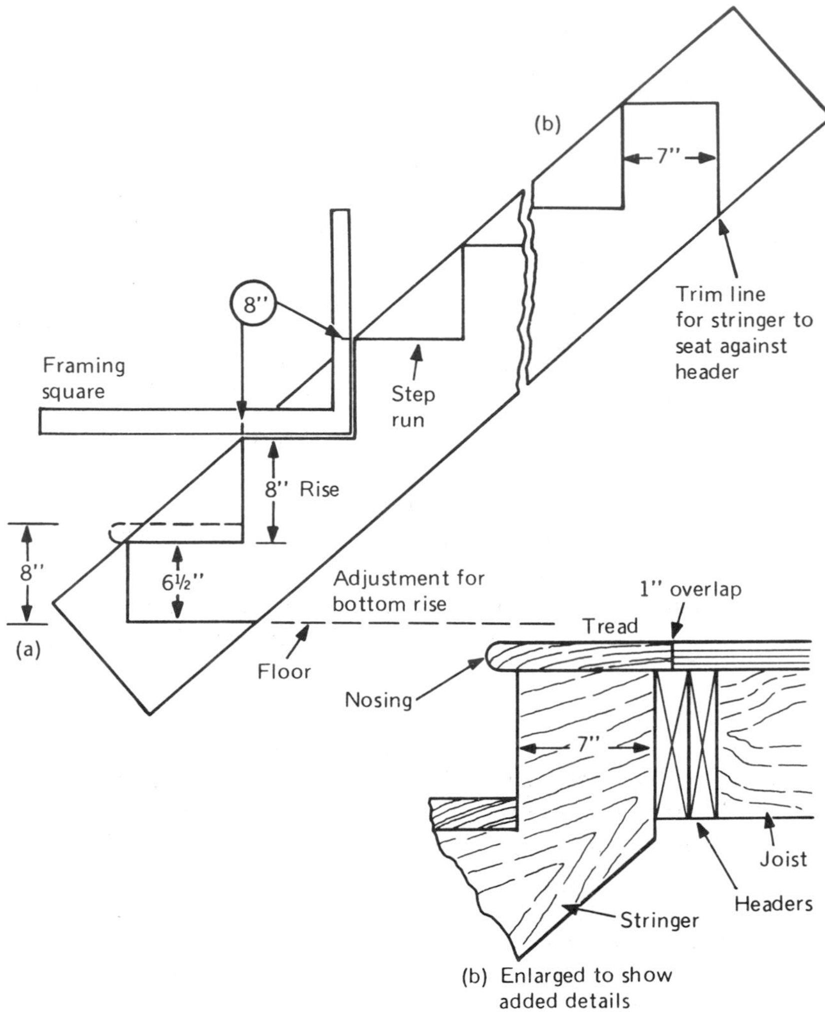

Figure 15–1 Basement stair stringer layout using an 8-in. riser and 9-in. tread with 1-in. nosing overhang.

The second stringer is set in place next. First it is set to the proper height and spaced properly for width at the headers. Then it is tack nailed with an 8d nail. Next, a straightedge is placed across the two stringers near the lower end as Figure 15–3 shows. The spirit level is laid on top of the straightedge and adjustments to the second stringer are made.

Once the adjustments have been made, spreader sticks cut to the exact width of the stairs should be nailed across the face of a lower rise and middle rise. These will help maintain alignment while the treads are being installed.

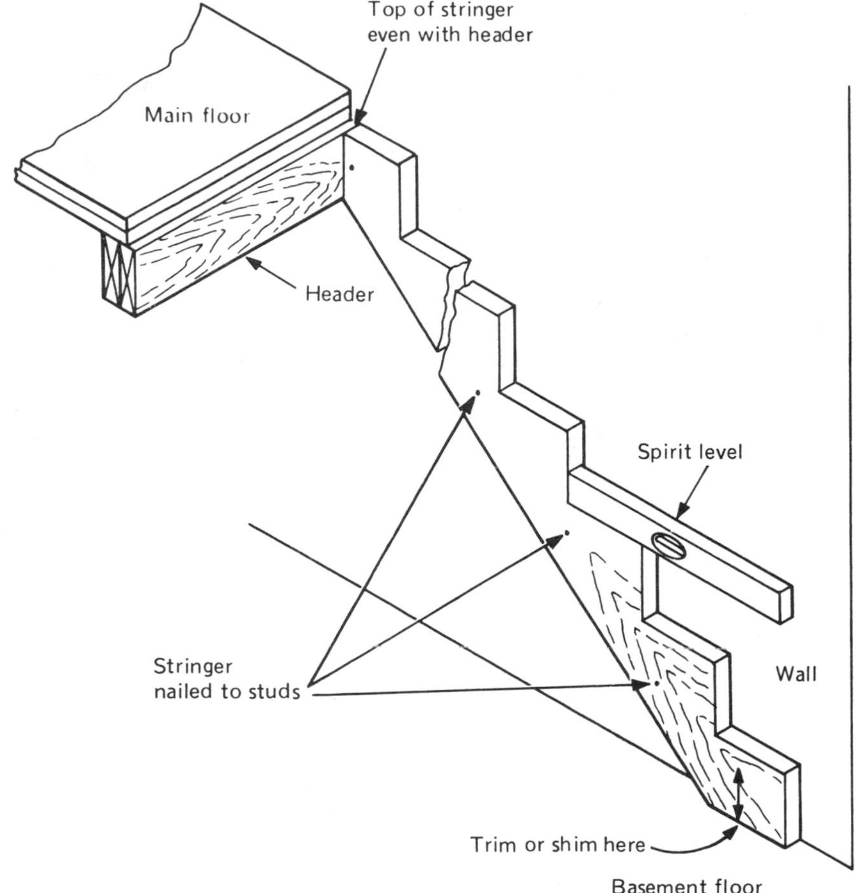

Figure 15–2 Installing stringer against wall after level adjustments are made.

Installing Treads and Risers

The sample set of stairs laid out in Figure 15–1 is designed for treads only, so the first operation after installing the stringers is to install the treads. Assuming that they are precut to length from 2 × 10 stock all noses should be rounded slightly with a block plane or router. Next, starting from the lowest step, each tread is positioned and nailed with face nails (casing type) through the tread into the stringer. Since no risers are used, each tread should butt to the rise cut on the stringer. The final or top tread should lap the header by 1 in. and be nailed to the header as well as the stringer.

The temporary spreaders can be removed once the treads are in place. For practical purposes the stairs are complete. However, if one side is open it must have

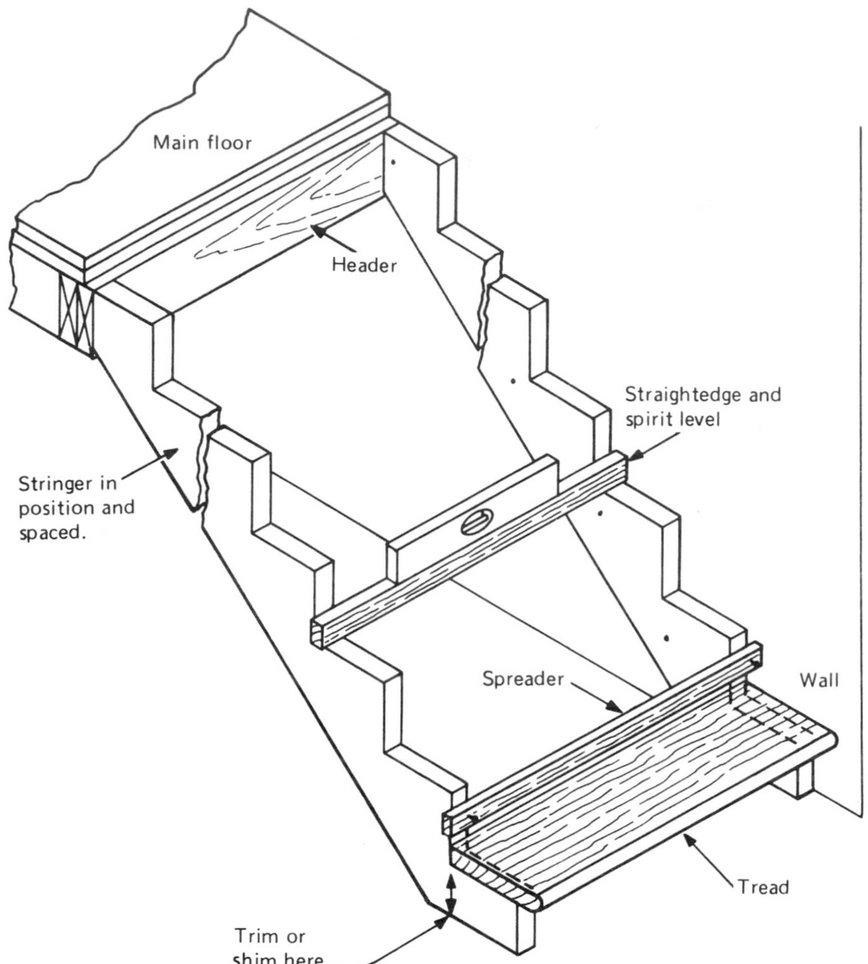

Figure 15-3 Installing the outer stringer, leveling and spacing it, and installing treads. Note spreader to aid in maintaining width.

a balustrade. If neither side is open a handrail mounted on brackets fastened to wall studs is used. For the description of installing a balustrade, skip to the end of the next section on formal stairs.

FORMAL STAIRS (FULL FLIGHT)

There are two distinctly different methods of constructing formal stairs for use between the main or first floor and second floor of a residential house. The first requires the use of the *dado* or more precisely the *blind dado* joint. Considerable skill is needed for layout as well as cutout and assembly. However, stairs constructed in this manner use the closed carriage and a minimum number of pieces.

The second method uses a variety of pieces and has as its foundation a pair of rough jacks. These are completely encased by trimmer boards, risers, and treads. Frequently, this method results in more work than the one previously identified because of all the precise cutting and fitting that must be done.

Laying Out and Dadoing the Closed Carriages

Rather than reiterate the formal description of determining total rise and run and performing all the calculations of how many step rises and runs, assume that a 7 ½-in. rise and 10-in. run plus 1½-in. nose are planned. Assume that 14 risers are needed.

The first requirement is to obtain clear stock 2 × 12s of fir or pine, preferably southern pine, for use as carriages. The grade must be the finest available for appearance and strength since no rough jacks are to be used. Layout begins with striking a line 1 to 1½ in. in from the board's edge, as Figure 15–4 shows. Then the framing square is used to lay off the rises and runs of all the steps. Next, the shape of the nosing is drawn so that its extreme edge is equal to the width of the tread (11½ in.). The one shown in Figure 15–4 (step 3) is half round. Then connecting lines are drawn as shown in step 4. The layout in step 4 also includes a layout for a wedge. The last step is shown in step 5, where the riser and wedge are laid out. If these series of steps are followed, an accurate layout is easily accomplished.

However, some simplification of the layout task can be made if templates are made for tread and wedge, and riser and wedge (Figure 15–4). These make the work much easier. If templates are used, steps 1 and 2 are performed first. Then the alignment mark on the tread template is held even with the rise mark and the pattern is drawn. The alignment mark on the riser template is held even with the tread (run) mark.

Once the layout is complete, the upper and lower markings are made and the housed stringer (Figure 15–5) is ready for cutting and dadoing. When completely cut and set in place it should resemble the segment shown in Figure 15–5b. Wedges should be cut and made ready.

Normally, the housed carriages are *not* installed before the stairs are assembled because of the difficulty involved. Rather, they are laid on the floor and assembled there. Figure 15–5c shows how the assembly is made from the underside of the stairs. Glue is applied to the dado, then a tread is set in place and a wedge is driven behind the tread. This action drives the tread firmly against the dado's edge. Next, the riser board is set in place and wedged and the lower edge of the riser is nailed to the tread. Once all pieces are installed into both carriages, glue blocks are installed as shown.

This type of staircase is usually used where both sides of a stair are enclosed with walls. Naturally, some small allowance, approximately ½ to ¾ in., is needed to make installation possible. This clearance can easily be covered with a small cove

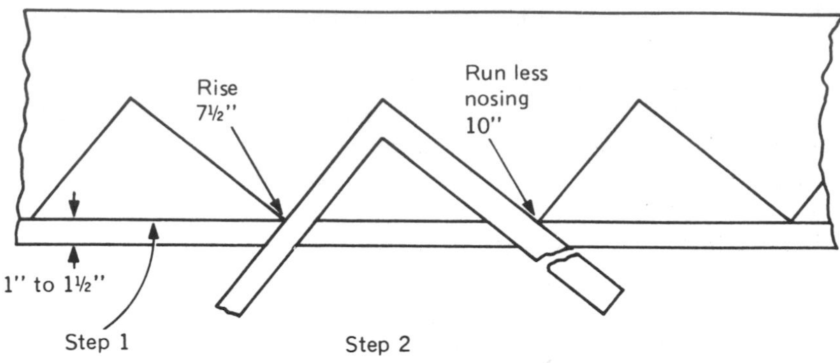

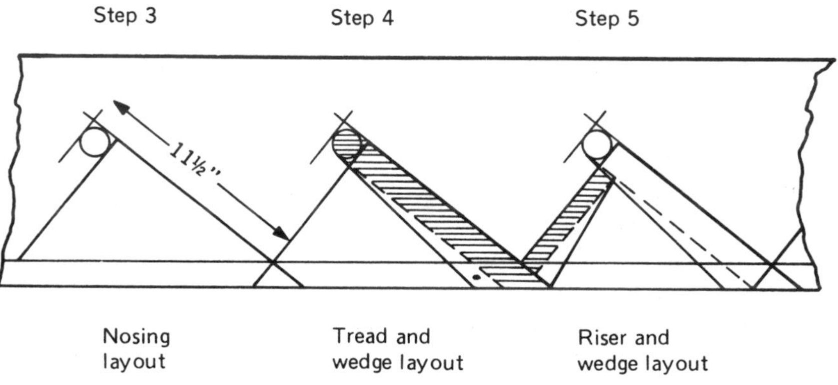

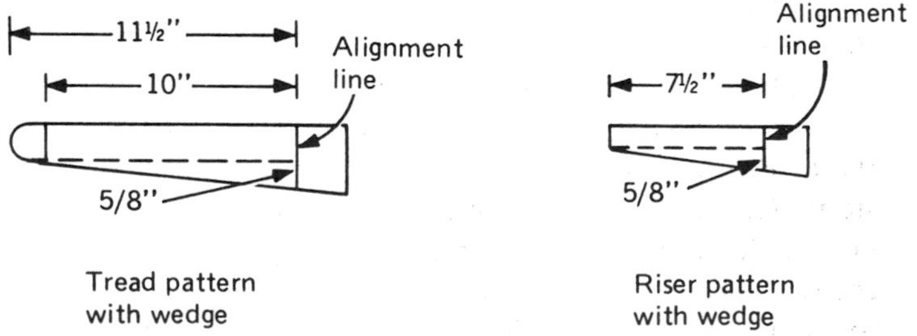

Figure 15-4 Layout sequence for closed or housed stringer templates aids and simplifies layout.

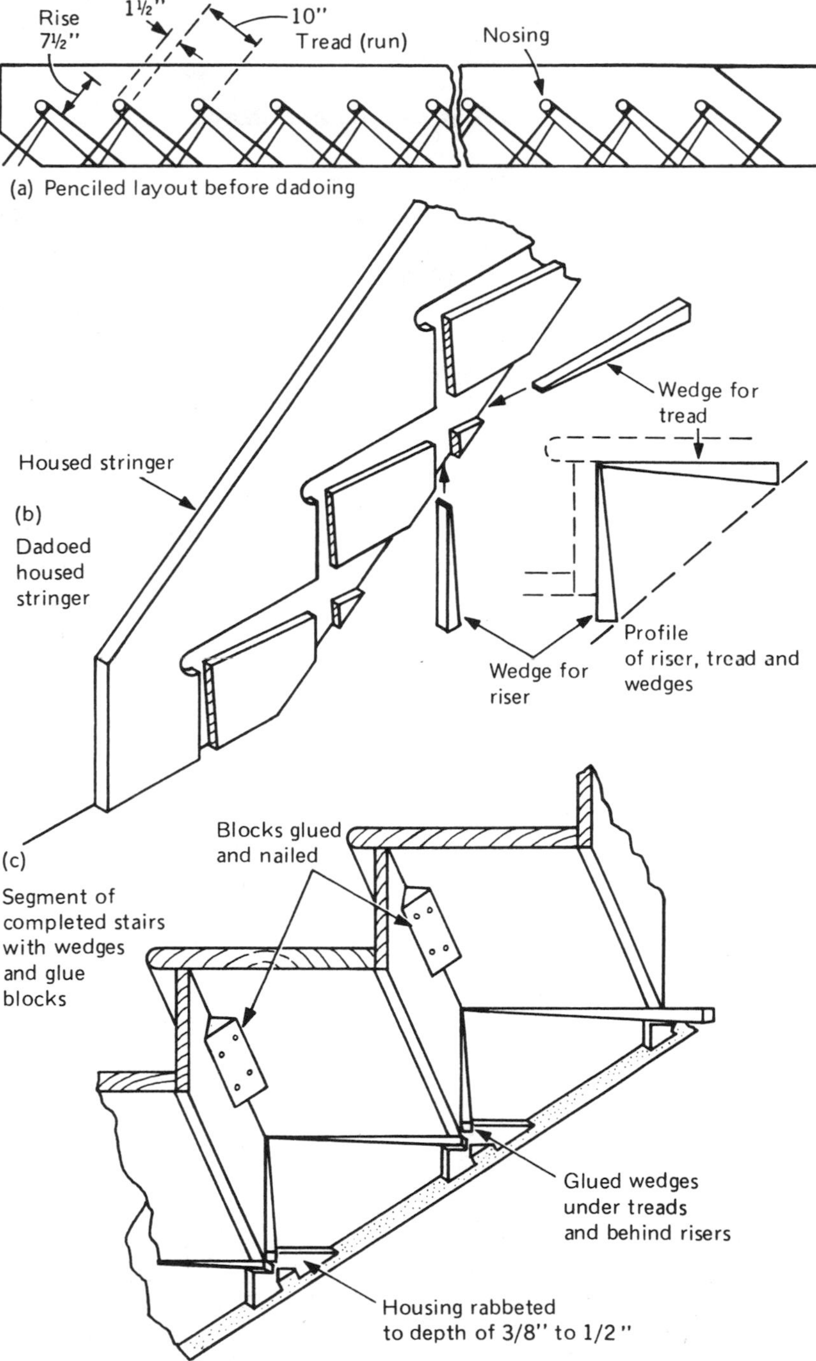

Figure 15-5 Layout, dadoed stringer and wedges, and a section of the glued and nailed stairs.

molding once the stairs are in place. Handrails are usually fastened to one wall or the other 30 to 34 in. above the stairs.

Stairs with Fitted Trimmer Boards and Mitered Risers

The other type of formal interior staircase is made with rough stringers, treads and fitted risers, and trimmer boards. Figure 15–6 shows such a staircase. This type of staircase is widely used when one side of the stairs is against a wall and the other side is open and uses a balustrade.

The construction of this type of staircase is straightforward. Rough stringers are laid out and cut. Next a trimmer board is cut and installed and then one of the stringers is nailed in place against it. Before nailing the second stringer in place it should be used to lay out the trimmer board that will be nailed to it. The line drawn for the tread run on the trimmer board is the cutting line; but the line drawn for the riser is the interior mark of a 45° miter. This means that a 45° layout must be made on each rise and then cut (Figure 15–7). The risers are cut with one end square and the other at 45° to match the trimmer board. This arrangement makes a professional job with no raw end wood showing when assembly is complete.

When the trimmer board is ready, assembly should begin by installing the rough stringer and trimmer board. The first riser needs to be cut to fit and trimmed

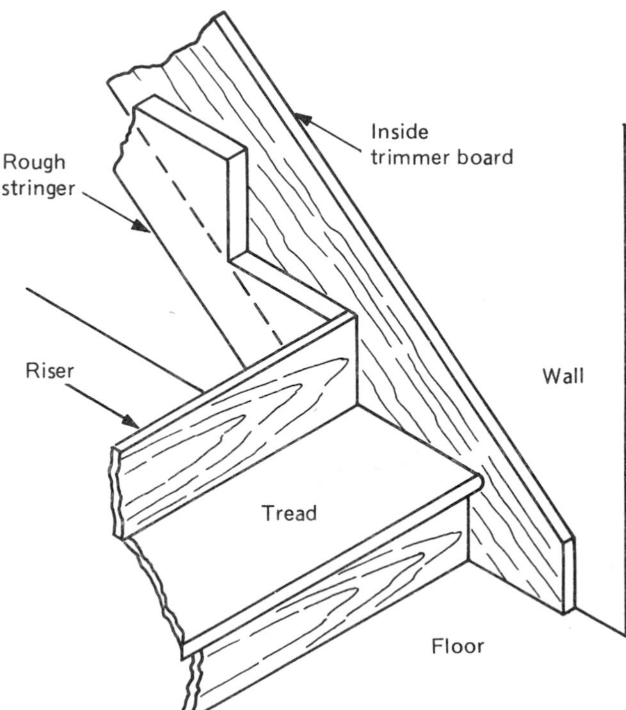

Figure 15–6 Using an inside trimmer board and rough stringer against a wall.

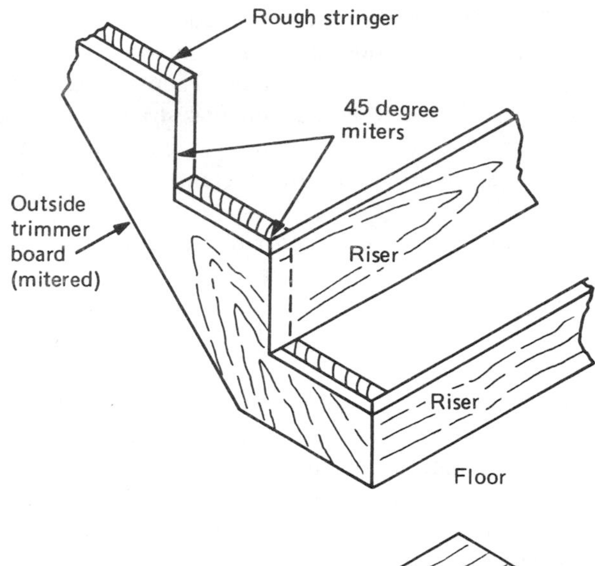

Figure 15–7 Using an outside trimmer cut to fit against a rough stringer. Mitered risers fit to 45° bevels made in trimmer.

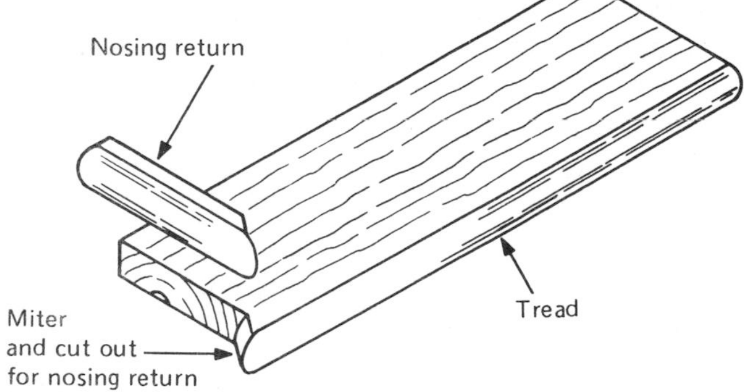

Figure 15–8 Self-nosing stair tread.

for width as well. Glue should be used on the miter joint and on the faces of the stringers for lasting hold and quality. Next, the tread needs to be prepared and installed. Since one end of the tread is to show, its raw edge must not be seen and a self-return is used (Figure 15–8). Actually, all treads should be prepared at one time. Notice also that the return stock extends about ¾ in. beyond the riser. The return stock should be glued as well as nailed.

Installing the Balustrade

Following the installation of all risers and treads, the balustrade is installed. Recall that this requires use of a handrail and balusters. Normally, two balusters are used on each tread, where they are toenailed and glued once their centering position is determined. The upper end of each baluster is tapered to fit under the handrail and there should be a short and long baluster per tread.

What is a simple way of defining the bevel cut needed for the top of each baluster? Line up a spirit level under or alongside the trimmer board and set a miter square against the level with the blade along the trimmer. Then tighten the wing nut. Or, set a framing square on the floor and let its tongue rest against the stair's trimmer board. Place the miter square against the square's tongue and tighten the wing nut when the miter square's blade is resting properly against the lower trimmer edge.

Applying the handrail to the balusters is the most difficult part of the task; how can it be done easily? One method makes use of a board ¾ × 1½ in. where the balusters are glued and nailed to it. Top nailing is used. Then the handrail is glued and screwed in place on top of the 1½ in. wide board. Screws are installed from beneath the 1 × 1½ in. stock.

Another method uses dowels, where a dowel is inserted into the top of each baluster and its other end is inserted into a hole drilled into the undersurface of the handrail. Since the lower end of the handrail and stairs opens onto the floor area a *newel* post is used to finish off the stairs. This post needs to be anchored very firmly. Its rigidity is absolutely essential. Therefore, toenailing is not appropriate by itself. The post should set firmly on the floor (Figure 15–9). A notch into the tread is surely called for or possibly a mortise needs to be made through the tread. The newel post is cut to fit into this opening and anchored to the stringer with lag screws, glue, and nails.

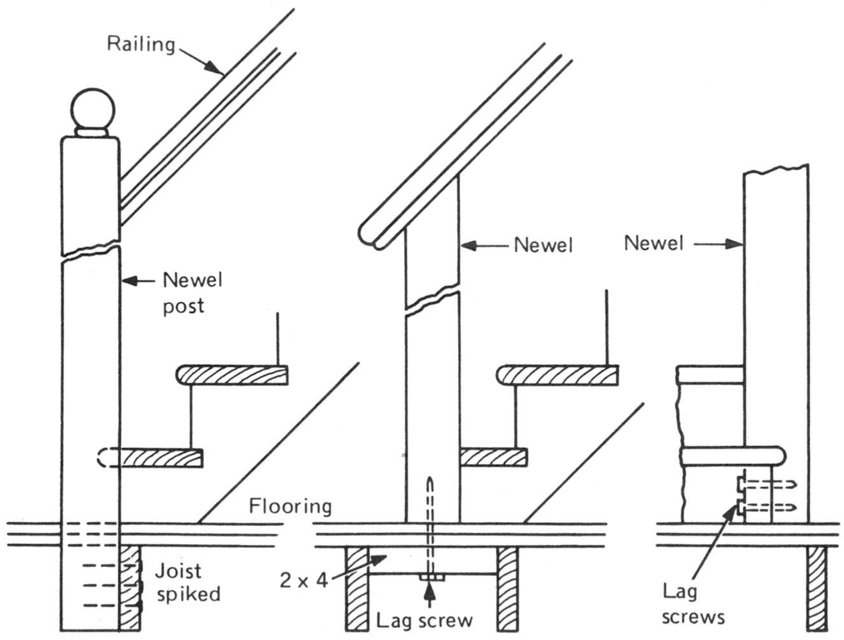

Figure 15–9 Several good methods of installing newel posts.

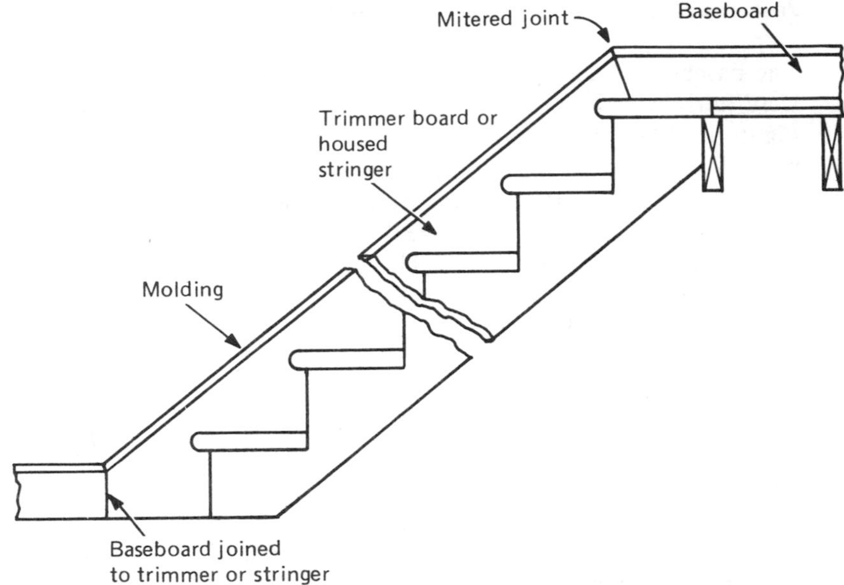

Figure 15-10 Properly joining baseboard to trimmer at the bottom of stairs and top of stairs. Baseboard is 1 × 5 or 1 × 6 variety.

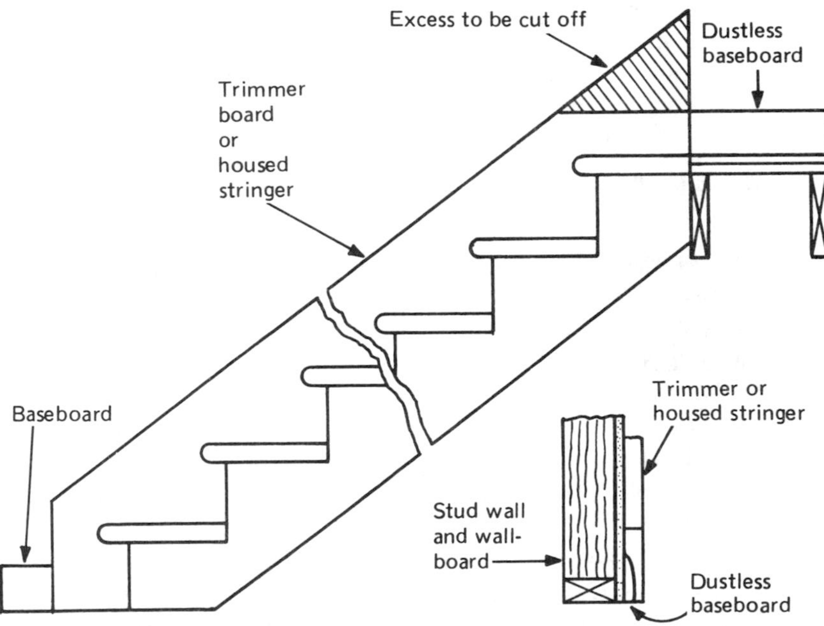

Figure 15-11 Joining baseboard to trimmer and trimming trimmer board when new dustless baseboard is used.

Joining Baseboard to Trimmer Board

The baseboard ends against the trimmer board as a rule. In the past, baseboards were fairly wide and the intersection of trimmer to baseboard was a special joint. Figure 15–10 shows this situation, and it should be noted that a miter is used so that no end wood of either the trimmer or baseboard shows. This technique is still used where baseboard is used on landings. The modern trend in baseboard design is the *dustless* shape, where the top edge is curved. Its size is relatively small and its thickness is reduced to less than ¾ in. This means that it is impossible to join the trimmer and baseboard in the same way. Figure 15–11 shows the modern trend. Unfortunately, some raw edge of the trimmer shows, making additional work for the painters. At the lower end a straight cut is used, but at the upper end the trimmer is trimmed to make the joint more pleasing.

SUMMARY

The building of a set of stairs, whether informal or formal, requires a great deal of fundamental understanding as well as skill. Formal stairs must integrate with walls and trim and be constructed accordingly. This presents many questions and problems for the carpenter. He or she must select the correct solutions from a variety of techniques. If both sides of a staircase are walls, a housed carriage design is appropriate, yet this design requires cabinetmaking skill. If one or both sides of a staircase are open, mitering of riser and trimmer is needed and a balustrade is employed. These applications require special skills and techniques.

PROBLEMS AND QUESTIONS

1. What is the maximum height for a set of stairs if the rise is exactly 8 in. and there are 13 risers?
2. Why should you sight a stringer 2 × 12 before beginning the stair layout?
3. What inch marks would you use to lay out the stringer if the rise is 7 in., the tread is 9 ½ in., there is no riser board, and the tread overhangs the riser cut by 1 in.?
   a. 7 and 8½ **b.** 7 and 9½
4. Why is the bottom rise on the stringer trimmed for the thickness of the tread?
5. Briefly explain what to do if when installing a stringer, the tread surface tilts up at the front edge.
6. What is the purpose of a *spreader* when installing a stringer?
7. Using Figure 15–4, explain what physical task is being performed in each step.
8. What is the purpose of putting a wedge under the tread and one behind the riser?
9. If one side of a formal staircase is open (not against the wall), what must be done to the riser cut of the trimmer?

10. What is a self-return on a tread used for?
11. Briefly explain how you would obtain the bevel angle that is necessary at the top of each baluster.
12. Explain one of the three ways to anchor a newel post soundly.
13. When should a *butt* joint be used when installing baseboard against trimmer or carriage? When should a miter joint be used?

PROJECT

In Chapter 14 we studied the design of stairs and a variety of subjects associated with stairs. If you completed the project at the chapter's end, you have a data base for completing this project. If not, refer to Figures 14–10 and 14–11 when necessary to complete this project. Our goal is to prepare a materials listing for a formal staircase that will be constructed by you on site.

A. Design Characteristics

Refer to Figures 15–3, 15–6, 15–7, 15–8, 15–9, 15–10, and, 15–11, and Figures 14–2, 14–10, 14–11, and 14–12. The overall height from floor to floor is 105 in. The stairs will have a rise of 7 in. and a step run of 11½ in. There will be a trimmer board against the wall and another on the outside of the outer stringer. There needs to be two rough stringers. Risers must be mitered into the outside trimmer board and butted against the inside trimmer board (Figures 15–6 and 15–7). Treads are to be made of oak and self-nosed. There will be a newel post and two balusters per step tread (Figure 14–2). The stair width will be 36 in. wide.

B. Materials Listing

Several materials can be defined once we know how many risers and treads will be needed, and thus the overall stair length. So let's determine this information first.

1. If you completed the project in Chapter 14, go there and extract the stair length from that listing and write it here. ____________
2. Otherwise, divide the 105-in. height by the riser height (7 in.) to obtain the number of steps. Then multiply the number of risers by the width of the step tread to find the overall length of the stairs.
   overall length = ____________ × ____________ in.
   = ____________ in. (____________ ft ____________ in.)
3. Now let's begin our list.
   a. Number of stair treads = ____________ × 36 in. = ____________ in. divided by 12 in. to obtain the number of linear feet = ____________. Then divide this number by 12 ft to find the number of 12-ft boards needed = ____________ 2 × 12s (rounded up).
   b. Tread materials: ____________ Oak 2 in. × 12 in. —____________ ft.
4. Next let's determine the riser materials needed. We know the number of risers and the rise in inches and the length of each riser since this information was given and calculated above. So let's install the numbers and get our results.

Number of risers × length/riser = total

$$\frac{\text{total}}{12 \text{ in.}} = \text{number of linear feet}$$

$$\frac{\text{Number of linear feet}}{12 \text{ ft}} = \text{number of 12-ft boards needed}$$

_________ × _________ = _____________

$$\frac{\rule{2cm}{0.4pt}}{12 \text{ in.}} = \rule{3cm}{0.4pt} \text{ lin. ft}$$

$$\frac{\rule{0.5cm}{0.4pt} \text{ lin. ft}}{12 \text{ ft}} = \rule{3cm}{0.4pt} \text{ 12-ft stock}$$

Since we know that the rise is 7 in., our stock must be a 1 × 8; thus we can now make the material entry:
Riser materials: _____________ oak 1 in. × 8 in. — _____________ft

5. Now let's determine the trimmer and stringer materials. Since the stairs are only 36 in. wide and we are using oak for risers and treads, only two stringers are needed. We already know from the design specifications that two trimmers are needed. From the text you know that the stringers are 2 × 12s and that trimmers are usually 1 × 12s. We shall plan to use these for the stairs. However, we do not know the length needed. This we will calculate by using the Pythagorean theorem (see pages 153 and 154, if necessary). Obtain the total rise and total stair length from the data above and determine the length of stringer and trimmer: _____________ ft long. You will need to round up to the next even length. Thus
   a. Stringer materials: 2 southern pine 2 × 12 — _____________ ft
   b. Trimmer materials: 2 oak 1 × 12 — _____________ ft
6. Next we define the balustrade and newel post requirements. One translation from the specifications may be needed. Since the outside trimmer and risers will be joined with a 45° miter, the implication is that a balustrade assembly is needed on the open side of the stair. Using the data from the specifications and the stringer length found in the last step, we will just compile a list of materials.
   a. Newel materials: _____________OAK NEWEL POST
   b. Baluster materials: _____________ (at 2/step) oak 36 in. balusters (refer to Figure 14–2)
   c. Handrail materials: _____________ oak handrail _____________ ft long
   d. Support materials: 2 lag bolts
      4d finish nails
      6d finish nails
7. a. Each stair tread must have a bullnose return on the open end. The top step tread may not require one, but without the plans we cannot identify this requirement. So we shall plan for a bullnose return on the topmost tread as well as the rest of the treads. Thus we must calculate the amount needed; then we can make the materials list item.

$$\frac{\text{Number of treads} \times \text{width of a tread}}{12 \text{ in.}} = \text{lin. ft.}$$

_____________ lin. ft

b. Since there will be mitering performed, it is desirable to allow for a miscut or two, so we should allow for 3 ft of waste. Now prepare the materials list item by using usual even-foot lengths.

Bullnose materials: ______________ 2-in. oak bullnose ______________ ft

and ______________ 2-in oak bullnose ______________ ft

C. Summary

Throughout this project we have made the decisions necessary to compile a materials list of materials needed to construct the stairs. There are some requirements for blocking and separators, but these are usually available around the job site, so no provisions are necessary in the list. If you actually do plan to construct a stair on-site, this project is very useful.

Answers:

1. 172.5 in. (14 ft 4 ½ in.)
2. 15 × 11 ½ in., 172 ½ in. (14 ft 4 ½ in.)
3. **a.** 15 × 36 in. = 540 in./ 12 in. = 45 lin. ft, 45 lin. ft/12 ft = 4 2 × 12s – 12 ft
   b. 4, 12
4. 15 × 36 in. = 540 in./12 in. = 45 lin. ft, 45 lin. ft, 4, 4 oak 1 × 8, 12 ft
5. 16.8 ft rounded up to 18 ft
   a. 18 ft
   b. 18 ft
6. **a.** 1
   b. 1 + 28 = 29, 1 on the first tread, thereafter two per tread
   c. 1, 18 ft
7. **a.** 15 × 12 in./12 in. = 15 lin. ft
   b. 1, 8 ft, 1, 10 ft

Chapter 16

Staircase for Deck to Ground

Constructing a set of exterior stairs employs the principles identified in Chapter 14 as well as some modifications to the stringers because of extended step runs and more gentle rises. In addition, wood decay becomes a critical factor. Special attention must be paid to this problem to minimize decay and extend the life of the stairs.

PREVENTING DECAY

Wood decay is caused by minute plants called *fungi*. Some merely discolor the wood of a stairs, but decay fungi destroy the fiber. Fungi can grow in wood only when it contains more than 20% moisture. Air-dried wood generally contains 19% moisture, which is below the danger point. As long as air freely circulates around the wood, the moisture content should remain below the danger point. However, other conditions, such as dew, moisture from humidity, and rain, all tend to raise the moisture content and contribute to the growth of fungi and thus to decay.

The conditions that cause fungi to grow and multiply can be eliminated or controlled by means other than air circulation. The heartwood of many species of trees is resistant to decay as well as to wood borers such as termites. Douglas fir and southern pine heartwood are classed as moderately resistant. Redwood, most cedars and a few species of red cypress are classed as highly resistant, and these can be in direct contact with soil. However, with time, even heartwood decays, and obtaining heartwoods is sometimes difficult.

Treating Lumber

The current trend in retarding decay is the use of treated lumber. Treating lumber by painting its surface does not as a rule prevent decay, since moisture may have been trapped under the paint. The application of one of several chemicals is more effective. Some of these chemicals, listed in Table 16–1, may be applied with brush or sprayer, while others are applied under high pressure, or used in vats where beams are dipped.

The salt preservatives leave the wood comparatively clean, paintable, and free from odor. Creosote, on the other hand, blackens the wood; it cannot be painted over and it retains a strong odor for many years. Copper naphthenate turns the wood green, and when air dried, the treated wood may be painted.

Drainage

Stairs exposed to the elements can last a long time if proper drainage around them is planned rather than left to chance. First, the slope of soil around the stairs should be uniform at a sufficient angle to afford complete drainage. This eliminates the chances of pools of water and the development of fungi that might occur. Light or thin shrubbery rather than dense kinds should be planted next to stairs so that air readily passes through and humidity is not trapped near stringers and treads. The bottom of the stringers should not be in contact with soil if possible. But if soil contact is unavoidable, pressure-treated stringers should be used.

One sure way to avoid contact with soil is to have each stringer rest on a block of concrete or pier made of stone, brick, or concrete as in Figure 16–1. When constructing the footing for the stringer to sit on, every effort should be made to slope the edges of the pier so that water sheds away from the stringer and does not become trapped between it and the footing. The one shown in Figure 16–1 has a *mortar* cap, which is shaped to provide the slopes needed for good drainage.

TABLE 16–1 PRESERVATIVES AND APPLICATION METHODS

Type of preservative	Method of application
Waterborne salt solutions	Pressure
Creosote and creosote coal tar solution	Brush, pressure
Pentachlorophenol	Pressure
Copper naphthenate	Brush, dipping

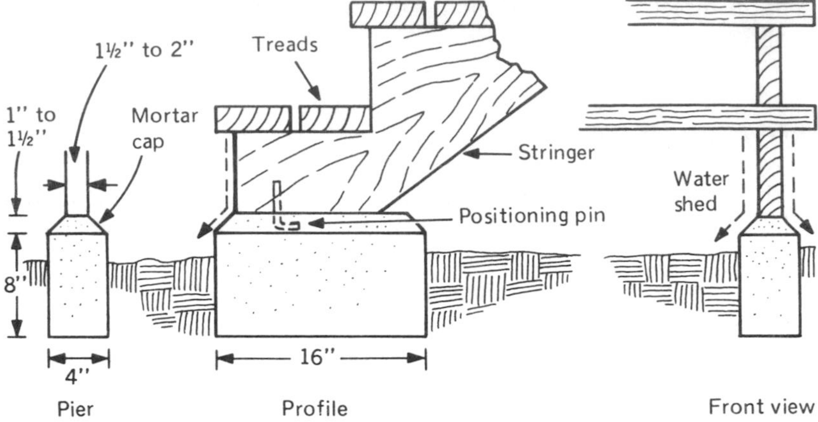

Figure 16–1 Preparing a pier for proper drainage and stringer anchoring.

Summary

Proper air circulation is fundamental to the prevention of fungi. Treated wood used for stringers and treads is another and is in common use since heartwood is hard to obtain. Even so, the lower ends of stringers should not be in contact with the soil. Piers are recommended to be used for stringers to rest on.

STAIRS WITH ROUGH STRINGERS AND 2 × 6 TREADS

The exterior staircase used as an example in this section is one that extends from the ground to a deck. The deck is made of a wood frame with 2 × 4 planking spaced ¼ in. to allow proper drainage and air circulation. The same principles must be applied to the stairs. In addition, the stringers and treads will be closer to ground level, so treated lumber must be used to resist decay. The use of a cement block as shown in Figure 16–1 is planned to be used as a pier for each stringer.

Example

Lay out and construct a set of stairs from ground to join a deck 3 ft above the ground. Each tread area is to be made from two 2 × 6s with a ½ in. space between them.

Solution—See Figure 16–2.

1. The step tread width is defined in the example as 2 times the width of a 2 × 6 plus ½-in. air space, which equals:

2 × 5½ in. =	11 in.
Air space =	½ in.
Total	11½ in.

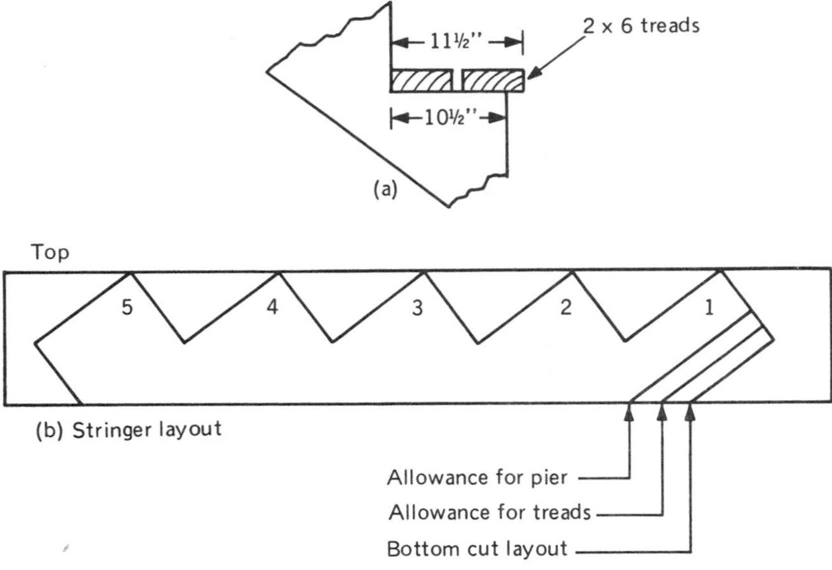

Figure 16–2 Stringer layout with bottom allowances shown.

2. Assume that the outer tread extends beyond the face of the rise by 1 in. This means that the tread run on the stringer is 10½ in. (Figure 16–2a).
3. Layout of the risers is next (Figure 16–2b).
   a. With a total rise of 3 ft and the greatest rise per step set for 8 in., an unequal distribution occurs (*not acceptable*).
   b. Using 5 five risers for a total rise of 36 in. results in a rise of 7¼ in. per step (*acceptable*).
      36 in. ÷ 5 = 7¼ in. L
   c. The layout is made as shown, then modified to account for the block and its cap, which extend above ground level (Figure 16–2).
4. Should the top rise be a part of the deck framing or part of the stringer?
   a. If the deck has a 2 × 8 header and joist framing with 2 × 4 decking, the total width equals 9 in. The option of using the framing as the fifth riser, as Figure 16–3a shows, means that *none* of the stringer makes contact. It is necessary to install a separate 2 × 8 header as shown so that the stringer can be fastened to it. If this option is selected, the final cut of stringer is made where the fourth riser ends.
   b. If a 2 × 10 frame with 2 × 4 decking is used, only 2 in. of bearing is obtained when the deck framing is used as the fifth riser. In this situation, a 2 × 6 header as shown in Figure 16–3b is added to provide more surface contact. As in the preceding solution, the stringer is cut at the end of the fourth riser.
   c. The third solution is shown in Figure 16–3c, where the fifth riser is a part of the stringer. Here no added framing is added to the deck. The stringer's top edge is held even with the deck's framing and the treads are flush with the 2 × 4 decking.

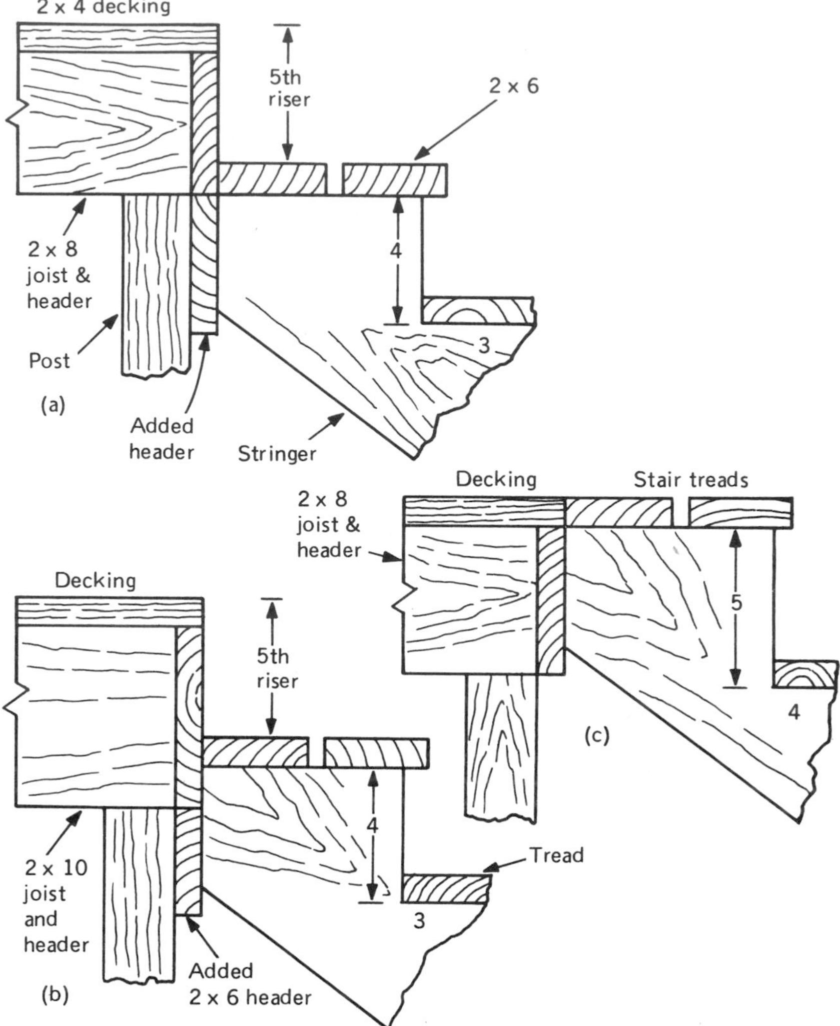

Figure 16–3 Headers added for supporting stringers (a and b) and deck framing as a support (c).

5. Install the stairs (Figure 16–4).
   a. With all stringers cut and treads precut for finished length, the assembly can begin.
   b. One pier block with mortar applied as shown in Figure 16–1 should be prepared for placement under each stringer. A bolt or bent pin can be set into the mortar before it dries.
   c. A small hole should be dug that is large enough to accommodate the block, and

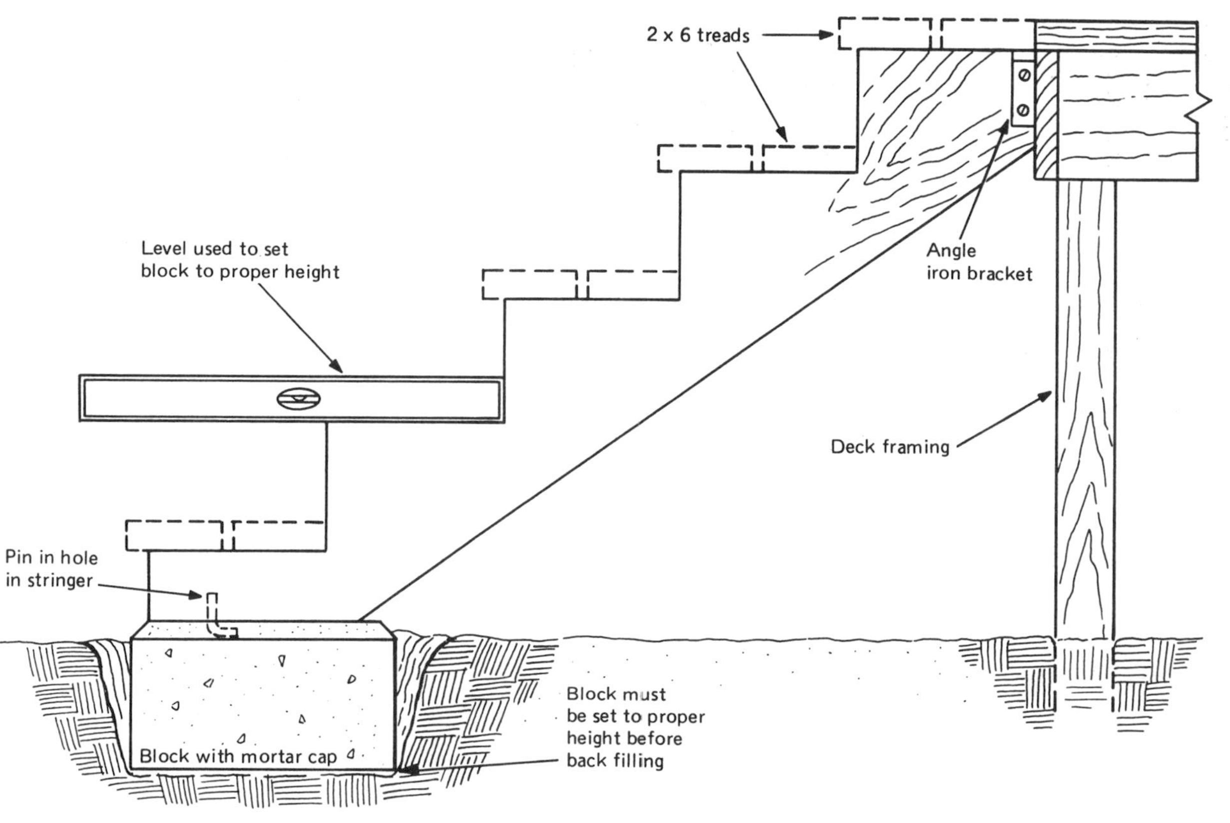

Figure 16–4 Installing a stringer and leveling with the pier height.

it should be set in place. Then the stringer is placed with its top edge properly in alignment against the deck's header.

d. Using a level, the tread (run) surface of the stringer is made level and adjustment to the block is made accordingly. At this time the drift-pin hole in the stringer bottom should be marked and drilled.
e. When ready, the stringer should be permanently fastened to the header. At this time there should also be a check made to verify that the stringer is at 90° to the header.
f. Repeating the previous steps is necessary to set each stringer in place.
g. All treads are installed with galvanized nails.

STAIRS WITH 2 × 4 CARRIAGES AND METAL TREAD SUPPORTS

It may be desirable in some situations to construct a set of exterior stairs using metal tread supports. If so, a 2 × 4 would suffice as a carriage onto which the metal tread supports would be lag bolted as in Figure 16–5.

Metal treads can either be ordered from a local metal shop or fabricated on-site. In either event the rise and run must be identified so that each tread support is bent to the proper angle and cut for the proper length. Then, too, small flanges must be made onto each end with holes bored for passage of the lag bolts. This detail is shown in Figure 16–5. Also, holes must be bored into the tread support surfaces to allow passage of the carriage bolts used to hold the treads in place.

When using this carriage arrangement the pier should be handled as explained in the preceding section, and an angle bracket could be used for securing the carriage to the headers of the deck framing.

BUILT-UP STRINGERS LAID HORIZONTALLY

The standard form of stairs does not always need to be used from deck to ground. Often a set of stairs can be a transition from deck to ground. This means that the stairs can be used as semidecks if each tread depth is made wide enough.

Figure 16–6 shows an example of this idea. Here, two tread surfaces are made 24 in. wide and act as semidecking surfaces. The standard inclined stringer would not be appropriate to use for tread support. Rather, a horizontal stringer is more appropriate.

As Figure 16–6 shows, 2 × 4s and 4 × 4s are used in built-up fashion to create the tread-bearing surfaces and rises. Blocks are installed under the lower 4 × 4 to provide sound support as well as elevation above ground level. Construction using this arrangement would be quite simple because the lower 4 × 4 could be leveled; then the short section of 2 × 4 could be nailed to the 4 × 4, and the short 4 × 4 could be toenailed to the 2 × 4. Once both or all stringer assemblies are in place, 2 × 4 treads cut to length are nailed to the stringers with 12d galvanized nails.

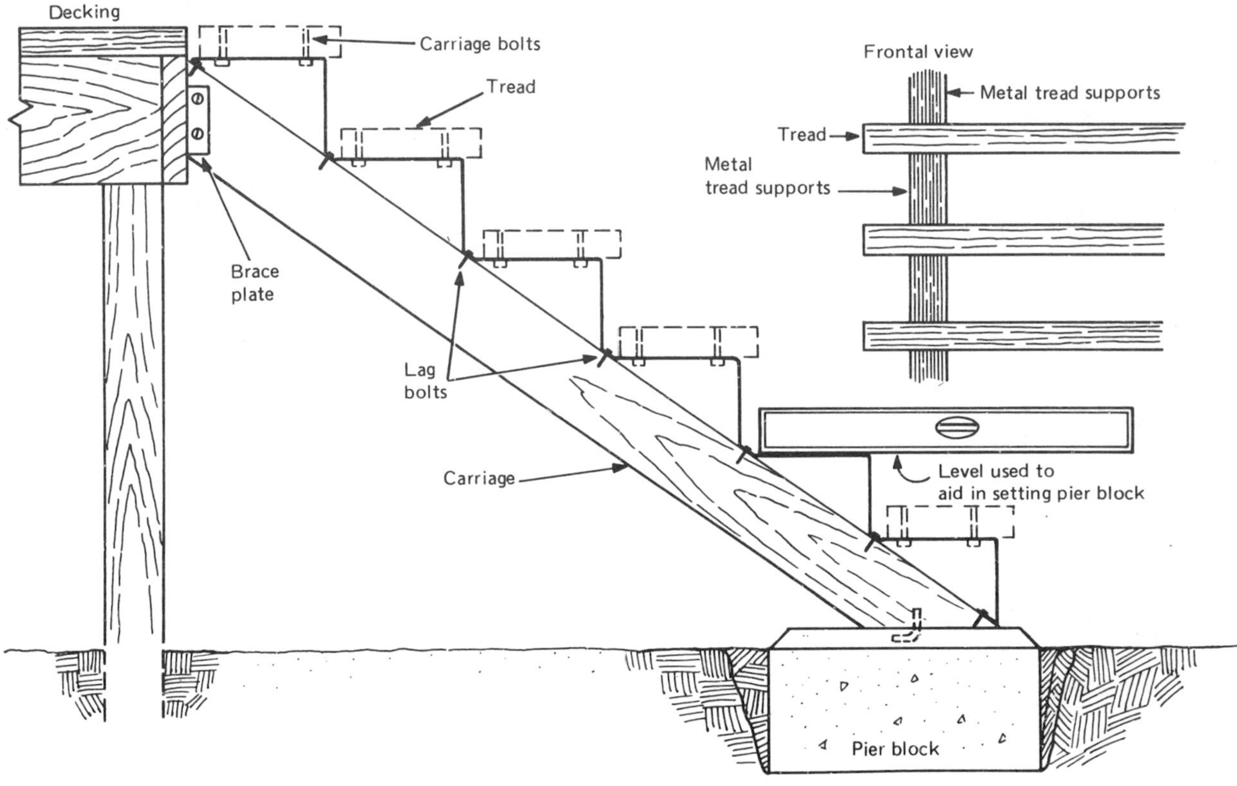

Figure 16–5 2 × 4 carriage and metal step supports.

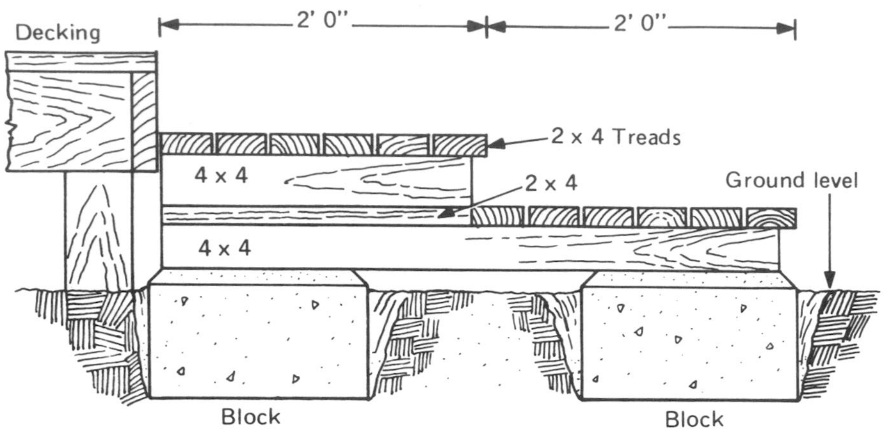

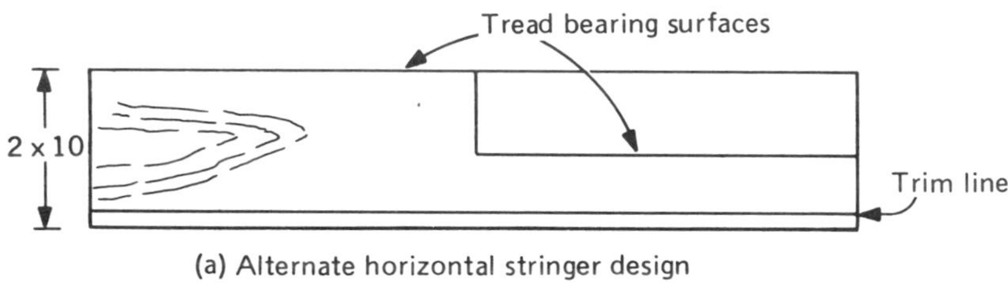

(a) Alternate horizontal stringer design

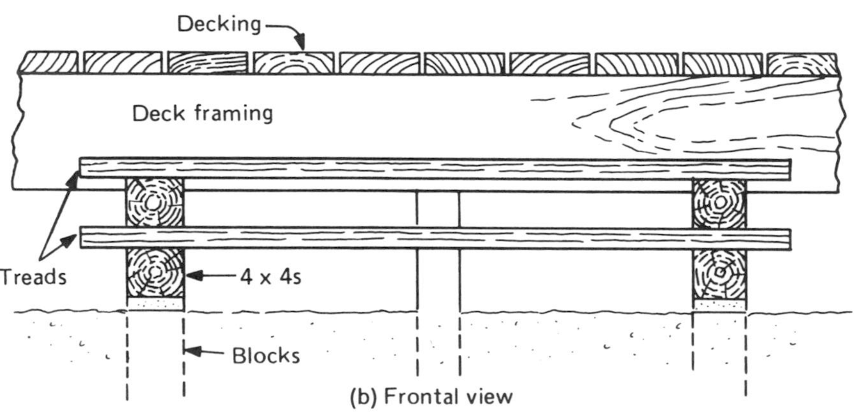

(b) Frontal view

Figure 16–6 Horizontal stringers for deck-style steps.

An alternative design to the horizontal stringer is shown in Figure 16–6a, where a 2 × 10 is laid out. A single member of this kind is easily prepared since all cuts are parallel and at right angles. Figure 16–6b shows the frontal view of the stairs. The ends of the 4 × 4s show through, but this is seldom a problem since end wood is expected to show on any open stringers. The openness, which includes spaces between tread pieces, no riser boards, and slight elevation above ground level, provides complete air circulation and therefore should provide lasting quality. However, it is still a good idea to use treated lumber for the exterior stairs.

PROBLEMS AND QUESTIONS

1. If fungi are to cause rot and decay in wood, what percentage of moisture content must be present?
   a. 10% **b.** 15% **c.** 20%
2. Which type of wood from a tree is best for exterior stringers?
   a. heartwood **b.** sapwood
3. Which species are classed as highly resistant to rot and termites?
4. Which type of wood preservative cannot be painted?
5. What might be a problem if brushing on a preservative is the treatment used on a stringer?
6. How can proper drainage of surface moisture extend the life of exterior stairs?
7. Assume that a deck is 48 in. above ground level and stairs are to be built to the ground; what would the average rise be if the maximum rise is 7½ in.?
   a. six 7½-in. risers **b.** seven 6¹³⁄₁₆-in. F risers
8. When the cut stringer is temporarily set in place, how is adjustment for a level tread surface made?
9. What is the purpose of a drift pin in the pier?
10. What size of carriage should be used for exterior stairs that are installed with metal tread supports?
11. Cite an example where the stringers would be made from 2 × 12s laid horizontally.

PROJECT

Throughout the book many end-of-chapter projects dealt with constructing a porch on a house. In this final chapter we have the opportunity to design a set of stairs that will lead from the porch floor to the ground. We can approach the project in two steps: (1) design the stairs, and (2) identify a materials list.

A. Design of the Stairs

Facts: The distance from the porch floor to the ground where the stairs will be built is 3 ft 6 in. The ground is firm and relatively level.

Specifications: The materials to be used must be treated. Every piece will be 2 in. stock pine. Step treads will be made from two 2 × 6s spaced ½ in. apart. A hollow 4 × 8 × 16 in. cement block will be used as a foundation for each stringer. One J bolt will be mortared into each block, the entire block filled with mortar, and a mortar cap made as shown in Figure 16–1. Because we anticipate closing in the porch with screen panels, we shall use the design approach shown in Figure 16–3c. The top of each stringer will be bolted to the porch framing with an angle bracket as shown in Figure 16–4. The length of each stair tread will be 6 ft, so three stringers and foundation blocks will be used. Each tread will overhang 4 in. outside the end stringer.

Since the cement block will be embedded in the ground so that just the mortar cap is above the ground, and we know that the total rise is 36 in., we should plan to use 36 in. Later, when laying out the stringer, we would cut 2 in. from the bottom to allow for the cap. Let's decide on the number of step rises to be used.

1. Divide the total rise by various rise values until an equal number of inches is calculated. Try 6½, 6¾, and 7 in., in ¼ in. increments, and so on. Your answer: ____________ in. per rise. See Figure 16–4; check the answer sheet to verify that we are on the same track.
2. Now calculate the step tread surface on the stringer. Allow a ¾-in. overhang of the outer 2 × 6 tread piece: ____________ in. Check your answer!
3. With five rises we will have five treads. Therefore, we can define the stair length as ____________ in. (____________ ft).
4. Now use the Pythagorean theorem to calculate the length of the stringer by using the total rise and total stair length: ____________ in. (____________ft ____________ in.). Rounded up to the next even foot: ____________-ft stringers.

B. Materials Listing

1. Stringer materials: ____________ 2 × 12, ____________ ft
2. Tread materials: ____________ 2 × 6, ____________ ft
3. Cement block: ____________ 4 × 8 × 16
4. J bolt: ____________
5. Mortar mix: ____________ bags
6. Sand = est. ____________ lb
7. Angle-iron materials: ____________, ____________ × ____________ in.
8. Lag bolts: ____________, ____________ × ____________
9. Galvanized common nails: ____________ lb 12d

Answers:

Part A: **1.** 7¼ in. = 35¼ in. The ¼ can be trimmed from the bottom when the allowance for the mortar cap is cut off.

2. 5½ in. + ½ in. + 5½ in. + 5½ in. − ¾ in. = 10¾ in.

3. 5 × 10¾ in. = 53¹³⁄₁₆ in. (4 ft 5¹³⁄₁₆ in)

4. 65 in. (5 ft 5 in.) rounded up to 6 ft

Part B:

1. 3, 6 ft
2. 10, 6 ft
3. 3
4. 3
5. 1
6. 50 lb
7. 3 (3 in. × 6 in. with 3 in. on each flange)
8. 12 (4 per bracket), ¼ in. × 2 in.
9. 4 lb

Appendix A

Design Values for Joists and Rafters

(Courtesy of National Forest Products Association, Washington, DC)

These species are representative of those used structurally in the United States. For additional species, contact the National Forest Products Association.

TABLE W-1 DESIGN VALUES FOR JOISTS AND RAFTERS—VISUAL GRADING

Species and grade	Size	Design value in bending F_b[a]			Modulus of elasticity E	Grading ru
		Normal duration	Snow loading	7-Day loading		
Douglas Fir-Larch (surfaced dry or surfaced green)[b]						
Dense Select Structural	2 × 4	2800	3220	3500	1,900,000	Western Wood Products Association
Select Structural		2400	2760	3000	1,800,000	
Dense No. 1		2400	2760	3000	1,900,000	
No. 1 and Appearance		2050	2360	2560	1,800,000	
Dense No. 2		1950	2240	2440	1,700,000	
No. 2		1650	1900	2060	1,700,000	
No. 3		925	1060	1160	1,500,000	
Stud		925	1060	1160	1,500,000	
Construction	2 × 4	1200	1380	1500	1,500,000	
Standard		675	780	840	1,500,000	
Utility		325	370	410	1,500,000	
Dense Select Structural	2 × 5 and wider	2400	2760	3000	1,900,000	West Coast Lumber Inspection Bureau
Select Structural		2050	2360	2560	1,800,000	
Dense No. 1		2050	2360	2560	1,900,000	
No. 1 and Appearance		1750	2010	2190	1,800,000	
Dense No. 2		1700	1960	2120	1,700,000	
No. 2		1450	1670	1810	1,700,000	
No. 3		850	980	1060	1,500,000	
Stud		850	980	1060	1,500,000	
Douglas Fir-Larch (North) (surfaced dry or surfaced green)[b]						
Select Structural	2 × 4	2400	2760	3000	1,800,000	National Lumb Grades Author (a Canadian agency)
No. 1 and Appearance		2050	2360	2560	1,800,000	
No. 2		1650	1900	2060	1,700,000	
No. 3		925	1060	1160	1,500,000	
Stud		925	1060	1160	1,500,000	
Construction	2 × 4	1200	1380	1500	1,500,000	
Standard		675	780	840	1,500,000	
Utility		325	370	410	1,500,000	
Select Structural	2 × 5 and wider	2050	2360	2560	1,800,000	
No. 1 and Appearance		1750	2010	2190	1,800,000	
No. 2		1450	1670	1810	1,700,000	
No. 3		850	980	1060	1,500,000	
Stud		850	980	1060	1,500,000	
Southern Pine (surfaced dry)[b]						
Select Structural	2 × 4	2300	2640	2880	1,700,000	
Dense Select Structural		2700	3100	3380	1,800,000	
No. 1		1950	2240	2440	1,700,000	
No. 1 Dense		2300	2640	2880	1,800,000	
No. 2		1650	1900	2060	1,600,000	
No. 2 Dense		1900	2180	2380	1,600,000	

TABLE W-1 DESIGN VALUES FOR JOISTS AND RAFTERS—VISUAL GRADING (*cont.*)

Species and grade	Size	Design value in bending F_b[a] Normal duration	Snow loading	7-Day loading	Modulus of elasticity E	Grading rules
No. 3		900	1040	1120	1,400,000	Southern Pine Inspection Bureau
No. 3 Dense		1050	1210	1310	1,500,000	
Stud		900	1040	1120	1,400,000	
Construction	2 × 4	1150	1320	1440	1,400,000	
Standard		675	780	840	1,400,000	
Utility		300	340	380	1,400,000	
Select Structural	2 × 5 and wider	2000	2300	2500	1,700,000	
Dense Select Structural		2350	2700	2940	1,800,000	
No. 1		1700	1960	2120	1,700,000	
No. 1 Dense		2000	2300	2500	1,800,000	
No. 2		1400	1610	1750	1,600,000	
No. 2 Dense		1650	1900	2060	1,600,000	
No. 3		800	920	1000	1,400,000	
No. 3 Dense		925	1060	1160	1,500,000	
Stud		850	980	1060	1,400,000	
Southern Pine (surfaced at 15% maximum moisture content—KD)						Southern Pine Inspection Bureau
Select Structural	2 × 4	2500	2880	3120	1,800,000	
Dense Select Structural		2900	3340	3620	1,900,000	
No. 1		2100	2420	2620	1,800,000	
No. 1 Dense		2450	2820	3060	1,900,000	
No. 2		1750	2010	2190	1,600,000	
No. 2 Dense		2050	2360	2560	1,700,000	
No. 3		975	1120	1220	1,500,000	
No. 3 Dense		1150	1320	1440	1,500,000	
Stud		975	1120	1220	1,500,000	
Construction	2 × 4	1250	1440	1560	1,500,000	
Standard		725	830	910	1,500,000	
Utility		300	340	380	1,500,000	
Select Structural	2 × 5 and wider	2150	2470	2690	1,800,000	
Dense Select Structural		2500	2880	3120	1,900,000	
No. 1		1850	2130	2310	1,800,000	
No. 1 Dense		2150	2470	2690	1,900,000	
No. 2		1500	1720	1880	1,600,000	
No. 2 Dense		1750	2010	2190	1,700,000	
No. 3		875	1010	1090	1,500,000	
No. 3 Dense		1000	1150	1250	1,500,000	
Stud		900	1040	1120	1,500,000	

[a]F_b values are for use where repetitive members are spaced not more than 24 in. For wider spacing, the F_b values should be reduced 13%.

[b]Values for surfaced dry or surfaced green lumber apply at 19% maximum moisture content in use.

Appendix B

Span Tables for Joists and Rafters

(Courtesy of National Forest Products Association, Washington, DC)

These species are representative of those used structurally in the United States. For additional species, contact the National Forest Products Association.

TABLE J–1 FLOOR JOISTS

40 psf live load

All rooms except those used for sleeping areas and attic floors

Design Criteria

Deflection: For 40 psf live load. Limited to span in inches divided by 360.

Strength: Live load of 40 psf plus dead load of 10 psf determines the required fiber stress value.

Joist size	spacing	Modulus of elasticity, E (1,000,000 psi)																	
(in.)	(in.)	0.4	0.5	0.6	0.7	0.8	0.9	1.0	1.1	1.2	1.3	1.4	1.5	1.6	1.7	1.8	1.9	2.0	2.2	2.4
2 × 6	12.0	6-9 450	7-3 520	7-9 590	8-2 660	8-6 720	8-10 780	9-2 830	9-6 890	9-9 940	10-0 990	10-3 1040	10-6 1090	10-9 1140	10-11 1190	11-2 1230	11-4 1280	11-7 1320	11-11 1410	12-3 1490
	13.7	6-6 470	7-0 550	7-5 620	7-9 690	8-2 750	8-6 810	8-9 870	9-1 930	9-4 980	9-7 1040	9-10 1090	10-0 1140	10-3 1190	10-6 1240	10-8 1290	10-10 1340	11-1 1380	11-5 1470	11-9 1560
	16.0	6-2 500	6-7 580	7-0 650	7-5 720	7-9 790	8-0 860	8-4 920	8-7 980	8-10 1040	9-1 1090	9-4 1150	9-6 1200	9-9 1250	9-11 1310	10-2 1360	10-4 1410	10-6 1460	10-10 1550	11-2 1640
	19.2	5-9 530	6-3 610	6-7 690	7-0 770	7-3 840	7-7 910	7-10 970	8-1 1040	8-4 1100	8-7 1160	8-9 1220	9-0 1280	9-2 1330	9-4 1390	9-6 1440	9-8 1500	9-10 1550	10-2 1650	10-6 1750
	24.0	5-4 570	5-9 660	6-2 750	6-6 830	6-9 900	7-0 980	7-3 1050	7-6 1120	7-9 1190	7-11 1250	8-2 1310	8-4 1380	8-6 1440	8-8 1500	8-10 1550	9-0 1610	9-2 1670	9-6 1780	9-9 1880
	32.0					6-2 1010	6-5 1090	6-7 1150	6-10 1230	7-0 1300	7-3 1390	7-5 1450	7-7 1520	7-9 1590	7-11 1660	8-0 1690	8-2 1760	8-4 1840	8-7 1950	8-10 2060
2 × 8	12.0	8-11 450	9-7 520	10-2 590	10-9 660	11-3 720	11-8 780	12-1 830	12-6 890	12-10 940	13-2 990	13-6 1040	13-10 1090	14-2 1140	14-5 1190	14-8 1230	15-0 1280	15-3 1320	15-9 1410	16-2 1490
	13.7	8-6 470	9-2 550	9-9 620	10-3 690	10-9 750	11-2 810	11-7 870	11-11 930	12-3 980	12-7 1040	12-11 1090	13-3 1140	13-6 1190	13-10 1240	14-1 1290	14-4 1340	14-7 1380	15-0 1470	15-6 1560
	16.0	8-1 500	8-9 580	9-3 650	9-9 720	10-2 790	10-7 850	11-0 920	11-4 980	11-8 1040	12-0 1090	12-3 1150	12-7 1200	12-10 1250	13-1 1310	13-4 1360	13-7 1410	13-10 1460	14-3 1550	14-8 1640
	19.2	7-7 530	8-2 610	8-9 690	9-2 770	9-7 840	10-0 910	10-4 970	10-8 1040	11-0 1100	11-3 1160	11-7 1220	11-10 1280	12-1 1330	12-4 1390	12-7 1440	12-10 1500	13-0 1550	13-5 1650	13-10 1750

	24.0	7-1 570	7-7 660	8-1 750	8-6 830	8-11 900	9-3 980	9-7 1050	9-11 1120	10-2 1190	10-6 1250	10-9 1310	11-0 1380	11-3 1440	11-5 1500	11-8 1550	11-11 1610	12-1 1670	12-6 1780	12-10 1880
	32.0					8-1 990	8-5 1080	8-9 1170	9-0 1230	9-3 1300	9-6 1370	9-9 1450	10-0 1520	10-2 1570	10-5 1650	10-7 1700	10-10 1790	11-0 1840	11-4 1950	11-8 2070
2 × 10	12.0	11-4 450	12-3 520	13-0 590	13-8 660	14-4 720	14-11 780	15-5 830	15-11 890	16-5 940	16-10 990	17-3 1040	17-8 1090	18-0 1140	18-5 1190	18-9 1230	19-1 1280	19-5 1320	20-1 1410	20-8 1490
	13.7	10-10 470	11-8 550	12-5 620	13-1 690	13-8 750	14-3 810	14-9 870	15-3 930	15-8 980	16-1 1040	16-6 1090	16-11 1140	17-3 1190	17-7 1240	17-11 1290	18-3 1340	18-7 1380	19-2 1470	19-9 1560
	16.0	10-4 500	11-1 580	11-10 650	12-5 720	13-0 790	13-6 850	14-0 920	14-6 980	14-11 1040	15-3 1090	15-8 1150	16-0 1200	16-5 1250	16-9 1310	17-0 1360	17-4 1410	17-8 1460	18-3 1550	18-9 1640
	19.2	9-9 530	10-6 610	11-1 690	11-8 770	12-3 840	12-9 910	13-2 970	13-7 1040	14-0 1100	14-5 1160	14-9 1220	15-1 1280	15-5 1330	15-9 1390	16-0 1440	16-4 1500	16-7 1550	17-2 1650	17-8 1750
	24.0	9-0 570	9-9 660	10-4 750	10-10 830	11-4 900	11-10 980	12-3 1050	12-8 1120	13-0 1190	13-4 1250	13-8 1310	14-0 1380	14-4 1440	14-7 1500	14-11 1550	15-2 1610	15-5 1670	15-11 1780	16-5 1880
	32.0					10-4 1000	10-9 1080	11-1 1150	11-6 1240	11-10 1310	12-2 1380	12-5 1440	12-9 1520	13-0 1580	13-3 1640	13-6 1700	13-9 1770	14-0 1830	14-6 1870	14-11 2080
2 × 12	12.0	13-10 450	14-11 520	15-10 590	16-8 660	17-5 720	18-1 780	18-9 830	19-4 890	19-11 940	20-6 990	21-0 1040	21-6 1090	21-11 1140	22-5 1190	22-10 1230	23-3 1280	23-7 1320	24-5 1410	25-1 1490
	13.7	13-3 470	14-3 550	15-2 620	15-11 690	16-8 750	17-4 810	17-11 870	18-6 930	19-1 980	19-7 1040	20-1 1090	20-6 1140	21-0 1190	21-5 1240	21-10 1290	22-3 1340	22-7 1380	23-4 1470	24-0 1560
	16.0	12-7 500	13-6 580	14-4 650	15-2 720	15-10 790	16-5 860	17-0 920	17-7 980	18-1 1040	18-7 1090	19-1 1150	19-6 1200	19-11 1250	20-4 1310	20-9 1360	21-1 1410	21-6 1460	22-2 1550	22-10 1640
	19.2	11-10 530	12-9 610	13-6 690	14-3 770	14-11 840	15-6 910	16-0 970	16-7 1040	17-0 1100	17-6 1160	17-11 1220	18-4 1280	18-9 1330	19-2 1390	19-6 1440	19-10 1500	20-2 1550	20-10 1650	21-6 1750
	24.0	11-0 570	11-10 660	12-7 750	13-3 830	13-10 900	14-4 980	14-11 1050	15-4 1120	15-10 1190	16-3 1250	16-8 1310	17-0 1380	17-5 1440	17-9 1500	18-1 1550	18-5 1610	18-9 1670	19-4 1780	19-11 1880
	32.0					12-7 1000	13-1 1080	13-6 1150	13-11 1220	14-4 1300	14-9 1380	15-2 1450	15-6 1520	15-10 1580	16-2 1650	16-5 1700	16-9 1770	17-0 1830	17-7 1950	18-1 2070

Note: The required extreme fiber stress in bending, F_b, in pounds per square inch is shown below each span.

TABLE J-4 CEILING JOISTS

20 psf live load

Limited attic storage where development of future rooms is not possible, Drywall ceiling

Design Criteria

Deflection: For 20 psf live load. Limited to span in inches divided by 240.

Strength: Live load of 20 psf plus dead load of 10 psf determines the required fiber stress value.

Joist size (in.)	spacing (in.)	0.4	0.5	0.6	0.7	0.8	0.9	1.0	1.1	1.2	1.3	1.4	1.5	1.6	1.7	1.8	1.9	2.0	2.2	2.4
		Modulus of elasticity, E (1,000,000 psi)																		
2 × 4	12.0	6-2 560	6-8 660	7-1 740	7-6 820	7-10 900	8-1 970	8-5 1040	8-8 1110	8-11 1170	9-2 1240	9-5 1300	9-8 1360	9-10 1420	10-0 1480	10-3 1540	10-5 1600	10-7 1650	10-11 1760	11-3 1860
	13.7	5-11 590	6-5 690	6-9 770	7-2 860	7-6 940	7-9 1010	8-1 1090	8-4 1160	8-7 1230	8-9 1300	9-0 1360	9-3 1420	9-5 1490	9-7 1550	9-9 1610	10-0 1670	10-2 1730	10-6 1840	10-9 1950
	16.0	5-8 620	6-1 720	6-5 810	6-9 900	7-1 990	7-5 1070	7-8 1140	7-11 1220	8-1 1290	8-4 1360	8-7 1430	8-9 1500	8-11 1570	9-1 1630	9-4 1690	9-6 1760	9-8 1820	9-11 1940	10-3 2050
	19.2	5-4 660	5-9 770	6-1 870	6-5 960	6-8 1050	6-11 1130	7-2 1220	7-5 1300	7-8 1370	7-10 1450	8-1 1520	8-3 1590	8-5 1660	8-7 1730	8-9 1800	8-11 1870	9-1 1930	9-4 2060	9-8 2180
	24.0	4-11 710	5-4 830	5-8 930	5-11 1030	6-2 1130	6-5 1220	6-8 1310	6-11 1400	7-1 1480	7-3 1560	7-6 1640	7-8 1720	7-10 1790	8-0 1870	8-1 1940	8-3 2010	8-5 2080	8-8 2220	8-11 2350
2 × 6	12.0	9-9 560	10-6 660	11-2 740	11-9 820	12-3 900	12-9 970	13-3 1040	13-8 1110	14-1 1170	14-5 1240	14-9 1300	15-2 1360	15-6 1420	15-9 1480	16-1 1540	16-4 1600	16-8 1650	17-2 1760	17-8 1860
	13.7	9-4 590	10-0 690	10-8 770	11-3 860	11-9 940	12-3 1010	12-8 1090	13-1 1160	13-5 1230	13-10 1300	14-2 1360	14-6 1420	14-9 1490	15-1 1550	15-5 1610	15-8 1670	15-11 1730	16-5 1840	16-11 1950
	16.0	8-10 620	9-6 720	10-2 810	10-8 900	11-2 990	11-7 1070	12-0 1140	12-5 1220	12-9 1290	13-1 1360	13-5 1430	13-9 1500	14-1 1570	14-4 1630	14-7 1690	14-11 1760	15-2 1820	15-7 1940	16-1 2050

	19.2	8-4 660	9-0 770	9-6 870	10-0 960	10-6 1050	10-11 1130	11-4 1220	11-8 1300	12-0 1370	12-4 1450	12-8 1520	12-11 1590	13-3 1660	13-6 1730	13-9 1800	14-0 1870	14-3 1930	14-8 2060	15-2 2180
	24.0	7-9 710	8-4 830	8-10 930	9-4 1030	9-9 1130	10-2 1220	10-6 1310	10-10 1400	11-2 1480	11-5 1560	11-9 1640	12-0 1720	12-3 1790	12-6 1870	12-9 1940	13-0 2010	13-3 2080	13-8 2220	14-1 2350
	12.0	12-10 560	13-10 660	14-8 740	15-6 820	16-2 900	16-10 970	17-5 1040	18-0 1110	18-6 1170	19-0 1240	19-6 1300	19-11 1360	20-5 1420	20-10 1480	21-2 1540	21-7 1600	21-11 1650	22-8 1760	23-4 1860
	13.7	12-3 590	13-3 690	14-1 770	14-10 860	15-6 940	16-1 1010	16-8 1090	17-2 1160	17-9 1230	18-2 1300	18-8 1360	19-1 1420	19-6 1490	19-11 1550	20-3 1610	20-8 1670	21-0 1730	21-8 1840	22-4 1950
2 × 8	16.0	11-8 620	12-7 720	13-4 810	14-1 900	14-8 990	15-3 1070	15-10 1140	16-4 1220	16-10 1290	17-3 1360	17-9 1430	18-2 1500	18-6 1570	18-11 1630	19-3 1690	19-7 1760	19-11 1820	20-7 1940	21-2 2050
	19.2	11-0 660	11-10 770	12-7 870	13-3 960	13-10 1050	14-5 1130	14-11 1220	15-5 1300	15-10 1370	16-3 1450	16-8 1520	17-1 1590	17-5 1660	17-9 1730	18-2 1800	18-5 1870	18-9 1930	19-5 2060	19-11 2180
	24.0	10-2 710	11-0 830	11-8 930	12-3 1030	12-10 1130	13-4 1220	13-10 1310	14-3 1400	14-8 1480	15-1 1560	15-6 1640	15-10 1720	16-2 1790	16-6 1870	16-10 1940	17-2 2010	17-5 2080	18-0 2220	18-6 2350
	12.0	16-5 560	17-8 660	18-9 740	19-9 820	20-8 900	21-6 970	22-3 1040	22-11 1110	23-8 1170	24-3 1240	24-10 1300	25-5 1360	26-0 1420	26-6 1480	27-1 1540	27-6 1600	28-0 1650	28-11 1760	29-9 1860
	13.7	15-8 590	16-11 690	17-11 770	18-11 860	19-9 940	20-6 1010	21-3 1090	21-11 1160	22-7 1230	23-3 1300	23-9 1360	24-4 1420	24-10 1490	25-5 1550	25-10 1610	26-4 1670	26-10 1730	27-8 1840	28-6 1950
2 × 10	16.0	14-11 620	16-0 720	17-0 810	17-11 900	18-9 990	19-6 1070	20-2 1140	20-10 1220	21-6 1290	22-1 1360	22-7 1430	23-2 1500	23-8 1570	24-1 1630	24-7 1690	25-0 1760	25-5 1820	26-3 1940	27-1 2050
	19.2	14-0 660	15-1 770	16-0 870	16-11 960	17-8 1050	18-4 1130	19-0 1220	19-7 1300	20-2 1370	20-9 1450	21-3 1520	21-9 1590	22-3 1660	22-8 1730	23-2 1800	23-7 1870	23-11 1930	24-9 2060	25-5 2180
	24.0	13-0 710	14-0 830	14-11 930	15-8 1030	16-5 1130	17-0 1220	17-8 1310	18-3 1400	18-9 1480	19-3 1560	19-9 1640	20-2 1720	20-8 1790	21-1 1870	21-6 1940	21-10 2010	22-3 2080	22-11 2220	23-8 2350

Note: The required extreme fiber stress in bending, F_b, in pounds per square inch is shown below each span.

TABLE R–7 FLAT OR LOW SLOPE RAFTERS

20 psf live load
No ceiling load Slope 3 in 12 or less

Design Criteria

Strength: 10 psf dead load plus 20 psf live load determines required fiber stress.

Deflection: For 20 psf live load, limited to span in inches divided by 240. Spans are measured along the horizontal projection and loads are considered as applied on the horizontal projection.

Rafter size	spacing	Extreme fiber stress in bending, F_b (psi)																				
(in.)	(in.)	300	400	500	600	700	800	900	1000	1100	1200	1300	1400	1500	1600	1700	1800	1900	2000	2100	2200	2400
	12.0	7-1 0.15	8-2 0.24	9-2 0.33	10-0 0.44	10-10 0.55	11-7 0.67	12-4 0.80	13-0 0.94	13-7 1.09	14-2 1.24	14-9 1.40	15-4 1.56	15-11 1.73	16-5 1.91	16-11 2.09	17-5 2.28	17-10 2.47				
	13.7	6-8 0.14	7-8 0.22	8-7 0.31	9-5 0.41	10-2 0.52	10-10 0.63	11-6 0.75	12-2 0.88	12-9 1.02	13-3 1.16	13-10 1.31	14-4 1.46	14-10 1.62	15-4 1.78	15-10 1.95	16-3 2.13	16-9 2.31	17-2 2.49			
2 × 6	16.0	6-2 0.13	7-1 0.21	7-11 0.29	8-8 0.38	9-5 0.48	10-0 0.58	10-8 0.70	11-3 0.82	11-9 0.94	12-4 1.07	12-10 1.21	13-3 1.35	13-9 1.50	14-2 1.65	14-8 1.81	15-1 1.97	15-6 2.14	15-11 2.31	16-3 2.48		
	19.2	5-7 0.12	6-6 0.19	7-3 0.26	7-11 0.35	8-7 0.44	9-2 0.53	9-9 0.64	10-3 0.75	10-9 0.86	11-3 0.98	11-8 1.10	12-2 1.23	12-7 1.37	13-0 1.51	13-4 1.65	13-9 1.80	14-2 1.95	14-6 2.11	14-10 2.27	15-2 2.43	
	24.0	5-0 0.11	5-10 0.17	6-6 0.24	7-1 0.31	7-8 0.39	8-2 0.48	8-8 0.57	9-2 0.67	9-7 0.77	10-0 0.88	10-5 0.99	10-10 1.10	11-3 1.22	11-7 1.35	11-11 1.48	12-4 1.61	12-8 1.75	13-0 1.89	13-3 2.03	13-7 2.18	14-2 2.48
	12.0	9-4 0.15	10-10 0.24	12-1 0.33	13-3 0.44	14-4 0.55	15-3 0.67	16-3 0.80	17-1 0.94	17-11 1.09	18-9 1.24	19-6 1.40	20-3 1.56	20-11 1.73	21-7 1.91	22-3 2.09	22-11 2.28	23-7 2.47				
	13.7	8-9 0.14	10-1 0.22	11-4 0.31	12-5 0.41	13-4 0.52	14-4 0.63	15-2 0.75	16-0 0.88	16-9 1.02	17-6 1.16	18-3 1.31	18-11 1.46	19-7 1.62	20-3 1.78	20-10 1.95	21-5 2.13	22-0 2.31	22-7 2.49			

2 × 8	16.0	8-1 0.13	9-4 0.21	10-6 0.29	11-6 0.38	12-5 0.48	13-3 0.58	14-0 0.70	14-10 0.82	15-6 0.94	16-3 1.07	16-10 1.21	17-6 1.35	18-2 1.50	18-9 1.65	19-4 1.81	19-10 1.97	20-5 2.14	20-11 2.31	21-5 2.48		
	19.2	7-5 0.12	8-7 0.19	9-7 0.26	10-6 0.35	11-4 0.44	12-1 0.53	12-10 0.64	13-6 0.75	14-2 0.86	14-10 0.98	15-5 1.10	16-0 1.23	16-7 1.37	17-1 1.51	17-7 1.65	18-2 1.80	18-7 1.95	19-1 2.11	19-7 2.27	20-0 2.43	
	24.0	6-7 0.11	7-8 0.17	8-7 0.24	9-4 0.31	10-1 0.39	10-10 0.48	11-6 0.57	12-1 0.67	12-8 0.77	13-3 0.88	13-9 0.99	14-4 1.10	14-10 1.22	15-3 1.35	15-9 1.48	16-3 1.61	16-8 1.75	17-1 1.89	17-6 2.03	17-11 2.18	18-9 2.48
	12.0	11-11 0.15	13-9 0.24	15-5 0.33	16-11 0.44	18-3 0.55	19-6 0.67	20-8 0.80	21-10 0.94	22-10 1.09	23-11 1.24	24-10 1.40	25-10 1.56	26-8 1.73	27-7 1.91	28-5 2.09	29-3 2.28	30-1 2.47				
	13.7	11-2 0.14	12-11 0.22	14-5 0.31	15-10 0.41	17-1 0.52	18-3 0.63	19-4 0.75	20-5 0.88	21-5 1.02	22-4 1.16	23-3 1.31	24-2 1.46	25-0 1.62	25-10 1.78	26-7 1.95	27-4 2.13	28-1 2.31	28-10 2.49			
2 × 10	16.0	10-4 0.13	11-11 0.21	13-4 0.29	14-8 0.38	15-10 0.48	16-11 0.58	17-11 0.70	18-11 0.82	19-10 0.94	20-8 1.07	21-6 1.21	22-4 1.35	23-2 1.50	23-11 1.65	24-7 1.81	25-4 1.97	26-0 2.14	26-8 2.31	27-4 2.48		
	19.2	9-5 0.12	10-11 0.19	12-2 0.26	13-4 0.35	14-5 0.44	15-5 0.53	16-4 0.64	17-3 0.75	18-1 0.86	18-11 0.98	19-8 1.10	20-5 1.23	21-1 1.37	21-10 1.51	22-6 1.65	23-2 1.80	23-9 1.95	24-5 2.11	25-0 2.27	25-7 2.43	
	24.0	8-5 0.11	9-9 0.17	10-11 0.24	11-11 0.31	12-11 0.39	13-9 0.48	14-8 0.57	15-5 0.67	16-2 0.77	16-11 0.88	17-7 0.99	18-3 1.10	18-11 1.22	19-6 1.35	20-1 1.48	20-8 1.61	21-3 1.75	21-10 1.89	22-4 2.03	22-10 2.18	23-11 2.48
	12.0	14-6 0.15	16-9 0.24	18-9 0.33	20-6 0.44	22-2 0.55	23-9 0.67	25-2 0.80	26-6 0.94	27-10 1.09	29-1 1.24	30-3 1.40	31-4 1.56	32-6 1.73	33-6 1.91	34-7 2.09	35-7 2.28	36-7 2.47				
	13-7	13-7 0.14	15-8 0.22	17-6 0.31	19-3 0.41	20-9 0.52	22-2 0.63	23-6 0.75	24-10 0.88	26-0 1.02	27-2 1.16	28-3 1.31	29-4 1.46	30-5 1.62	31-4 1.78	32-4 1.95	33-3 2.13	34-2 2.31	35-1 2.49			
2 × 12	16.0	12-7 0.13	14-6 0.21	16-3 0.29	17-9 0.38	19-3 0.48	20-6 0.58	21-9 0.70	23-0 0.82	24-1 0.94	25-2 1.07	26-2 1.21	27-2 1.35	28-2 1.50	29-1 1.65	29-11 1.81	30-10 1.97	31-8 2.14	32-6 2.31	33-3 2.48		
	19.2	11-6 0.12	13-3 0.19	14-10 0.26	16-3 0.35	17-6 0.44	18-9 0.53	19-11 0.64	21-0 0.75	22-0 0.86	23-0 0.98	23-11 1.10	24-10 1.23	25-8 1.37	26-6 1.51	27-4 1.65	28-2 1.80	28-11 1.95	29-8 2.11	30-5 2.27	31-1 2.43	
	24.0	10-3 0.11	11-10 0.17	13-3 0.24	14-6 0.31	15-8 0.39	16-9 0.48	17-9 0.57	18-9 0.67	19-8 0.77	20-6 0.88	21-5 0.99	22-2 1.10	23-0 1.22	23-9 1.35	24-5 1.48	25-2 1.61	25-10 1.75	26-6 1.89	27-2 2.03	27-10 2.18	29-1 2.48

Note: The required modulus of elasticity, *E*, in 1,000,000 psi is shown below each span.

TABLE R-10 MEDIUM OR HIGH SLOPE RAFTERS

20 psf live load

No ceiling load Slope over 3 in 12 Heavy roof covering

Design Criteria

Strength: 15 psf dead load plus 20 psf live load determines required fiber stress.

Deflection: For 20 psf live load, limited to span in inches divided by 180. Spans are measured along the horizontal projection and loads are considered as applied on the horizontal projection.

Rafter size spacing		Extreme fiber stress in bending, F_b (psi)																					
(in.)	(in.)	200	300	400	500	600	700	800	900	1000	1100	1200	1300	1400	1500	1600	1700	1800	1900	2000	2100	2200	2400	2700	3000
	12.0	3-5 0.05	4-2 0.09	4-10 0.14	5-5 0.20	5-11 0.26	6-5 0.33	6-10 0.40	7-3 0.48	7-8 0.56	8-0 0.65	8-4 0.74	8-8 0.83	9-0 0.93	9-4 1.03	9-8 1.14	9-11 1.24	10-3 1.36	10-6 1.47	10-10 1.59	11-1 1.71	11-4 1.83	11-10 2.09	12-7 2.49	
	13.7	3-2 0.05	3-11 0.09	4-6 0.13	5-1 0.19	5-6 0.24	6-0 0.31	6-5 0.38	6-9 0.45	7-2 0.52	7-6 0.61	7-10 0.69	8-2 0.78	8-5 0.87	8-9 0.96	9-0 1.06	9-4 1.16	9-7 1.27	9-10 1.37	10-1 1.48	10-4 1.60	10-7 1.71	11-1 1.95	11-9 2.33	
2 × 4	16.0	2-11 0.04	3-7 0.08	4-2 0.12	4-8 0.17	5-1 0.23	5-6 0.28	5-11 0.35	6-3 0.41	6-7 0.49	6-11 0.56	7-3 0.64	7-6 0.72	7-10 0.80	8-1 0.89	8-4 0.98	8-7 1.08	8-10 1.17	9-1 1.27	9-4 1.37	9-7 1.48	9-10 1.59	10-3 1.81	10-10 2.16	11-5 2.53
	19.2	2-8 0.04	3-4 0.07	3-10 0.11	4-3 0.16	4-8 0.21	5-1 0.26	5-5 0.32	5-9 0.38	6-0 0.44	6-4 0.51	6-7 0.58	6-11 0.66	7-2 0.73	7-5 0.81	7-8 0.90	7-10 0.98	8-1 1.07	8-4 1.16	8-6 1.25	8-9 1.35	8-11 1.45	9-4 1.65	9-11 1.97	10-5 2.31
	24.0	2-5 0.04	2-11 0.07	3-5 0.10	3-10 0.14	4-2 0.18	4-6 0.23	4-10 0.28	5-1 0.34	5-5 0.40	5-8 0.46	5-11 0.52	6-2 0.59	6-5 0.66	6-7 0.73	6-10 0.80	7-0 0.88	7-3 0.96	7-5 1.04	7-8 1.12	7-10 1.21	8-0 1.29	8-4 1.48	8-10 1.76	9-4 2.06
	12.0	5-4 0.05	6-7 0.09	7-7 0.14	8-6 0.20	9-4 0.26	10-0 0.33	10-9 0.40	11-5 0.48	12-0 0.56	12-7 0.65	13-2 0.74	13-8 0.83	14-2 0.93	14-8 1.03	15-2 1.14	15-8 1.24	16-1 1.36	16-7 1.47	17-0 1.59	17-5 1.71	17-10 1.83	18-7 2.09	19-9 2.49	
	13.7	5-0 0.05	6-2 0.09	7-1 0.13	7-11 0.19	8-8 0.24	9-5 0.31	10-0 0.38	10-8 0.45	11-3 0.52	11-9 0.61	12-4 0.69	12-10 0.78	13-3 0.87	13-9 0.96	14-2 1.06	14-8 1.16	15-1 1.27	15-6 1.37	15-11 1.48	16-3 1.60	16-8 1.71	17-5 1.95	18-5 2.33	

2 × 6	16.0	4-8 0.04	5-8 0.08	6-7 0.12	7-4 0.17	8-1 0.23	8-8 0.28	9-4 0.35	9-10 0.41	10-5 0.49	10-11 0.56	11-5 0.64	11-10 0.72	12-4 0.80	12-9 0.89	13-2 0.98	13-7 1.08	13-11 1.17	14-4 1.27	14-8 1.37	15-1 1.48	15-5 1.59	16-1 1.81	17-1 2.16	18-0 2.53
	19.2	4-3 0.04	5-2 0.07	6-0 0.11	6-9 0.16	7-4 0.21	7-11 0.26	8-6 0.32	9-0 0.38	9-6 0.44	9-11 0.51	10-5 0.58	10-10 0.66	11-3 0.73	11-7 0.81	12-0 0.90	12-4 0.98	12-9 1.07	13-1 1.16	13-5 1.25	13-9 1.35	14-1 1.45	14-8 1.65	15-7 1.97	16-5 2.31
	24.0	3-10 0.04	4-8 0.07	5-4 0.10	6-0 0.14	6-7 0.18	7-1 0.23	7-7 0.28	8-1 0.34	8-6 0.40	8-11 0.46	9-4 0.52	9-8 0.59	10-0 0.66	10-5 0.73	10-9 0.80	11-1 0.88	11-5 0.96	11-8 1.04	12-0 1.12	12-4 1.21	12-7 1.29	13-2 1.48	13-11 1.76	14-8 2.06
	12.0	7-1 0.05	8-8 0.09	10-0 0.14	11-2 0.20	12-3 0.26	13-3 0.33	14-2 0.40	15-0 0.48	15-10 0.56	16-7 0.65	17-4 0.74	18-0 0.83	18-9 0.93	19-5 1.03	20-0 1.14	20-8 1.24	21-3 1.36	21-10 1.47	22-4 1.59	22-11 1.71	23-6 1.83	24-6 2.09	26-0 2.49	
	13.7	6-7 0.05	8-1 0.09	9-4 0.13	10-6 0.19	11-6 0.24	12-5 0.31	13-3 0.38	14-0 0.45	14-10 0.52	15-6 0.61	16-3 0.69	16-10 0.78	17-6 0.87	18-2 0.96	18-9 1.06	19-4 1.16	19-10 1.27	20-5 1.37	20-11 1.48	21-5 1.60	21-11 1.71	22-11 1.95	24-4 2.33	
2 × 8	16.0	6-2 0.04	7-6 0.08	8-8 0.12	9-8 0.17	10-7 0.23	11-6 0.28	12-3 0.35	13-0 0.41	13-8 0.49	14-4 0.56	15-0 0.,64	15-7 0.72	16-3 0.80	16-9 0.89	17-4 0.98	17-10 1.08	18-5 1.17	18-11 1.27	19-5 1.37	19-10 1.48	20-4 1.59	21-3 1.81	22-6 2.16	23-9 2.53
	19.2	5-7 0.04	6-10 0.07	7-11 0.11	8-10 0.16	9-8 0.21	10-6 0.26	11-2 0.32	11-10 0.38	12-6 0.44	13-1 0.51	13-8 0.58	14-3 0.66	14-10 0.73	15-4 0.81	15-10 0.90	16-4 0.98	16-9 1.07	17-3 1.16	17-8 1.25	18-2 1.35	18-7 1.45	19-5 1.65	20-7 1.97	21-8 2.31
	24.0	5-0 0.04	6-2 0.07	7-1 0.10	7-11 0.14	8-8 0.18	9-4 0.23	10-0 0.28	10-7 0.34	11-2 0.40	11-9 0.46	12-3 0.52	12-9 0.59	13-3 0.66	13-8 0.73	14-2 0.80	14-7 0.88	15-0 0.96	15-5 1.04	15-10 1.12	16-3 1.21	16-7 1.29	17-4 1.48	18-5 1.76	19-5 2.06
	12.0	9-0 0.05	11-1 0.09	12-9 0.14	14-3 0.20	15-8 0.26	16-11 0.33	18-1 0.40	19-2 0.48	20-2 0.56	21-2 0.65	22-1 0.74	23-0 0.83	23-11 0.93	24-9 1.03	25-6 1.14	26-4 1.24	27-1 1.36	27-10 1.47	28-7 1.59	29-3 1.71	29-11 1.83	31-3 2.09	33-2 2.49	
	13.7	8-5 0.05	10-4 0.09	11-11 0.13	13-4 0.19	14-8 0.24	15-10 0.31	16-11 0.38	17-11 0.45	18-11 0.52	19-10 0.61	20-8 0.69	21-6 0.78	22-4 0.87	23-2 0.96	23-11 1.06	24-7 1.16	25-4 1.27	26-0 1.37	26-8 1.48	27-4 1.60	28-0 1.71	29-3 1.95	31-0 2.33	
2 × 10	16.0	7-10 0.04	9-7 0.08	11-1 0.12	12-4 0.17	13-6 0.23	14-8 0.28	15-8 0.35	16-7 0.41	17-6 0.49	18-4 0.56	19-2 0.64	19-11 0.72	20-8 0.80	21-5 0.89	22-1 0.98	22-10 1.08	23-5 1.17	24-1 1.27	24-9 1.37	25-4 1.48	25-11 1.59	27-1 1.81	28-9 2.16	30-3 2.53
	19.2	7-2 0.04	8-9 0.07	10-1 0.11	11-3 0.16	12-4 0.21	13-4 0.26	14-3 0.32	15-2 0.38	15-11 0.44	16-9 0.51	17-6 0.58	18-2 0.66	18-11 0.73	19-7 0.81	20-2 0.90	20-10 0.98	21-5 1.07	22-0 1.16	22-7 1.25	23-2 1.35	23-8 1.45	24-9 1.65	26-3 1.97	27-8 2.31
	24.0	6-5 0.04	7-10 0.07	9-0 0.10	10-1 0.14	11-1 0.18	11-11 0.23	12-9 0.28	13-6 0.34	14-3 0.40	15-0 0.46	15-8 0.52	16-3 0.59	16-11 0.66	17-6 0.73	18-1 0.80	18-7 0.88	19-2 0.96	19-8 1.04	20-2 1.12	20-8 1.21	21-2 1.29	22-1 1.48	23-5 1.76	24-9 2.06

Note: The required modulus of elasticity, *E*, in 1,000,000 psi is shown below each span.

Appendix C

Maximum Spans for Joists and Rafters

(Courtesy of Southern Forest Products Association)

TABLE NO. 1 FLOOR JOISTS

30 psf live load.
Sleeping rooms and attic floors.

Size (in.)	Spacing (in. o.c.)	Grade[a]: Dense Sel Str KD and No. 1 Dense KD	Dense Sel Str, Sel Str KD, No. 1 Dense, and No. 1 KD	Sel Str, No. 1 and No. 2 Dense KD	No. 2 Dense, No. 2 KD, and No. 2	No. 3 Dense KD[b]	No. 3 Dense[b]	No. 3 KD	No. 3
2 × 6	12.0	12-6	12-3	12-0	11-10	**11-3**	**10-11**	**10-5**	**10-1**
	13.7	11-11	11-9	11-6	11-3	**10-6**	**10-3**	**9-9**	**9-5**
	16.0	11-4	11-2	10-11	10-9	**9-9**	**9-6**	**9-0**	**8-9**
	19.2	10-8	10-6	10-4	10-1	**8-11**	**8-8**	**8-3**	**8-0**
	24.0	9-11	9-9	9-7	9-4	**8-0**	**7-9**	**7-4**	**7-1**
2 × 8	12.0	16-6	16-2	15-10	15-7	**14-10**	**14-5**	**13-9**	**13-3**
	13.7	15-9	15-6	15-2	14-11	**13-11**	**13-6**	**12-10**	**12-5**
	16.0	15-0	14-8	14-5	14-2	**12-10**	**12-6**	**11-11**	**11-6**
	19.2	14-1	13-10	13-7	13-4	**11-9**	**11-5**	**10-10**	**10-6**
	24.0	13-1	12-10	12-7	12-4	**10-6**	**10-2**	**9-9**	**9-5**
2 × 10	12.0	21-0	20-8	20-3	19-10	**18-11**	**18-5**	**17-6**	**16-11**
	13.7	20-1	19-9	19-4	19-0	**17-9**	**17-2**	**16-5**	**15-10**
	16.0	19-1	18-9	18-5	18-0	**16-5**	**15-11**	**15-2**	**14-8**
	19.2	18-0	17-8	17-4	17-0	**15-0**	**14-6**	**13-10**	**13-5**
	24.0	16-8	16-5	16-1	15-9[c]	**13-5**	**13-0**	**12-5**	**12-0**
2 × 12	12.0	25-7	25-1	24-8	24-2	**23-0**	**22-4**	**21-4**	**20-7**
	13.7	24-5	24-0	23-7	23-1	**21-7**	**20-11**	**19-11**	**19-3**
	16.0	23-3	22-10	22-5	21-11	**19-11**	**19-4**	**18-6**	**17-10**
	19.2	21-10	21-6	21-1	20-8	**18-3**	**17-8**	**16-10**	**16-3**
	24.0	20-3	19-11	19-7	19-2[c]	**16-3**	**15-10**	**15-1**	**14-7**

Note: Spans shown in lightface type are based on a deflection limitation of $l/360$. Spans shown in boldface type are limited by the recommended extreme fiber stress in bending value of the grade and includes a 10-psf dead load.

[a]Terms and abbreviations: Sel. Str. means select structural; KD means KD15, dried to a moisture content of 15% or less; where KD is not shown, the material is dried to a moisture content of 19% or less. Lumber dried to 19% or less will be stamped S-DRY or KD S-DRY.

[b]These grades may not be commonly available.

[c]The span for No. 2 grade, 24 in. o.c. spacing is: 2 × 10, **15-8;** 2 × 12, **19-1.**

TABLE NO. 2 FLOOR JOISTS

40 psf live load.
All rooms except sleeping rooms and attic floors.

		Grade[a]							
Size (in.)	Spacing (in. o.c.)	Dense Sel Str KD and No. 1 Dense KD	Dense Sel Str, Sel Str KD, No. 1 Dense, and No. 1 KD	Sel Str, No. 1 and No. 2 Dense KD	No. 2 Dense, No. 2 KD and No. 2	No. 3 Dense KD[b]	No. 3 Dense[b]	No. 3 KD	No. 3
2 × 6	12.0	11-4	11-2	10-11	10-9	**10-1**	**9-9**	**9-4**	**9-0**
	13.7	10-10	10-8	10-6	10-3	**9-5**	**9-2**	**8-9**	**8-5**
	16.0	10-4	10-2	9-11	9-9	**8-9**	**8-6**	**8-1**	**7-10**
	19.2	9-8	9-6	9-4	9-2	**8-0**	**7-9**	**7-4**	**7-1**
	24.0	9-0	8-10	8-8	8-6[c]	**7-1**	**6-11**	**6-7**	**6-4**
2 × 8	12.0	15-0	14-8	14-5	14-2	**13-3**	**12-11**	**12-4**	**11-11**
	13.7	14-4	14-1	13-10	13-6	**12-5**	**12-1**	**11-6**	**11-1**
	16.0	13-7	13-4	13-1	12-10	**11-6**	**11-2**	**10-8**	**10-3**
	19.2	12-10	12-7	12-4	12-1	**10-6**	**10-2**	**9-9**	**9-5**
	24.0	11-11	11-8	11-5	11-3[c]	**9-5**	**9-1**	**8-8**	**8-5**
2 × 10	12.0	19-1	18-9	18-5	18-0	**16-11**	**16-5**	**15-8**	**15-2**
	13.7	18-3	17-11	17-7	17-3	**15-10**	**15-5**	**14-8**	**14-2**
	16.0	17-4	17-0	16-9	16-5	**14-8**	**14-3**	**13-7**	**13-1**
	19.2	16-4	16-0	15-9	15-5	**13-5**	**13-0**	**12-5**	**12-0**
	24.0	15-2	14-11	14-7	14-4[c]	**12-0**	**11-8**	**11-1**	**10-9**
2 × 12	12.0	23-3	22-10	22-5	21-11	**20-7**	**20-0**	**19-1**	**18-5**
	13.7	22-3	21-10	21-5	21-0	**19-3**	**18-9**	**17-10**	**17-3**
	16.0	21-1	20-9	20-4	19-11	**17-10**	**17-4**	**16-6**	**16-0**
	19.2	19-10	19-6	19-2	18-9	**16-3**	**15-10**	**15-1**	**14-7**
	24.0	18-5	18-1	17-9	17-5[c]	**14-7**	**14-2**	**13-6**	**13-0**

Note: Spans shown in lightface type are based on a deflection limitation of $l/360$. Spans shown in boldface type are limited by the recommended extreme fiber stress in bending value of the grade and includes a 10-psf dead load.

[a]Terms and abbreviations: Sel. Str. means select structural; KD means KD15, dried to a moisture content of 15% or less; where KD is not shown the material is dried to a moisture content of 19% or less. Lumber dried to 19% or less will be stamped S-DRY or KD S-DRY.

[b]These grades may not be commonly available.

[c]The span for No. 2 grade, 24 in. o.c. spacing is: 2 × 6, **8-4**; 2 × 8, **11-0**; 2 × 10, **14-0**; 2 × 12, **17-1**.

TABLE NO. 8 CEILING JOISTS

Drywall ceiling
20 psf live load
No future sleeping rooms but limited storage available

		Grade[a]													
Size (in.)	Spacing (in. o.c.)	Dense Sel Str KD and No. 1 Dense KD	Dense Sel Str, Sel Str KD, No. 1 Dense, and No. 1 KD	Sel Str, No. 1 and No. 2 Dense KD	No. 2 Dense	No. 2 KD	No. 2	No. 3 Dense KD[b]	No. 3 Dense[b]	No. 3 KD	No. 3	Con-struc-tion KD	Con-struc-tion	Stan-dard KD	Stan-dard
2 × 4	12.0	10-5	10-3	10-0	9-10	9-10	9-10	**8-10**	**8-6**	**8-2**	**7-9**	**9-3**	**8-10**	**7-0**	**6-9**
	13.7	10-0	9-9	9-7	9-5	9-5	9-5	**8-3**	**8-0**	**7-8**	**7-3**	**8-8**	**8-3**	**6-7**	**6-3**
	16.0	9-6	9-4	9-1	8-11	8-11	8-11	**7-8**	**7-4**	**7-1**	**6-9**	**8-0**	**7-8**	**6-1**	**5-10**
	19.2	8-11	8-9	8-7	8-5	8-5	**8-3**	**7-0**	**6-9**	**6-5**	**6-2**	**7-4**	**7-0**	**5-6**	**5-4**
	24.0	8-3	8-1	8-0	7-10	**7-9**	**7-5**	**6-3**	**6-0**	**5-9**	**5-6**	**6-7**	**6-3**	**4-11**	**4-9**
2 × 6	12.0	16-4	16-1	15-9	15-6	**15-6**	**15-3**	**13-0**	**12-8**	**12-0**	**11-8**				
	13.7	15-8	15-5	15-1	14-9	14-9	**14-3**	**12-2**	**11-10**	**11-3**	**10-11**				
	16.0	14-11	14-7	14-4	14-1	**13-9**	**13-2**	**11-3**	**10-11**	**10-5**	**10-1**				
	19.2	14-0	13-9	13-6[c]	**13-0**	**12-6**	**12-0**	**10-3**	**10-0**	**9-6**	**9-2**				
	24.0	13-0	12-9[d]	12-6	**11-8**	**11-2**	**10-9**	**9-2**	**8-11**	**8-6**	**8-3**				

	12.0	21-7	21-2	20-10	20-5	20-5	**20-1**	**17-2**	**16-8**	**15-10**	**15-4**				
	13.7	20-8	20-3	19-11	19-6	19-6	**18-9**	**16-0**	**15-7**	**14-10**	**14-4**				
2 × 8	16.0	19-7	19-3	18-11	18-6	**18-1**	**17-5**	**14-10**	**14-5**	**13-9**	**13-3**				
	19.2	18-5	18-2	17-9[c]	**17-2**	**16-6**	**15-10**	**13-7**	**13-2**	**12-7**	**12-1**				
	24.0	17-2	16-10[d]	16-6	**15-4**	**14-9**	**14-2**	**12-1**	**11-9**	**11-3**	**10-10**				
	12.0	27-6	27-1	26-6	26-0	26-0	**25-7**	**21-10**	**21-3**	**20-3**	**19-7**				
	13.7	26-4	25-10	25-5	24-11	24-11	**24-0**	**20-6**	**19-10**	**18-11**	**18-4**				
2 × 10	16.0	25-0	24-7	24-1	23-8	**23-1**	**22-2**	**18-11**	**18-5**	**17-6**	**16-11**				
	19.2	23-7	23-2	22-8[c]	**21-10**	**21-1**	**20-3**	**17-3**	**16-9**	**16-0**	**15-6**				
	24.0	21-10	21-6[d]	21-1	**19-7**	**18-10**	**18-1**	**15-6**	**15-0**	**14-4**	**13-10**				

Note: Spans shown in lightface type are based on a deflection limitation of *l*/240. Spans shown in boldface type are limited by the recommended extreme fiber stress in bending value of the grade and includes a 10-psf dead load.

[a]Terms and abbreviations: Sel. Str. means selected structural; KD means KD15, dried to a moisture content of 15% or less; where KD is not shown the material is dried to a moisture content of 19% or less. Lumber dried to 19% or less will be stamped S-DRY or KD S-DRY.

[b]These grades may not be commonly available.

[c]The span for No. 1 grade is 2 × 6, 19.2 o.c., **13-3;** 24 o.c., **11-10;** 2 × 8, 19.2 o.c., **17-5;** 24 o.c., **15-7;** 2 × 10, 19.2 o.c., **22-3;** 24 o.c., **19-11.**

[d]The span for No. 1 KD grade, 24 in. o.c. is: 2 × 6, **12-5;** 2 × 8, **16-5;** 2 × 10, **20-11.**

[e]The span for No. 2 Dense KD grade, 24 in. o.c. is: 2 × 6, **12-3;** 2 × 8, **16-2;** 2 × 10, **20-7.**

TABLE NO. 6 CEILING JOISTS

Drywall ceiling
10 psf live load.
No future sleeping rooms and no attic storage, roof slopes 3 in 12 or less

		Grade[a]											
Size (in.)	Spacing (in. o.c.)	Dense Sel Str KD and No. 1 Dense KD	Dense Sel Str, Sel Str KD, No. 1 Dense, and No. 1 KD	Sel Str, No. 1 and No. 2 Dense KD	No. 2 Dense, No. 2 KD, and No. 2	No. 3 Dense KD[b]	No. 3 Dense[b]	No. 3 KD	No. 3	Con-struction KD	Con-struction	Standard KD	Standard
2 × 4	12.0	13-2	12-11	12-8	12-5	12-2	**12-0**	**11-6**	**11-0**	12-2	11-10	**9-11**	**9-6**
	13.7	12-7	12-4	12-1	11-10	11-7	**11-3**	**10-9**	**10-4**	11-7	11-4	**9-3**	**8-11**
	16.0	11-11	11-9	11-6	11-3	**10-10**	**10-5**	**10-0**	**9-6**	11-0	10-9	**8-7**	**8-3**
	19.2	11-3	11-0	10-10	10-7	**9-11**	**9-6**	**9-1**	**8-8**	**10-4**	**9-11**	**7-10**	**7-6**
	24.0	10-5	10-3	10-0	9-10	**8-10**	**8-6**	**8-2**	**7-9**	**9-3**	**8-10**	**7-0**	**6-9**
2 × 6	12.0	20-8	20-3	19-11	19-6	**18-5**	**17-10**	**17-0**	**16-5**				
	13.7	19-9	19-5	19-0	18-8	**17-2**	**16-8**	**15-11**	**15-5**				
	16.0	18-9	18-5	18-1	17-8	**15-11**	**15-6**	**14-9**	**14-3**				
	19.2	17-8	17-4	17-0	16-8	**14-6**	**14-1**	**13-6**	**13-0**				
	24.0	16-4	16-1	15-9	15-6[c]	**13-0**	**12-8**	**12-0**	**11-8**				

	12.0	27-2	26-9	26-2	25-8	**24-3**	**23-6**	**22-5**	**21-8**				
	13.7	26-0	25-7	25-1	24-7	**22-8**	**22-0**	**21-0**	**20-3**				
2 × 8	16.0	24-8	24-3	23-10	23-4	**21-0**	**20-5**	**19-5**	**18-9**				
	19.2	23-3	22-10	22-5	21-11	**19-2**	**18-7**	**17-9**	**17-2**				
	24.0	21-7	21-2	20-10	20-5[c]	**17-2**	**16-8**	**15-10**	**15-4**				
	12.0	34-8	34-1	33-5	32-9	**30-11**	**30-0**	**28-8**	**27-8**				
	13.7	33-2	32-7	32-0	31-4	**28-11**	**28-1**	**26-9**	**25-11**				
2 × 10	16.0	31-6	31-0	30-5	29-9	**26-9**	**26-0**	**24-10**	**23-11**				
	19.2	29-8	29-2	28-7	28-0	**24-5**	**23-9**	**22-8**	**21-10**				
	24.0	27-6	27-1	26-6	26-0[c]	**21-10**	**21-3**	**20-3**	**19-7**				

Note: Spans shown in lightface type are based on a deflection limitation of $l/240$. Spans shown in boldface type are limited by the recommended extreme fiber stress in bending value of the grade and includes a 5-psf dead load.

[a]Terms and abbreviations: Sel. Str. means select structural; KD means KD15, dried to a moisture content of 15% or less; where KD is not shown the material is dried to a moisture content of 19% or less. Lumber dried to 19% or less will be stamped S-DRY or KD S-DRY.

[b]These grades may not be commonly available.

[c]The span for No. 2 grade, 24 in. o.c. spacing is: 2 × 6, **15-3;** 2 × 8, **20-1;** 2 × 10, **25-7.**

TABLE NO. 12 RAFTERS

Any slope
With drywall ceiling
20 psf live load

		Grade[a]														
Size (in.)	Spacing (in. o.c.)	Dense Sel Str KD	No. 1 Dense KD	Dense Sel Str, and Sel Str KD	No. 1 Dense	Sel Str	No. 1 KD	No. 2 Dense KD	No. 1	No. 2 Dense	No. 2 KD	No. 2	No. 3 Dense KD[b]	No. 3 Dense[b]	No. 3 KD	No. 3
	12.0	16-4	16-4	16-1	16-1	15-9	16-1	15-9	15-6	15-3	14-8	14-1	12-0	11-8	11-2	10-9
	13.7	15-8	15-8	15-5	15-5	15-1	15-3	15-0	14-6	14-3	13-9	13-2	11-3	10-11	10-5	10-1
2 × 6	16.0	14-11	14-11	14-7	14-6	14-4	14-1	13-11	13-5	13-2	12-9	12-3	10-5	10-1	9-8	9-4
	19.2	14-0	13-10	13-9	13-3	13-6	12-10	12-8	12-3	12-0	11-7	11-2	9-6	9-3	8-10	8-6
	24.0	13-0	12-5	12-9[c]	11-10	12-0	11-6	11-4	11.0	10-9	10-5	10-0	8-6	8-3	7-11	7-7
	12.0	21-7	21-7	21-2	21-2	20-10	21-2	20-10	20-5	20-1	19-4	18-7	15-10	15-5	14-8	14-2
	13.7	20-8	20-8	20-3	20-3	19-11	20-1	19-9	19-1	18-9	18-1	17-5	14-10	14-5	13-9	13-3
2 × 8	16.0	19-7	19-7	19-3	19-2	18-11	18-7	18-4	17-8	17-5	16-9	16-1	13-9	13-4	12-9	12-4
	19.2	18-5	18-3	18-2	17-6	17-9	17.0	16-8	16-2	15-10	15-4	14-8	12-7	12-2	11-7	11-3
	24.0	17-2	16-4	16-10[c]	15-8[c]	15-10	15-2	14-11	14-5	14-2	13-8	13-2	11-3	10-11	10-5	10-0

	12.0	27-6	27-6	27-1	27-1	26-6	27-1	26-6	**26-1**	**25-7**	**24-8**	**23-9**	**20-3**	**19-8**	**18-9**	**18-1**
	13.7	26-4	26-4	25-10	25-10	25-5	**25-7**	**25-3**	**24-5**	**24-0**	**23-1**	**22-2**	**18-11**	**18-5**	**17-6**	**16-11**
2 × 10	16.0	25-0	25.0	24-7	**24-5**	24-1	**23-9**	**23-4**	**22-7**	**22-2**	**21-4**	**20-6**	**17-6**	**17-0**	**16-3**	**15-8**
	19.2	23-7	**23-3**	23-2	**22-4**	**22-8**	**21-8**	**21-4**	**20-7**	**20-3**	**19-6**	**18-9**	**16-0**	**15-7**	**14-10**	**14-4**
	24.0	21-10	**20-10**	21-6[c]	**19-11**	**20-3**	**19-4**	**19-1**	**18-5**	**18-1**	**17-5**	**16-9**	**14-4**	**13-11**	**13-3**	**12-10**
	12.0	33-6	33-6	32-11	32.11	32-3	32-11	32-3	**31-8**	**31-2**	**30-0**	**28-10**	**24-8**	**23-11**	**22-10**	**22-0**
	13.7	32-0	32-0	31-6	31-6	30-10	**31-2**	**30-8**	**29-8**	**29-2**	**28-1**	**27-0**	**23-0**	**22-4**	**21-4**	**20-7**
2 × 12	16.0	30-5	30-5	29-11	**29-9**	29-4	**28-10**	**28-5**	**27-5**	**27-0**	**26-0**	**25-0**	**21-4**	**20-9**	**19-9**	**19-1**
	19.2	28-8	**28-4**	28-2	**27-2**	**27-6**	**26-4**	**25-11**	**25-1**	**24-8**	**23-9**	**22-10**	**19-6**	**18-11**	**18-0**	**17-5**
	24.0	26-7	**25-4**	26-1[c]	**24-3**	**24-8**	**23-7**	**23-2**	**22-5**	**22-0**	**21-3**	**20-5**	**17-5**	**16-11**	**16-1**	**15-7**

Note: Spans shown in lightface type are based on a deflection limitation of $l/240$. Spans shown in boldface type are limited by the recommended extreme fiber stress in bending value of the grade and includes a 15-psf dead load.

[a]Terms and abbreviations: Sel. Str. means select structural; KD means KD15, dried to a moisture content of 15% or less; where KD is not shown the material is dried to a moisture content of 19% or less. Lumber dried to 19% or less will be stamped S-DRY or KD S-DRY.

[b]These grades may not be commonly available.

[c]The span for Select Structural KD, 24 inches o.c. spacing is: 2 × 6, **12-5;** 2 × 8, **16-4;** 2 × 10, **20-10;** 2 × 12, **25-4.**

TABLE NO. 13 RAFTERS

Any slope
With drywall ceiling
30 psf live load

Size (in.)	Spacing (in. o.c.)	Grade[a]: Dense Sel Str KD	No. 1 Dense KD	Dense Sel Str and Sel Str KD	No. 1 Dense	Sel Str	No. 1 KD	No. 2 Dense KD	No. 1	No. 2 Dense	No. 2 KD	No. 2	No. 3 Dense KD[b]	No. 3 Dense[b]	No. 3 KD	No. 3
2 × 6	12.0	14-4	14-4	14-1	14-1	13-9	14-1	13-9	13-8	13-5	12-11	12-5	10-7	10-4	9-10	9-6
	13.7	13-8	13-8	13-5	13-5	13-2	13-5	13-2	12-9	12-7	12-1	11-8	9-11	9-8	9-2	8-11
	16.0	13-0	13-0	12-9	12-9	12-6	12-5	12-3	11-10	11-8	11-2	10-9	9-2	8-11	8-6	8-3
	19.2	12-3	12-2	12-0	11-8	11-9	11-4	11-2	10-10	10-7	10-3	9-10	8-5	8-2	7-9	7-6
	24.0	11-4	10-11	11-2[c]	10-6	10-7	10-2	10-0	9-8	9-6	9-2	8-10	7-6	7-3	6-11	6-9
2 × 8	12.0	18-10	18-10	18-6	18-6	18-2	18-6	18-2	18-0	17-8	17-1	16-5	14-0	13-7	12-11	12-6
	13.7	18-0	18-0	17-9	17-9	17-5	17-8	17-5	16-10	16-7	16-0	15-4	13-1	12-9	12-1	11-9
	16.0	17-2	17-2	16-10	16-10	16-6	16-5	16-2	15-7	15-4	14-9	14-2	12-1	11-9	11-3	10-10
	19.2	16-1	16-1	15-10	15-5	15-6	15-0	14-9	14-3	14-0	13-6	12-11	11-1	10-9	10-3	9-11
	24.0	15-0	14-5	14-8[c]	13-10	14-0	13-5	13-2	12-9	12-6	12-1	11-7	9-11	9-7	9-2	8-10

	12.0	24-1	24-1	23-8	23-8	23-2	23-8	23-2	**23-0**	**22-7**	**21-9**	**20-11**	**17-10**	**17-4**	**16-6**	**16-0**
	13.7	23-0	23-0	22-7	22-7	22-2	**22-7**	22-2	**21-6**	**21-2**	**20-4**	**19-7**	**16-8**	**16-3**	**15-6**	**14-11**
2 × 10	16.0	21-10	21-10	21-6	21-6	21-1	**20-11**	**20-7**	**19-11**	**19-7**	**18-10**	**18-1**	**15-6**	**15-0**	**14-4**	**13-10**
	19.2	20-7	**20-6**	20-2	**19-8**	19-10	**19-1**	**18-9**	**18-2**	**17-10**	**17-2**	**16-6**	**14-1**	**13-8**	**13-1**	**12-8**
	24.0	19-1	**18-4**	18-9[c]	**17-7**	**17-10**	**17-1**	**16-10**	**16-3**	**16-0**	**15-5**	**14-9**	**12-8**	**12-3**	**11-8**	**11-4**
	12.0	29-3	29-3	28-9	28-9	28-2	28-9	28-2	**27-11**	**27-6**	**26-6**	**25-5**	**21-9**	**21-1**	**20-1**	**19-5**
	13.7	28-0	28-0	27-6	27-6	27-0	**27-6**	27-0	**26-2**	**25-8**	**24-9**	**23-10**	**20-4**	**19-9**	**18-10**	**18-2**
2 × 12	16.0	26-7	26-7	26-1	26-1	25-7	**25-5**	**25-0**	**24-3**	**23-9**	**22-11**	**22-0**	**18-10**	**18-3**	**17-5**	**16-10**
	19.2	25.0	**25-0**	24-7	**23-11**	24-1	**23-3**	**22-10**	**22-1**	**21-9**	**20-11**	**20-1**	**17-2**	**16-8**	**15-11**	**15-4**
	24.0	23-3	**22-4**	22-10[c]	**21-5**	**21-9**	**20-9**	**20-5**	**19-9**	**19-5**	**18-9**	**18-0**	**15-4**	**14-11**	**14-3**	**13-9**

Note: Spans shown in lightface type are based on a deflection limitation of $l/240$. Spans shown in boldface type are limited by the recommended fiber stress in bending value of the grade and includes a 15-psf dead load.

[a]Terms and abbreviations: Sel. Str. means select structural; KD means KD15, dried to a moisture content of 15% or less; where KD is not shown the material is dried to a moisture content of 19% or less. Lumber dried to 19% or less will be stamped S-DRY or KD S-DRY.

[b]These grades may not be commonly available.

[c]The span for Select Structural KD, 24 in. o.c. spacing is: 2 × 6, **10-11;** 2 × 8, **14-5;** 2 × 10, **18-4;** 2 × 12, **22-4.**

TABLE NO. 14 RAFTERS

Any slope
With drywall ceiling
40 psf live load.

Size (in.)	Spacing (in. o.c.)	Grade[a] Dense Sel Str KD	No. 1 Dense KD	Dense Sel Str and Sel Str KD	No. 1 Dense	Sel Str	No. 1 KD	No. 2 Dense KD	No. 1	No. 2 Dense	No. 2 KD	No. 2	No. 3 Dense KD[b]	No. 3 Dense[b]	No. 3 KD	No. 3
	12.0	13-0	13-0	12-9	12-9	12-6	12-9	12-6	**12-4**	**12-2**	**11-8**	**11-3**	**9-7**	**9-4**	**8-11**	**8-7**
	13.7	12-5	12-5	12-3	12-3	12-0	**12-2**	**12-0**	**11-7**	**11-4**	**10-11**	**10-6**	**9-0**	**8-9**	**8-4**	**8-0**
2 × 6	16.0	11-10	11-10	11-7	**11-7**	11-5	**11-3**	**11-1**	**10-8**	**10-6**	**10-2**	**9-9**	**8-4**	**8-1**	**7-8**	**7-5**
	19.2	11-1	**11-0**	10-11	**10-7**	10-8	**10-3**	**10-1**	**9-9**	**9-7**	**9-3**	**8-11**	**7-7**	**7-4**	**7-0**	**6-9**
	24.0	10-4	**9-10**	10-2[c]	**9-6**	**9-7**	**9-2**	**9-0**	**8-9**	**8-7**	**8-3**	**7-11**	**6-9**	**6-7**	**6-3**	**6-1**
	12.0	17-2	17-2	16-10	16-10	16-6	16-10	16-6	**16-4**	**16-0**	**15-5**	**14-10**	**12-8**	**12-4**	**11-9**	**11-4**
	13.7	16-5	16-5	16-1	16-1	15-9	**16-0**	**15-9**	**15-3**	**15-0**	**14-5**	**13-10**	**11-10**	**11-6**	**11-0**	**10-7**
2 × 8	16.0	15-7	15-7	15-3	**15-3**	15-0	**14-10**	**14-7**	**14-1**	**13-10**	**13-4**	**12-10**	**11-0**	**10-8**	**10-2**	**9-10**
	19.2	14-8	**14-7**	14-5	**13-11**	14-1	**13-6**	**13-4**	**12-11**	**12-8**	**12-2**	**11-9**	**10-0**	**9-9**	**9-3**	**8-11**
	24.0	13-7	**13-0**	13-4[c]	**12-6**	**12-8**	**12-1**	**11-11**	**11-6**	**11-4**	**10-11**	**10-6**	**8-11**	**8-8**	**8-3**	**8-0**

2 × 10	12.0	21-10	21-10	21-6	21-6	21-1	21-6	21-1	**20-10**	**20-5**	**19-8**	**18-11**	**16-2**	**15-8**	**14-11**	**14-5**
	13.7	20-11	20-11	20-6	20-6	20-2	**20-5**	**20-1**	**19-5**	**19-1**	**18-5**	**17-8**	**15-1**	**14-8**	**14-0**	**13-6**
	16.0	19-10	19-10	19-6	**19-6**	19-2	**18-11**	**18-7**	**18-0**	**17-8**	**17-1**	**16-5**	**14-0**	**13-7**	**12-11**	**12-6**
	19.2	18-8	**18-7**	18-4	**17-10**	18-0	**17-3**	**17-0**	**16-5**	**16-2**	**15-7**	**14-11**	**12-9**	**12-5**	**11-10**	**11-5**
	24.0	17-4	**16-7**	**16-7**[c]	**15-11**	**16-2**	**15-5**	**15-2**	**14-8**	**14-5**	**13-11**	**13-5**	**11-5**	**11-1**	**10-7**	**10-3**
2 × 12	12.0	26-7	26-7	26-1	26-1	25-7	26-1	25-7	**25-3**	**24-10**	**23-11**	**23-0**	**19-8**	**19-1**	**18-2**	**17-7**
	13.7	25-5	25-5	25-0	25-0	24-6	**24-10**	24-6	**23-8**	**23-3**	**22-5**	**21-6**	**18-5**	**17-10**	**17-0**	**16-5**
	16.0	24-2	24-2	23-9	**23-9**	23-3	**23-0**	**22-8**	**21-11**	**21-6**	**20-9**	**19-11**	**17-0**	**16-6**	**15-9**	**15-3**
	19.2	22-9	**22-7**	22-4	**21-8**	21-11	**21-0**	**20-8**	**20-0**	**19-8**	**18-11**	**18-2**	**15-6**	**15-1**	**14-5**	**13-11**
	24.0	21-1	**20-2**	20-9[c]	**19-4**	**19-8**	**18-9**	**18-6**	**17-11**	**17-7**	**16-11**	**16-3**	**13-11**	**13-6**	**12-10**	**12-5**

Note: Spans shown in lightface type are based on a deflection limitation of $l/240$. Spans shown in boldface type are limited by the recommended fiber stress in bending value of the grade and includes a 15-psf dead load.

[a]Terms and abbreviations: Sel. Str. means select structural; KD means KD15, dried to a moisture content of 15% or less; where KD is not shown the material is dried to a moisture content of 19% or less. Lumber dried to 19% or less will be stamped S-DRY or KD S-DRY.

[b]These grades may not be commonly available.

[c]The span for Select Structural KD, 24 in. o.c. spacing is: 2 × 6, **9-10;** 2 × 8, **13-0;** 2 × 10, **16-2;** 2 × 12, **20-2.**

TABLE NO. 15 RAFTERS

Low slope (3 in 12 or less)
With no finished ceiling
20 psf live load

Size (in.)	Spacing (in. o.c.)	Grade[a]										
		Dense Sel Str KD and No. 1 Dense KD	Dense Sel Str, Sel Str KD, No. 1 Dense, and No. 1 KD	Sel Str and No. 2 Dense KD	No. 1	No. 2 Dense	No. 2 KD	No. 2	No. 3 Dense KD[b]	No. 3 Dense[b]	No. 3 KD	No. 3
2 × 6	12.0	16-4	16-1	15-9	15-9	15-6	15-6	**15-3**	**13-0**	**12-8**	**12-0**	**11-8**
	13.7	15-8	15-5	15-1	15-1	14-9	14-9	**14-3**	**12-2**	**11-10**	**11-3**	**10-11**
	16.0	14-11	14-7	14-4	14-4	14-1	**13-9**	**13-2**	**11-3**	**10-11**	**10-5**	**10-1**
	19.2	14-0	13-9	13-6	**13-3**	**13-0**	**12-6**	**12-0**	**10-3**	**10-0**	**9-6**	**9-2**
	24.0	13-0	12-9[c]	12-6[d]	**11-10**	**11-8**	**11-2**	**10-9**	**9-2**	**8-11**	**8-6**	**8-3**
2 × 8	12.0	21-7	21-2	20-10	20-10	20-5	20-5	**20-1**	**17-2**	**16-8**	**15-10**	**15-4**
	13.7	20-8	20-3	19-11	19-11	19-6	19-6	**18-9**	**16-0**	**15-7**	**14-10**	**14-4**
	16.0	19-7	19-3	18-11	18-11	18-6	**18-1**	**17-5**	**14-10**	**14-5**	**13-9**	**13-3**
	19.2	18-5	18-2	17-9	**17-5**	**17-2**	**16-6**	**15-10**	**13-7**	**13-2**	**12-7**	**12-1**
	24.0	17-2	16-10[c]	16-6[d]	**15-7**	**15-4**	**14-9**	**14-2**	**12-1**	**11-9**	**11-3**	**10-10**

	12.0	27-6	27-1	26-6	26-6	26-0	26-0	**25-7**	**21-10**	**21-3**	**20-3**	**19-7**
	13.7	26-4	25-10	25-5	25-5	24-11	24-11	**24-0**	**20-6**	**19-10**	**18-11**	**18-4**
2 × 10	16.0	25-0	24-7	24-1	24-1	23-8	**23-1**	**22-2**	**18-11**	**18-5**	**17-6**	**16-11**
	19.2	23-7	23-2	22-8	**22-3**	**21-10**	**21-1**	**20-3**	**17-3**	**16-9**	**16-0**	**15-6**
	24.0	21-10	21-6[c]	21-1[d]	**19-11**	**19-7**	**18-10**	**18-1**	**15-6**	**15-0**	**14-4**	**13-10**
	12.0	33-6	32-11	32-3	32-3	31-8	31-8	**31-2**	**26-7**	**25-10**	**24-8**	**23-9**
	13.7	32.0	31-6	30-10	30-10	30-3	30-3	**29-2**	**24-11**	**24-2**	**23-0**	**22-3**
2 × 12	16.0	30-5	29-11	29-4	29-4	28-9	**28-1**	**27-0**	**23-0**	**22-4**	**21-4**	**20-7**
	19.2	28-8	28-2	27-7	**27-1**	**26-7**	**25-8**	**24-8**	**21-0**	**20-5**	**19-6**	**18-10**
	24.0	26-7	26-1[c]	25-7[d]	**24-3**	**23-9**	**22-11**	**22-0**	**18-10**	**18-3**	**17-5**	**16-10**

Note: Spans shown in lightface type are based on a deflection limitation of $l/240$. Spans shown in boldface type are limited by the recommended extreme fiber stress in bending value of the grade and includes a 10-psf dead load.

[a]Terms and abbreviations: Sel. Str. means select structural; KD means KD15, dried to a moisture content of 15% or less; where KD is not shown the material is dried to a moisture content of 19% or less. Lumber dried to 19% or less will be stamped S-DRY or KD S-DRY.

[b]These grades may not be commonly available.

[c]The span for No. 1 KD, 24 in. o.c. is: 2 × 6, **12-5;** 2 × 8, **16-5;** 2 × 10, **20-11;** 2 × 12, **25-5.**

[d]The span for No. 2 Dense KD, 24 in. o.c. is: 2 × 6, **12-3;** 2 × 8, **16-2;** 2 × 10, **20-7;** 2 × 12, **25-0.**

TABLE NO. 16 RAFTERS

Low slope (3 in 12 or less)
With no finished ceiling
30 psf live load

Size (in.)	Spacing (in. o.c.)	Grade[a]: Dense Sel Str KD and No. 1 Dense KD	Dense Sel Str and Sel Str KD	No. 1 Dense and No. 1 KD	Sel Str	No. 2 Dense KD	No. 1	No. 2 Dense	No. 2 KD	No. 2	No. 3 Dense KD[b]	No. 3 Dense[b]	No. 3 KD	No. 3
2 × 6	12.0	14-4	14-1	14-1	13-9	13-9	13-9	13-6	13-6	13-2	11-3	10-11	10-5	10-1
	13.7	13-8	13-5	13-5	13-2	13-2	13-2	12-11	12-10	12-4	10-6	10-3	9-9	9-5
	16.0	13-0	12-9	12-9	12-6	12-6	12-6	12-3	11-11	11-5	9-9	9-6	9-0	8-9
	19.2	12-3	12-0	12-0	11-9	11-9	11-6	11-3	10-10	10-5	8-11	8-8	8-3	8-0
	24.0	11-4	11-2	11-1[c]	10-11	10-7	10-3	10-1	9-8	9-4	8-0	7-9	7-4	7-1
2 × 8	12.0	18-10	18-6	18-6	18-2	18-2	18-2	17-10	17-10	17-5	14-10	14-5	13-9	13-3
	13.7	18-0	17-9	17-9	17-5	17-5	17-5	17-0	16-11	16-3	13-11	13-6	12-10	12-5
	16.0	17-2	16-10	16-10	16-6	16-6	16-6	16-2	15-8	15-1	12-10	12-6	11-11	11-6
	19.2	16-1	15-10	15-10	15-6	15-6	15-1	14-10	14-4	13-9	11-9	11-5	10-10	10-6
	24.0	15-0	14-8	14-8[c]	14-5	14-0	13-6	13-3	12-10	12-4	10-6	10-2	9-9	9-5

	12.0	24-1	23-8	23-8	23-2	23-2	23-2	22-9	22-9	**22-2**	**18-11**	**18-5**	**17-6**	**16-11**
	13.7	23-0	22-7	22-7	22-2	22-2	22-2	21-9	**21-7**	**20-9**	**17-9**	**17-2**	**16-5**	**15-10**
2 × 10	16.0	21-10	21-6	21-6	21-1	21-1	21-1	20-8	**20-0**	**19-3**	**16-5**	**15-11**	**15-2**	**14-8**
	19.2	20-7	20-2	20-2	19-10	19-10	**19-3**	**18-11**	**18-3**	**17-6**	**15-0**	**14-6**	**13-10**	**13-5**
	24.0	19-1	18-9	**18-8**[c]	18-5	**17-10**	**17-3**	**16-11**	**16-4**	**15-8**	**13-5**	**13-0**	**12-5**	**12-0**
	12.0	29-3	28-9	28-9	28-2	28-2	28-2	27-8	27-8	**27-0**	**23-0**	**22-4**	**21-4**	**20-7**
	13.7	28-0	27-6	27-6	27-0	27-0	27-0	26-5	**26-3**	**25-3**	**21-7**	**20-11**	**19-11**	**19-3**
2 × 12	16.0	26-7	26-1	26-1	25-7	25-7	25-7	25-1	**24-4**	**23-4**	**19-11**	**19-4**	**18-6**	**17-10**
	19.2	25-0	24-7	24-7	24-1	24-1	**23-5**	**23-0**	**22-2**	**21-4**	**18-3**	**17-8**	**16-10**	**16-3**
	24.0	23-3	22-10	**22-8**[c]	22-5	**21-8**	**21-0**	**20-7**	**19-10**	**19-1**	**16-3**	**15-10**	**15-1**	**14-7**

Note: Spans shown in lightface type are based on a deflection limitation of $l/240$. Spans shown in boldface type are limited by the recommended extreme fiber stress in bending value of the grade and includes a 10-psf dead load.

[a]Terms and abbreviations: Sel. Str. means select structural; KD means KD15, dried to a moisture content of 15% or less; where KD is not shown the material is dried to a moisture content of 19% or less. Lumber dried to 19% or less will be stamped S-DRY or KD S-DRY.

[b]These grades may not be commonly available.

[c]The span for No. 1 KD, 24 inches o.c. spacing is: 2 × 6, **10-9;** 2 × 8, **14-2;** 2 × 10, **18-1;** 2 × 12, **22-0.**

TABLE NO. 18 RAFTERS

High slope (over 3 in 12)
With no finished ceiling
20 psf live load + 15 psf dead load—heavy roofing

Size (in.)	Spacing (in. o.c.)	Grade[a] Dense Sel Str KD	Dense Sel Str	No. 1 Dense KD and Sel Str KD	Sel Str	No. 1 Dense	No. 1 KD	No. 2 Dense KD	No. 1	No De
	12.0	11-6	11-3	11-6[c]	11-1	11-3	**11-2**	**11-0**	**10-8**	**10-**
	13.7	11-0	10-9	11-0[c]	10-7	10-9	**10-5**	**10-3**	**10-0**	**9-**
2 × 4	16.0	10-5	10-3	**10-5**[c]	**10-0**	**10-0**	**9-8**	**9-6**	**9-3**	**9-**
	19.2	9-10	9-8	**9-6**	**9-2**	**9-2**	**8-10**	**8-8**	**8-5**	**8-**
	24.0	9-1	**8-11**	**8-6**	**8-2**	**8-2**	**7-11**	**7-9**	**7-7**	**7-**
	12.0	18-0	17-8	**17-6**	**17-0**	**16-9**	**16-3**	**16-0**	**15-6**	**15-**
	13.7	17-3	16-11	**16-5**	**15-11**	**15-8**	**15-3**	**15-0**	**14-6**	**14-**
2 × 6	16.0	16-4	**16-0**	**15-2**	**14-9**	**14-6**	**14-1**	**13-11**	**13-5**	**13-**
	19.2	**15-1**	**14-7**	**13-10**	**13-6**	**13-3**	**12-10**	**12-8**	**12-3**	**12-**
	24.0	**13-6**	**13-0**	**12-5**	**12-0**	**11-10**	**11-6**	**11-4**	**11-0**	**10-**
	12.0	23-9	23-4	**23-1**	**22-5**	**22-1**	**21-6**	**21-1**	**20-5**	**20-**
	13.7	22-9	22-4	**21-7**	**21-0**	**20-8**	**20-1**	**19-9**	**19-1**	**18-**
2 × 8	16.0	21-7	**21-0**	**20-0**	**19-5**	**19-2**	**18-7**	**18-4**	**17-8**	**17-**
	19.2	**19-11**	**19-2**	**18-3**	**17-9**	**17-6**	**17-0**	**16-8**	**16-2**	**15-**
	24.0	**17-10**	**17-2**	**16-4**	**15-10**	**15-8**	**15-2**	**14-11**	**14-5**	**14-**
	12.0	30-4	29-9	**29-5**	**28-8**	**28-3**	**27-5**	**26-11**	**26-1**	**25-**
	13.7	29-0	28-6	**27-7**	**26-9**	**26-5**	**25-7**	**25-3**	**24-5**	**24-**
2 × 10	16.0	27-6	**26-10**	**25-6**	**24-10**	**24-5**	**23-9**	**23-4**	**22-7**	**22**
	19.2	**25-5**	**24-6**	**23-3**	**22-8**	**22-4**	**21-8**	**21-4**	**20-7**	**20-**
	24.0	**22-8**	**21-11**	**20-10**	**20-3**	**19-11**	**19-4**	**19-1**	**18-5**	**18-**

Note: Spans shown in lightface type are based on a deflection limitation of *l*/180. Spans shown in boldface are limited by the recommended extreme fiber stress in bending value of the grade and includes a 15-psf dead l

[a]Terms and abbreviations: Sel. Str. means select structural; KD means KD15, dried to moisture content of or less; where KD is not shown the material is dried to a moisture content of 19% or less. Lumber dried to or less will be stamped S-DRY or KD S-DRY.

No. 2 KD	No. 2	No. 3 Dense KD[b]	No. 3 Dense[b]	No. 3 KD	No. 3	Con-struc-tion KD	Con-struc-tion	Stan-dard KD	Stan-dard
10-2	**9-8**	**8-2**	**7-11**	**7-7**	**7-3**	**8-7**	**8-2**	**6-6**	**6-3**
9-6	**9-1**	**7-8**	**7-4**	**7-1**	**6-9**	**8-0**	**7-8**	**6-1**	**5-10**
8-10	**8-5**	**7-1**	**6-10**	**6-6**	**6-3**	**7-5**	**7-1**	**5-7**	**5-5**
8-1	**7-8**	**6-6**	**6-3**	**6-0**	**5-8**	**6-9**	**6-6**	**5-1**	**4-11**
7-3	**6-10**	**5-9**	**5-7**	**5-4**	**5-1**	**6-1**	**5-9**	**4-7**	**4-5**
14-8	**14-1**	**12-0**	**11-8**	**11-2**	**10-9**				
13-9	**13-2**	**11-3**	**10-11**	**10-5**	**10-1**				
12-9	**12-3**	**10-5**	**10-1**	**9-8**	**9-4**				
11-7	**11-2**	**9-6**	**9-3**	**8-10**	**8-6**				
10-5	**10-0**	**8-6**	**8-3**	**7-11**	**7-7**				
19-4	**18-7**	**15-10**	**15-5**	**14-8**	**14-2**				
18-1	**17-5**	**14-10**	**14-5**	**13-9**	**13-3**				
16-9	**16-1**	**13-9**	**13-4**	**12-9**	**12-4**				
15-4	**14-8**	**12-7**	**12-2**	**11-7**	**11-3**				
13-8	**13-2**	**11-3**	**10-11**	**10-5**	**10-0**				
24-8	**23-9**	**20-3**	**19-8**	**18-9**	**18-1**				
23-1	**22-2**	**18-11**	**18-5**	**17-6**	**16-11**				
21-4	**20-6**	**17-6**	**17-0**	**16-3**	**15-8**				
19-6	**18-9**	**16-0**	**15-7**	**14-10**	**14-4**				
17-5	**16-9**	**14-4**	**13-11**	**13-3**	**12-10**				

[b]These grades may not be commonly available.

[c]The span for Select Structural KD, 2 × 4, 12 inches o.c. is 11-3; 13.7 inches o.c., 10-9, and 16 inches o.c., 10-3.

TABLE NO. 19 RAFTERS

High slope (over 3 in 12)
With no finished ceiling.
30 psf live load + 15 psf dead load—heavy roofing

		Grade[a]								
Size (in.)	Spacing (in. o.c.)	Dense Sel Str KD	Dense Sel Str	No. 1 Dense KD and Sel Str KD	Sel Str	No. 1 Dense	No. 1 KD	No. 2 Dense KD	No. 1	No. Der
	12.0	10-0	9-10	10-0[c]	9-8	9-10	**9-10**	9-8	**9-5**	9-
	13.7	9-7	9-5	9-7[c]	9-3	9-5	**9-2**	**9-1**	**8-10**	8-
2 × 4	16.0	9-1	8-11	9-1[c]	8-9	**8-10**	**8-6**	**8-5**	**8-2**	8-
	19.2	8-7	8-5	**8-4**	**8-1**	**8-1**	**7-9**	**7-8**	**7-5**	7-
	24.0	7-11	7-10	**7-6**	**7-3**	**7-3**	**6-11**	**6-10**	**6-8**	6-
	12.0	15-9	15-6	**15-5**	**15-0**	**14-10**	**14-4**	**14-2**	**13-8**	13-
	13.7	15-1	14-9	**14-5**	**14-1**	**13-10**	**13-5**	**13-3**	**12-9**	12-
2 × 6	16.0	14-4	14-1	**13-4**	**13-0**	**12-10**	**12-5**	**12-3**	**11-10**	11-
	19.2	**13-4**	**12-10**	**12-2**	**11-10**	**11-8**	**11-4**	**11-2**	**10-10**	10-
	24.0	**11-11**	**11-6**	**10-11**	**10-7**	**10-6**	**10-2**	**10-0**	**9-8**	9-
	12.0	20-9	20-5	**20-4**	**19-10**	**19-6**	**18-11**	**18-8**	**18-0**	17-
	13.7	19-10	19-6	**19-0**	**18-6**	**18-3**	**17-8**	**17-5**	**16-10**	16-
2 × 8	16.0	18-10	18-6	**17-7**	**17-2**	**16-11**	**16-5**	**16-2**	**15-7**	15-
	19.2	**17-7**	**16-11**	**16-1**	**15-8**	**15-5**	**15-0**	**14-9**	**14-3**	14-
	24.0	**15-8**	**15-2**	**14-5**	**14-0**	**13-10**	**13-5**	**13-2**	**12-9**	12-
	12.0	26-6	26-0	**26-0**	**25-3**	**24-11**	**24-2**	**23-9**	**23-0**	22-
	13.7	25-4	24-11	**24-4**	**23-8**	**23-3**	**22-7**	**22-3**	**21-6**	21-
2 × 10	16.0	24-1	23-8	**22-6**	**21-10**	**21-7**	**20-11**	**20-7**	**19-11**	19-
	19.2	**22-5**	**21-7**	**20-6**	**20-0**	**19-8**	**19-1**	**18-9**	**18-2**	17-
	24.0	**20-0**	**19-4**	**18-4**	**17-10**	**17-7**	**17-1**	**16-10**	**16-3**	16-

Note: Spans shown in lightface type are based on a deflection limitation of $l/180$. Spans shown in boldface are limited by the recommended extreme fiber stress in bending value of the grade and includes a 15-psf dead l

[a]Terms and abbreviations: Sel. Str. means select structural; KD means KD15, dried to a moisture content of or less; where KD is not shown the material is dried to a moisture content of 19% or less. Lumber dried to or less will be stamped S-DRY or KD S-DRY.

No. 2 KD	No. 2	No. 3 Dense KD[b]	No. 3 Dense[b]	No. 3 KD	No. 3	Construction KD	Construction	Standard KD	Standard
9-0	8-7	7-3	6-11	6-8	6-4	7-7	7-3	5-9	5-6
8-5	8-0	6-9	6-6	6-3	5-11	7-1	6-9	5-4	5-1
7-9	7-5	6-3	6-0	5-9	5-6	6-7	6-3	4-11	4-9
7-1	6-9	5-9	5-6	5-3	5-0	6-0	5-9	4-6	4-4
6-4	6-1	5-1	4-11	4-9	4-6	5-4	5-1	4-0	3-10
12-11	12-5	10-7	10-4	9-10	9-6				
12-1	11-8	9-11	9-8	9-2	8-11				
11-2	10-9	9-2	8-11	8-6	8-3				
10-3	9-10	8-5	8-2	7-9	7-6				
9-2	8-10	7-6	7-3	6-11	6-9				
17-1	16-5	14-0	13-7	12-11	12-6				
16-0	15-4	13-1	12-9	12-1	11-9				
14-9	14-2	12-1	11-9	11-3	10-10				
13-6	12-11	11-1	10-9	10-3	9-11				
12-1	11-7	9-11	9-7	9-2	8-10				
21-9	20-11	17-10	17-4	16-6	16-0				
20-4	19-7	16-8	16-3	15-6	14-11				
18-10	18-1	15-6	15-0	14-4	13-10				
17-2	16-6	14-1	13-8	13-1	12-8				
15-5	14-9	12-8	12-3	11-8	11-4				

[b]These grades may not be commonly available.

[c]The span for Select Structural KD, 2 × 4, 12 in. o.c. is 9-10; 13-7 in. o.c., 9-5, and 16 in. o.c., 8-11.

TABLE NO. 21 RAFTERS

High slope (over 3 in 12)
With no finished ceiling
20 psf live load + 7 psf dead load—light roofing

Size (in.)	Spacing (in. o.c.)	Grade[a] Dense Sel Str KD	Sel Str KD and Dense Sel Str	No. 1 Dense KD	Sel Str	No. 1 Dense	No. 1 KD	No. 2 Dense KD	No. 1	No Den
2 × 4	12.0	11-6	11-3	11-6	11-1	11-3	11-3	11-1	11-1	10-
	13.7	11-0	10-9	11-0	10-7	10-9	10-9	10-7	10-7	10-
	16.0	10-5	10-3	10-5	10-0	10-3	10-3	10-0	10-0	9-
	19.2	9-10	9-8	9-10	9-5	9-8	9-8	9-5	9-5	9-
	24.0	9-1	8-11	9-1	8-9	8-11	8-11	8-9	**8-7**	**8-**
2 × 6	12.0	18-0	17-8	18-0	17-4	17-8	17-8	17-4	17-4	17-
	13.7	17-3	16-11	17-3	16-7	16-11	16-11	16-7	**16-6**	**16-**
	16.0	16-4	16-1	16-4	15-9	16-1	**16-1**	15-9	**15-3**	**15-**
	19.2	15-5	15-2	15-5	14-10	**15-1**	**14-8**	**14-5**	**13-11**	**13-**
	24.0	14-4	14-1	**14-1**	**13-8**	**13-6**	**13-1**	**12-11**	**12-6**	**12-**
2 × 8	12.0	23-9	23-4	23-9	22-11	23-4	23-4	22-11	22-11	22-
	13.7	22-9	22-4	22-9	21-11	22-4	22-4	21-11	**21-9**	**21-**
	16.0	21-7	21-2	21-7	20-10	21-2	**21-2**	20-10	**20-2**	**19-**
	19.2	20-4	19-11	20-4	19-7	**19-11**	**19-4**	**19-0**	**18-5**	**18-**
	24.0	18-10	18-6	**18-7**	**18-1**	**17-10**	**17-3**	**17-0**	**16-5**	**16-**
2 × 10	12.0	30-4	29-9	30-4	29-2	29-9	29-9	29-2	29-2	28-
	13.7	29.0	28.6	29.0	27-11	28-6	28-6	27-11	**27-9**	**27-**
	16.0	27-6	27-1	27-6	26-6	27-1	**27-0**	26-6	**25-8**	**25-**
	19.2	25-11	25-5	25-11	25-0	**25-5**	**24-8**	**24-3**	**23-6**	**23-**
	24.0	24-1	23-8	**23-8**	**23-1**	**22-9**	**22-1**	**21-8**	**21-0**	**20-**

Note: Spans shown in lightface type are based on a deflection limitation of $l/180$. Spans shown in boldface are limited by the recommended extreme fiber stress in bending value of the grade and includes a 7-psf dead l

[a]Terms and abbreviations: Sel. Str. means select structural; KD means KD15, dried to a moisture content of

No. 2 KD	No. 2	No. 3 Dense KD[b]	No. 3 Dense[b]	No. 3 KD	No. 3	Con-struc-tion KD	Con-struc-tion	Stan-dard KD	Stan-dard
10-10	10-10	9-4	9-0	8-7	8-3	9-9	9-4	7-4	7-1
10-4	10-4	8-9	8-5	8-1	7-8	9-2	8-9	6-11	6-7
9-10	9-7	8-1	7-9	7-5	7-1	8-6	8-1	6-5	6-1
9-2	8-9	7-4	7-1	6-10	6-6	7-9	7-4	5-10	5-7
8-3	7-10	6-7	6-4	6-1	5-10	6-11	6-7	5-3	5-0
16-8	16-1	13-8	13-4	12-8	12-3				
15-8	15-0	12-10	12-5	11-10	11-6				
14-6	13-11	11-10	11-6	11-0	10-7				
13-3	12-8	10-10	10-6	10-0	9-8				
11-10	11-4	9-8	9-5	9-0	8-8				
22-0	21-2	18-1	17-7	16-9	16-2				
20-7	19-10	16-11	16-5	15-8	15-1				
19-1	18-4	15-8	15-2	14-6	14-0				
17-5	16-9	14-3	13-10	13-3	12-9				
15-7	15-0	12-9	12-5	11-10	11-5				
28-1	27-0	23-1	22-5	21-4	20-7				
26-3	25-3	21-7	20-11	20-0	19-3				
24-4	23-5	20-0	19-5	18-6	17-10				
22-3	21-4	18-3	17-8	16-10	16-4				
19-10	19-1	16-4	15-10	15-1	14-7				

or less; where KD is not shown the material is dried to a moisture content of 19% or less. Lumber dried to 19% or less will be stamped S-DRY or KD S-DRY.

[b]These grades may not be commonly available.

TABLE NO. 22 RAFTERS

High slope (over 3 in 12)
With no finished ceiling
30 psf live load + 7 psf dead load—light roofing

		Grade[a]								
Size (in.)	Spacing (in. o.c.)	Dense Sel Str KD	Sel Str KD and Dense Sel Str	No. 1 Dense KD	Sel Str	No. 1 Dense	No. 1 KD	No. 2 Dense KD	No. 1	No. Den
2 × 4	12.0	10-0	9-10	10-0	9-8	9-10	9-10	9-8	9-8	9-6
	13.7	9-7	9-5	9-7	9-3	9-5	9-5	9-3	9-3	9-1
	16.0	9-1	8-11	9-1	8-9	8-11	8-11	8-9	8-9	8-7
	19.2	8-7	8-5	8-7	8-3	8-5	8-5	8-3	**8-3**	8-1
	24.0	7-11	7-10	7-11	7-8	7-10	**7-8**	**7-7**	**7-4**	**7-3**
2 × 6	12.0	15-9	15-6	15-9	15-2	15-6	15-6	15-2	**15-1**	**14-1**
	13.7	15-1	14-9	15-1	14-6	14-9	14-9	14-6	**14-1**	**13-1**
	16.0	14-4	14-1	14-4	13-9	14-1	**13-9**	**13-6**	**13-1**	**12-1**
	19.2	13-6	13-3	**13-6**	13-0	**12-11**	**12-6**	**12-4**	**11-11**	**11-9**
	24.0	12-6	12-3[c]	**12-0**	**11-9**	**11-6**	**11-2**	**11-0**	**10-8**	**10-6**
2 × 8	12.0	20-9	20-5	20-9	20-0	20-5	20-5	20-0	**19-10**	**19-6**
	13.7	19-10	19-6	19-10	19-2	19-6	19-6	19-2	**18-7**	**18-3**
	16.0	18-10	18-6	18-10	18-2	18-6	**18-1**	**17-9**	**17-2**	**16-1**
	19.2	17-9	17-5	**17-9**	17-1	**17-0**	**16-6**	**16-3**	**15-9**	**15-5**
	24.0	16.6	16-2[c]	**15-10**	**15-5**	**15-3**	**14-9**	**14-6**	**14-1**	**13-1**
2 × 10	12.0	26-6	26-0	26-6	25-6	26-0	26-0	25-6	**25-4**	**24-1**
	13.7	25-4	24-11	25-4	24-5	24-11	24-11	24-5	**23-9**	**23-4**
	16.0	24-1	23-8	24-1	23-2	23-8	**23-1**	**22-8**	**21-11**	**21-7**
	19.2	22-8	22-3	**22-8**	21-10	**21-8**	**21-1**	**20-9**	**20-1**	**19-**
	24.0	21-0	20-8[c]	**20-3**	**19-8**	**19-5**	**18-10**	**18-6**	**17-11**	**17-7**

Note: Spans shown in lightface type are based on a deflection limitation of $l/180$. Spans shown in boldface t are limited by the recommended extreme fiber stress in bending value of the grade and includes a 7-psf dead lo

[a]Terms and abbreviations: Sel. Str. means select structural; KD means KD15, dried to a moisture content of 1 or less; where KD is not shown the material is dried to a moisture content of 19% or less. Lumber dried to 1 or less will be stamped S-DRY or KD S-DRY.

No. 2 KD	No. 2	No. 3 Dense KD[b]	No. 3 Dense[b]	No. 3 KD	No. 3	Construction KD	Construction	Standard KD	Standard
9-6	**9-5**	**8-0**	**7-8**	**7-4**	**7-0**	**8-4**	**8-0**	**6-4**	**6-0**
9-1	**8-10**	**7-5**	**7-2**	**6-10**	**6-7**	**7-10**	**7-5**	**5-11**	**5-8**
8-7	**8-2**	**6-11**	**6-8**	**6-4**	**6-1**	**7-3**	**6-11**	**5-5**	**5-3**
7-10	**7-5**	**6-4**	**6-1**	**5-10**	**5-7**	**6-7**	**6-4**	**5-0**	**4-9**
7-0	**6-8**	**5-8**	**5-5**	**5-2**	**5-0**	**5-11**	**5-8**	**4-5**	**4-3**
14-3	**13-9**	**11-9**	**11-4**	**10-10**	**10-6**				
13-4	**12-10**	**10-11**	**10-8**	**10-2**	**9-10**				
12-4	**11-11**	**10-2**	**9-10**	**9-5**	**9-1**				
11-3	**10-10**	**9-3**	**9-0**	**8-7**	**8-3**				
10-1	**9-8**	**8-3**	**8-0**	**7-8**	**7-5**				
18-10	**18-1**	**15-5**	**15-0**	**14-3**	**13-10**				
17-7	**16-11**	**14-5**	**14-0**	**13-4**	**12-11**				
16-4	**15-8**	**13-4**	**13-0**	**12-5**	**11-11**				
14-10	**14-3**	**12-2**	**11-10**	**11-4**	**10-11**				
13-4	**12-9**	**10-11**	**10-7**	**10-1**	**9-9**				
24-0	**23-1**	**19-8**	**19-1**	**18-3**	**17-7**				
22-5	**21-7**	**18-5**	**17-11**	**17-1**	**16-6**				
20-9	**20-0**	**17-1**	**16-7**	**15-9**	**15-3**				
19-0	**18-3**	**15-7**	**15-1**	**14-5**	**13-11**				
17-0	**16-4**	**13-11**	**13-6**	**12-11**	**12-5**				

[b]These grades may not be commonly available.

[c]The span for Select Structural KD, 24 inches o.c. is: 2 × 6, **12-0;** 2 × 8, **15-10;** 2 × 10, **20-3.**

Appendix D

Plank and Beam Framing for Residential Buildings

(Courtesy of National Forest Products Association, Washington, DC)

Tables 1, 4, 7, and 10 are representative of all tables of this kind. For a complete set of tables, contact the National Forest Products Association.

DESIGN DATA FOR PLANKS

Design data for plank floors and roofs are included in **Table 1.** Computations for bending are based on the live load indicated, plus 10 psf of dead load. Computations for deflection are based on the live load only. The table shows four general arrangements of planks as follows:

Type A: extending over a single span
Type B: continuous over two equal spans
Type C: continuous over three equal spans
Type D: a combination of types A and B

On the basis of a section of planking 12 in. wide, the following formulas were used in making the computations:

For type A:

$$M = \frac{wL^2}{8} \quad \text{and} \quad D = \frac{5wL^4(12)^3}{384EI}$$

For type B:

$$M = \frac{wL^2}{8} \quad \text{and} \quad D = \frac{wL^4(12)^3}{185EI}$$

For type C:

$$M = \frac{wL^2}{10} \quad \text{and} \quad D = \frac{4wL^4(12)^3}{581EI}$$

For type D:

$$M = \frac{wL^2}{8} \quad \text{and} \quad D = \frac{1}{2}\left[\frac{5wL^4(12)^3}{384EI} + \frac{wL^4(12)^3}{185EI}\right]$$

To use Table 1, first determine the plank arrangement (type A, B, C, or D the span, the live load to be supported, and the deflection limitation. Then sele from the table the corresponding required values for fiber stress in bending (*f*) a modulus of elasticity (*E*). The plank to be used should be of a grade and speci that meets these minimum values. The maximum span for a specific grade and sp cies of plank may be determined by reversing these steps.

For those who prefer to use random-length planks (instead of arrangeme type A, B, C, or D), similar technical information concerning it is included in *Ra dom Length Wood Decking,* a publication of the National Forest Products Ass ciation.

DESIGN DATA FOR BEAMS

Design data for beams are included in **Tables 2 through 10**. Computations for ben ing are based on the live load indicated plus 10 psf of dead load. Computations f deflection are based on the live load only. All beams in the table were designed extend over a single span and the following formulas were used:

For type A:

$$M = \frac{wL^2}{8} \quad \text{and} \quad D = \frac{5wL^4(12)^3}{384EI}$$

To use the tables, first determine the span, the live load to be supported, a the deflection limitation. Then select from the tables the proper size of beam wi the corresponding required values for fiber stress in bending (*f*) and modulus elasticity (*E*). The beam used should be of a grade and species that meets the minimum values. The maximum span for a beam of specific size, grade, and speci can be determined by reversing these steps.

NOTATIONS

In the preceding formulas and in the tables the symbols have the following meanings:

w = load in pounds per linear foot
L = span in feet
M = induced bending moment in pound-feet
f = fiber stress in bending in pounds per square inch
E = modulus of elasticity in pounds per square inch
I = moment of inertia in inches to the fourth power
D = deflection in inches

LUMBER SIZES

Tabular data provided herein are based on net dimensions (S4S) as listed in American Softwood Lumber Standard, VPS 20-70.

TABLE 1 NOMINAL 2-INCH PLANK

Live load of 20, 30, or 40 psf
Deflection limitation of $l/240$, $l/300$, or $l/360$

Plank span (ft)	Live load (psf)	Deflection limitation	Type A		Type B		Type C		Type D	
			f (psi)	*E* (psi)	*f* (psi)	*E* (psi)	*f* (psi)	*E* (psi)	*f* (psi)	*E* (psi)
6	20	$\frac{l}{240}$	360	576,000	360	239,000	288	305,000	360	408,000
		$\frac{l}{300}$	360	720,000	360	299,000	288	381,000	360	509,000
		$\frac{l}{360}$	360	864,000	360	359,000	288	457,000	360	611,000
	30	$\frac{l}{240}$	480	864,000	480	359,000	384	457,000	480	611,000
		$\frac{l}{300}$	480	1,080,000	480	448,000	384	571,000	480	764,000
		$\frac{l}{360}$	480	1,296,000	480	538,000	384	685,000	480	917,000
	40	$\frac{l}{240}$	600	1,152,000	600	478,000	480	609,000	600	815,000
		$\frac{l}{300}$	600	1,440,000	600	598,000	480	762,000	600	1,019,000
		$\frac{l}{360}$	600	1,728,000	600	717,000	480	914,000	600	1,223,000
	20	$\frac{l}{240}$	490	915,000	490	380,000	392	484,000	490	647,000
		$\frac{l}{300}$	490	1,143,000	490	475,000	392	605,000	490	809,000
		$\frac{l}{360}$	490	1,372,000	490	570,000	392	726,000	490	971,000

7	30	$\frac{l}{240}$	653	1,372,000	653	570,000	522	726,000	653	971,000
		$\frac{l}{300}$	653	1,715,000	653	712,000	522	907,000	653	1,213,000
		$\frac{l}{360}$	653	2,058,000	653	854,000	522	1,088,000	653	1,456,000
	40	$\frac{l}{240}$	817	1,829,000	817	759,000	653	968,000	817	1,294,000
		$\frac{l}{300}$	817	2,287,000	817	949,000	653	1,209,000	817	1,618,000
		$\frac{l}{360}$	817	2,744,000	817	1,139,000	653	1,451,000	817	1,941,000
8	20	$\frac{l}{240}$	640	1,365,000	640	567,000	512	722,000	640	966,000
		$\frac{l}{300}$	640	1,707,000	640	708,000	512	903,000	640	1,208,000
		$\frac{l}{360}$	640	2,048,000	640	850,000	512	1,083,000	640	1,449,000
	30	$\frac{l}{240}$	853	2,048,000	853	850,000	682	1,083,000	853	1,449,000
		$\frac{l}{300}$	853	2,560,000	853	1,063,000	682	1,354,000	853	1,811,000
		$\frac{l}{360}$	853	3,072,000	853	1,275,000	682	1,625,000	853	2,174,000
	40	$\frac{l}{240}$	1067	2,731,000	1067	1,134,000	853	1,444,000	1067	1,932,000
		$\frac{l}{300}$	1067	3,413,000	1067	1,417,000	853	1,805,000	1067	2,415,000
		$\frac{l}{360}$	1067	4,096,000	1067	1,700,000	853	2,166,000	1067	2,898,000

TABLE 4 FLOOR AND ROOF BEAMS

Live load of 20 psf
Deflection limitation of $l/360$

Beam span (ft)	Nominal size of beam	Minimum f and E (psi) for beams spaced: 6′-0″		7′-0″		8′-0″	
		f	E	f	E	f	E
10	2-3x6	1070	1170000	1250	1365000	1430	1560000
	1-3x8	1235	1020000	1440	1192000	1645	1359000
	2-2x8	1030	855000	1200	997000	1370	1140000
	1-4x8	880	727000	1030	847000	1175	969000
	3-2x8	685	570000	800	667000	915	759000
	2-3x8	615	510000	720	600000	820	679000
	1-6x8	525	419000	615	489000	700	558000
11	2-3x6	1295	1555000	1510	1815000	1730	2073000
	1-3x8	1490	1357000	1740	1584000	1990	1809000
	2-2x8	1245	1131000	1450	1320000	1660	1507000
	1-4x8	1065	970000	1245	1132000	1420	1293000
	3-2x8	830	754000	970	880000	1105	1005000
	2-3x8	745	679000	870	793000	995	906000
	1-6x8	635	558000	740	651000	845	744000
12	1-3x8	1775	1762000	2070	2056000	2370	2349000
	2-2x8	1480	1470000	1725	1716000	1970	1959000
	1-4x8	1270	1260000	1480	1470000	1690	1680000
	3-2x8	985	979000	1150	1143000	1315	1305000
	2-3x8	890	882000	1035	1029000	1185	1176000
	1-6x8	775	724000	880	846000	1005	966000
	2-2x10	910	708000	1060	826000	1210	943000
	1-3x10	1090	849000	1275	991000	1455	1132000
13	2-2x8	1740	1867000	2025	2179000	2315	2490000
	1-4x8	1490	1600000	1735	1867000	1985	2133000
	3-2x8	1160	1245000	1350	1453000	1545	1659000
	2-3x8	1045	1120000	1215	1308000	1390	1494000
	1-6x8	885	921000	1040	1074000	1185	1227000
	2-2x10	1070	900000	1245	1050000	1420	1200000
	1-3x10	1280	1078000	1495	1258000	1710	1437000
	1-4x10	915	771000	1070	900000	1220	1028000

Beam span (ft)	Nominal size of beam	Minimum f and E (psi) for beams spaced: 6′-0″		7′-0″		8′-0″	
		f	E	f	E	f	E
17	2-2x10	1825	2011000	2125	2347000	2430	2680000
	1-4x10	1565	1723000	1825	2011000	2085	2298000
	3-2x10	1215	1341000	1420	1564000	1620	1788000
	2-3x10	1095	1206000	1280	1407000	1460	1608000
	1-6x10	945	1012000	1100	1182000	1260	1350000
	4-2x10	910	1005000	1065	1173000	1215	1341000
	1-8x10	690	742000	805	867000	910	990000
	1-3x12	1480	1341000	1725	1564000	1975	1788000
	2-2x12	1235	1117000	1440	1303000	1645	1489000
	1-4x12	1060	958000	1230	1119000	1410	1278000
	3-2x12	820	745000	960	870000	1095	994000
	2-3x12	740	671000	865	782000	990	894000
18	3-2x10	1365	1591000	1590	1857000	1815	2122000
	2-3x10	1270	1492000	1480	1671000	1695	1909000
	1-6x10	1060	1201000	1235	1402000	1415	1602000
	4-2x10	1020	1194000	1195	1393000	1365	1593000
	1-8x10	780	882000	910	1029000	1040	1176000
	1-3x12	1660	1593000	1935	1858000	2210	2124000
	2-2x12	1380	1327000	1615	1549000	1845	1770000
	1-4x12	1185	1137000	1385	1327000	1580	1516000
	3-2x12	920	885000	1075	1032000	1230	1179000
	2-3x12	830	796000	970	930000	1105	1062000
	1-6x12	720	678000	840	790000	960	904000
	4-2x12	690	663000	805	774000	920	884000
19	3-2x10	1520	1872000	1775	2184000	2025	2496000
	2-3x10	1365	1684000	1595	1965000	1825	2245000
	1-6x10	1170	1414000	1365	1650000	1560	1885000
	4-2x10	1140	1404000	1330	1638000	1520	1872000
	2-4x10	975	1203000	1140	1404000	1300	1605000
	1-8x10	860	1036000	1005	1209000	1145	1381000
	2-2x12	1540	1561000	1800	1822000	2055	2082000
	1-4x12	1320	1338000	1540	1561000	1760	1785000
	3-2x12	1025	1041000	1200	1215000	1370	2055000
	2-3x12	925	936000	1080	1092000	1230	1248000
	1-6x12	805	796000	940	930000	1070	1062000
	4-2x12	770	780000	900	910000	1025	1040000

14	3-2x8	1340	1555000	1570	1815000	1790	2073000
	2-3x8	1210	1399000	1410	1633000	1610	1866000
	1-6x8	1025	1149000	1200	1341000	1370	1531000
	1-3x10	1485	1348000	1730	1573000	1980	1797000
	2-2x10	1235	1123000	1445	1311000	1650	1497000
	1-4x10	1060	963000	1240	1123000	1415	1284000
	3-2x10	825	748000	965	873000	1100	997000
	2-3x10	740	673000	865	786000	990	897000
	1-6x10	640	565000	745	660000	850	753000
	4-2x10	620	561000	720	655000	825	749000
	2-2x12	835	624000	975	728000	1115	832000
15	3-2x8	1540	1912000	1800	2232000	2055	2548000
	2-3x8	1390	1722000	1620	2010000	1850	2295000
	1-6x8	1180	1414000	1375	1650000	1570	1885000
	1-3x10	1705	1657000	1990	1933000	2270	2209000
	2-2x10	1420	1381000	1660	1612000	1895	1842000
	1-4x10	1220	1183000	1420	1381000	1625	1578000
	3-2x10	950	921000	1105	1075000	1265	1227000
	2-3x10	850	829000	995	967000	1135	1105000
	1-6x10	735	696000	855	811000	980	927000
	4-2x10	710	691000	830	807000	945	921000
	2-2x12	960	768000	1120	895000	1280	1023000
	1-4x12	825	659000	960	769000	1100	878000
16	2-2x10	1615	1677000	1890	1957000	2155	2235000
	1-4x10	1385	1437000	1615	1677000	1845	1915000
	3-2x10	1075	1117000	1260	1303000	1435	1489000
	2-3x10	970	1006000	1130	1174000	1290	1341000
	1-6x10	835	844000	975	985000	1130	1125000
	4-2x10	810	838000	945	978000	1080	1117000
	1-8x10	615	619000	715	723000	815	825000
	1-3x12	1310	1119000	1530	1306000	1750	1491000
	2-2x12	1090	931000	1275	1087000	1455	1242000
	1-4x12	935	799000	1090	932000	1250	1066000
	3-2x12	730	622000	850	725000	970	828000
	2-3x12	655	559000	765	652000	875	745000

20	2-3x10	1515	1965000	1770	2293000	2020	2620000
	1-6x10	1300	1648000	1515	1923000	1735	2197000
	4-2x10	1260	1638000	1475	1911000	1685	2184000
	2-4x10	1080	1404000	1265	1638000	1445	1872000
	1-8x10	960	1209000	1120	1411000	1280	1612000
	2-2x12	1705	1821000	1990	2125000	2275	2428000
	1-4x12	1465	1560000	1710	1821000	1950	2080000
	3-2x12	1140	1213000	1330	1416000	1520	1618000
	2-3x12	1025	1092000	1195	1275000	1365	1456000
	1-6x12	970	930000	1130	1085000	1295	1239000
	4-2x12	855	911000	995	1063000	1135	1214000
	2-4x12	730	780000	850	910000	975	1040000
21	4-2x10	1390	1896000	1625	2212000	1855	2529000
	2-4x10	1195	1624000	1390	1896000	1590	2166000
	1-8x10	1050	1399000	1225	1633000	1400	1866000
	1-4x12	1615	1806000	1880	2107000	2150	2409000
	3-2x12	1255	1405000	1465	1639000	1670	1873000
	2-3x12	1130	1264000	1320	1476000	1505	1686000
	1-6x12	970	1075000	1130	1255000	1295	1434000
	4-2x12	940	1054000	1100	1230000	1255	1404000
	2-4x12	805	903000	940	1053000	1075	1203000
	3-3x12	750	834000	875	973000	1000	1112000
	1-8x12	720	789000	840	921000	960	1052000
	1-10x12	570	623000	665	727000	760	830000
22	2-4x10	1310	1867000	1530	2179000	1745	2490000
	1-8x10	1160	1609000	1355	1878000	1545	2146000
	3-2x12	1375	1615000	1605	1885000	1835	2154000
	2-3x12	1240	1453000	1445	1695000	1655	1936000
	1-6x12	1080	1237000	1260	1444000	1440	1650000
	4-2x12	1035	1212000	1205	1414000	1375	1617000
	2-4x12	885	1038000	1035	1210000	1180	1383000
	5-2x12	825	969000	965	1131000	1105	1293000
	3-3x12	825	958000	965	1119000	1105	1278000
	1-8x12	790	907000	920	1058000	1055	1209000
	1-10x12	625	716000	730	835000	835	954000

TABLE 7 FLOOR AND ROOF BEAMS

Live load of 30 psf
Deflection limitation of $l/360$

Plank span (ft)	Nominal size of beam	Minimum f and E (psi) for beams spaced:					
		6′-0″		7′-0″		8′-0″	
		f	E	f	E	f	E
10	2-3x6	1430	1754000	1670	2047000	1905	2338000
	1-3x8	1645	1530000	1920	1785000	2195	2039000
	1-4x8	1175	1091000	1370	1273000	1565	1454000
	3-2x8	915	854000	1070	997000	1220	1138000
	2-3x8	820	764000	955	891000	1095	1018000
	2-4x8	590	546000	690	637000	785	728000
	2-2x10	840	613000	980	715000	1120	817000
11	1-3x8	1990	2035000	2320	2375000	2655	2713000
	1-4x8	1420	1454000	1660	1697000	1895	1938000
	3-2x8	1105	1130000	1290	1319000	1475	1506000
	2-3x8	995	1019000	1160	1189000	1325	1358000
	2-4x8	710	727000	830	848000	945	969000
	2-2x10	1020	816000	1190	952000	1360	1088000
	1-3x10	1220	980000	1425	1144000	1625	1306000
12	1-4x8	1690	1890000	1970	2206000	2255	2519000
	3-2x8	1315	1469000	1535	1714000	1755	1958000
	2-3x8	1185	1322000	1385	1543000	1580	1762000
	2-4x8	845	944000	985	1102000	1125	1258000
	1-6x8	1005	1086000	1175	1267000	1340	1448000
	2-2x10	1210	1062000	1410	1239000	1615	1416000
	3-2x10	810	708000	945	826000	1080	944000
	2-3x10	725	636000	845	742000	965	848000
	1-4x10	1040	909000	1215	1061000	1385	1212000
13	3-2x8	1545	1867000	1805	2179000	2060	2489000
	2-3x8	1390	1680000	1620	1960000	1855	2239000
	2-4x8	990	1201000	1155	1401000	1320	1600000
	1-6x8	1180	1381000	1375	1612000	1575	1841000
	2-2x10	1425	1350000	1665	1575000	1900	1799000
	3-2x10	950	900000	1110	1050000	1265	1200000
	2-3x10	855	810000	1000	945000	1140	1080000
	1-4x10	1220	1385000	1425	1616000	1625	1846000
	1-6x10	735	679000	855	792000	980	905000

Plank span (ft)	Nominal size of beam	Minimum f and E (psi) for beams spaced:					
		6′-0″		7′-0″		8′-0″	
		f	E	f	E	f	E
17	2-3x10	1460	1808000	1705	2110000	1945	2410000
	1-6x10	1255	1518000	1465	1771000	1675	2023000
	2-4x10	1040	1292000	1215	1508000	1385	1722000
	4-2x10	1215	1507000	1420	1758000	1620	2009000
	1-8x10	920	1112000	1075	1298000	1225	1482000
	2-2x12	1645	1675000	1920	1955000	2195	2233000
	1-4x12	1410	1436000	1645	1676000	1880	1914000
	3-2x12	1095	1117000	1280	1303000	1460	1498000
	2-3x12	985	1007000	1150	1175000	1315	1342000
	4-2x12	820	839000	955	979000	1095	1118000
	2-4x12	705	719000	820	839000	940	958000
	5-2x12	655	671000	765	783000	875	894000
18	2-4x10	1170	1535000	1365	1791000	1560	2046000
	4-2x10	1360	1790000	1590	2089000	1815	2386000
	1-8x10	1040	1322000	1215	1543000	1385	1762000
	2-2x12	1840	1991000	2150	2323000	2455	2654000
	1-4x12	1580	1705000	1845	1990000	2105	2273000
	3-2x12	1230	1327000	1435	1549000	1640	1769000
	2-3x12	1105	1194000	1290	1393000	1475	1591000
	4-2x12	920	995000	1075	1161000	1225	1326000
	2-4x12	790	853000	920	995000	1055	1137000
	5-2x12	735	797000	860	930000	980	1062000
	1-6x12	960	1016000	1120	1186000	1280	1354000
	3-3x12	740	787000	865	918000	985	1049000
19	2-4x10	1300	1805000	1515	2106000	1735	2406000
	1-8x10	1145	1554000	1335	1813000	1525	2071000
	3-2x12	1370	1561000	1600	1822000	1825	2081000
	2-3x12	1230	1404000	1435	1638000	1640	1871000
	4-2x12	1025	1170000	1195	1365000	1365	1560000
	2-4x12	880	1003000	1025	1170000	1175	1337000
	5-2x12	820	936000	955	1092000	1095	1248000
	1-6x12	1070	1194000	1250	1393000	1425	1592000
	3-3x12	820	925000	955	1079000	1095	1233000
	1-8x12	785	877000	915	1023000	1045	1169000
	1-10x12	620	692000	725	807000	825	922000
	4-3x12	615	702000	715	819000	820	936000

14	2-4x8	1150	1500000	1340	1750000	1535	1999000
	1-6x8	1370	1723000	1600	2011000	1825	2297000
	2-2x10	1650	1684000	1925	1965000	2200	2245000
	3-2x10	1100	1122000	1285	1309000	1465	1496000
	2-3x10	990	1009000	1155	1177000	1320	1345000
	1-4x10	1415	1445000	1650	1686000	1885	1926000
	1-6x10	915	1415000	1070	1651000	1220	1886000
	2-4x10	705	721000	825	841000	940	961000
	4-2x10	825	842000	960	983000	1100	1122000
	1-8x10	625	622000	730	726000	835	829000
15	2-2x10	1895	2071000	2210	2417000	2525	2760000
	3-2x10	1260	1381000	1470	1612000	1680	1841000
	2-3x10	1135	1243000	1325	1450000	1515	1657000
	1-4x10	1620	1775000	1890	2071000	2160	2366000
	1-6x10	980	1044000	1145	1218000	1305	1392000
	2-4x10	810	888000	945	1036000	1080	1184000
	4-2x10	945	1037000	1105	1210000	1260	1382000
	1-8x10	720	764000	840	891000	960	1018000
	2-2x12	1280	1152000	1495	1344000	1705	1536000
	1-4x12	1095	986000	1280	1151000	1460	1314000
	3-2x12	855	768000	1000	896000	1140	1024000
	2-3x12	770	691000	900	806000	1025	921000
16	3-2x10	1435	1675000	1675	1955000	1915	2233000
	2-3x10	1290	1508000	1505	1760000	1720	2010000
	1-6x10	1115	1266000	1300	1477000	1485	1687000
	2-4x10	925	1079000	1080	1259000	1235	1438000
	4-2x10	1075	1256000	1255	1466000	1435	1674000
	1-8x10	815	929000	950	1084000	1085	1238000
	2-2x12	1455	1397000	1700	1630000	1940	1862000
	1-4x12	1250	1199000	1460	1399000	1665	1598000
	3-2x12	970	931000	1130	1086000	1295	1241000
	2-3x12	875	839000	1020	979000	1165	1118000
	4-2x12	730	699000	850	816000	975	932000
	2-4x12	625	599000	730	699000	835	798000

20	3-2x12	1520	1819000	1775	2123000	2025	2425000
	2-3x12	1365	1638000	1595	1911000	1820	2183000
	4-2x12	1025	1364000	1195	1592000	1365	1818000
	2-4x12	975	1170000	1140	1365000	1300	1560000
	5-2x12	910	1092000	1060	1274000	1215	1456000
	1-6x12	1295	1394000	1510	1627000	1725	1858000
	3-3x12	910	1080000	1060	1260000	1215	1440000
	1-8x12	870	1022000	1015	1193000	1160	1362000
	1-10x12	690	806000	805	941000	920	1074000
	4-3x12	680	818000	795	955000	905	1090000
	2-3x14	985	1003000	1150	1170000	1315	1337000
	1-6x14	860	862000	1005	1006000	1145	1149000
21	2-3x12	1505	1896000	1755	2213000	2005	2527000
	4-2x12	1255	1580000	1465	1844000	1675	2106000
	2-4x12	1075	1355000	1255	1581000	1435	1806000
	5-2x12	1005	1265000	1175	1476000	1340	1686000
	1-6x12	1295	1613000	1510	1882000	1725	2150000
	3-3x12	1005	1249000	1175	1457000	1340	1665000
	1-8x12	960	1183000	1120	1380000	1280	1577000
	1-10x12	760	935000	885	1091000	1015	1246000
	4-3x12	750	948000	875	1106000	1000	1264000
	2-3x14	1085	1160000	1265	1354000	1445	1546000
	1-6x14	950	998000	1110	1165000	1265	1330000
	3-3x14	725	769000	845	897000	965	1025000
22	4-2x12	1375	1818000	1605	2122000	1835	2423000
	2-4x12	1180	1556000	1380	1816000	1575	2074000
	5-2x12	1100	1453000	1285	1696000	1465	1937000
	3-3x12	1100	1436000	1285	1676000	1465	1914000
	1-8x12	1055	1361000	1230	1588000	1405	1814000
	1-10x12	830	1074000	970	1253000	1105	1432000
	4-3x12	825	1091000	965	1273000	1100	1454000
	2-3x14	1190	1334000	1390	1557000	1585	1778000
	1-6x14	1045	1147000	1220	1338000	1395	1529000
	3-3x14	795	883000	930	1030000	1060	1177000
	2-4x14	820	901000	955	1051000	1095	1201000
	4-3x14	595	667000	695	778000	795	889000

TABLE 10 FLOOR AND ROOF BEAMS

Live load of 40 psf

Deflection limitation of $l/360$

Plank span (ft)	Nominal size of beam	Minimum f and E (psi) for beams spaced: 6′-0″ f	6′-0″ E	7′-0″ f	7′-0″ E	8′-0″ f	8′-0″ E
10	1-3x8	2055	2040000	2400	2381000	2740	2719000
	2-2x8	1710	1701000	1995	1985000	2280	2267000
	1-4x8	1470	1453000	1715	1696000	1960	1937000
	1-6x8	875	837000	1020	977000	1165	1116000
	2-2x10	1050	817000	1225	953000	1400	1089000
	1-3x10	1260	982000	1470	1146000	1680	1309000
	1-4x10	900	702000	1050	819000	1200	936000
11	1-4x8	1775	1939000	2070	2263000	2365	2585000
	1-6x8	1055	1114000	1230	1300000	1405	1485000
	2-2x10	1275	1087000	1490	1268000	1700	1449000
	1-3x10	1525	1308000	1780	1526000	2030	1743000
	1-4x10	1090	934000	1270	1090000	1455	1245000
	3-2x10	850	726000	990	847000	1135	968000
	2-3x10	765	654000	890	763000	1020	872000
12	1-6x8	1255	1447000	1465	1689000	1670	1929000
	3-2x8	1645	1957000	1920	2284000	2190	2609000
	2-2x10	1510	1416000	1760	1652000	2010	1887000
	1-3x10	1820	1698000	2125	1981000	2425	2263000
	1-4x10	1300	1212000	1515	1414000	1735	1615000
	2-3x10	905	847000	1055	988000	1205	1129000
	3-2x10	1010	943000	1180	1100000	1345	1257000
	1-6x10	785	713000	915	832000	1045	950000
	2-4x10	650	606000	760	707000	865	808000
13	2-4x8	1235	1602000	1440	1869000	1645	2135000
	2-2x10	1780	1800000	2075	2100000	2370	2400000
	3-2x10	1185	1200000	1380	1400000	1580	1600000
	2-3x10	1070	1080000	1250	1260000	1425	1440000
	1-4x10	1525	1845000	1780	2153000	2035	2459000
	2-4x10	760	771000	890	900000	1015	1028000
	3-3x10	710	719000	830	839000	945	958000
	1-6x10	920	906000	1075	1057000	1225	1208000
	4-2x10	890	899000	1040	1049000	1185	1198000

Plank span (ft)	Nominal size of beam	Minimum f and E (psi) for beams spaced: 6′-0″ f	6′-0″ E	7′-0″ f	7′-0″ E	8′-0″ f	8′-0″ E
17	2-4x10	1300	1723000	1520	2011000	1735	2297000
	3-3x10	1215	1609000	1420	1878000	1620	2145000
	1-8x10	1150	1483000	1340	1731000	1535	1977000
	3-2x12	1370	1489000	1600	1738000	1825	1985000
	4-2x12	1025	1117000	1195	1303000	1365	1489000
	5-2x12	820	894000	955	1043000	1095	1192000
	2-3x12	1230	1342000	1435	1566000	1640	1789000
	3-3x12	820	885000	955	1033000	1095	1180000
	2-4x12	880	958000	1025	1118000	1175	1277000
	1-6x12	1070	1141000	1250	1331000	1425	1521000
	1-8x12	785	837000	915	977000	1045	1116000
	3-4x12	585	639000	680	746000	780	852000
18	3-3x10	1365	1909000	1595	2228000	1820	2545000
	1-8x10	1300	1764000	1515	2058000	1730	2351000
	3-2x12	1540	1770000	1800	2065000	2050	2359000
	4-2x12	1150	1326000	1340	1547000	1530	1767000
	5-2x12	920	1062000	1075	1239000	1225	1416000
	2-3x12	1380	1591000	1610	1857000	1840	2121000
	3-3x12	920	1050000	1075	1225000	1225	1400000
	2-4x12	990	1138000	1155	1328000	1320	1517000
	1-6x12	1200	1354000	1400	1580000	1600	1805000
	1-8x12	880	994000	1025	1160000	1175	1325000
	3-4x12	660	758000	770	884000	880	1010000
	4-3x12	690	796000	805	929000	920	1061000
19	4-2x12	1280	1560000	1495	1820000	1705	2079000
	5-2x12	1025	1248000	1195	1456000	1365	1663000
	2-3x12	1540	1872000	1795	2185000	2050	2495000
	3-3x12	1025	1234000	1195	1440000	1365	1645000
	2-4x12	1100	1338000	1280	1561000	1465	1783000
	1-6x12	1335	1591000	1560	1857000	1780	2121000
	1-8x12	980	1168000	1145	1363000	1305	1557000
	3-4x12	735	892000	860	1041000	980	1189000
	4-3x12	770	936000	900	1092000	1025	1248000
	2-6x12	670	1594000	780	186000	895	2125000
	1-10x12	775	923000	905	1077000	1035	1230000
	2-3x14	1110	1146000	1295	1337000	1480	1528000

14	2-4x8	1435	1999000	1675	2333000	1915	2665000
	3-2x10	1375	1495000	1605	1745000	1830	1993000
	2-3x10	1235	1345000	1440	1570000	1645	1793000
	2-4x10	880	961000	1025	1121000	1175	1281000
	3-3x10	825	898000	960	1048000	1100	1197000
	1-6x10	1145	1885000	1335	2200000	1525	2513000
	1-8x10	780	829000	910	967000	1040	1105000
	4-2x10	1030	1123000	1200	1310000	1375	1497000
	2-2x12	1395	1248000	1630	1456000	1860	1663000
	3-2x12	930	832000	1085	971000	1240	1109000
15	3-2x10	1575	1842000	1840	2150000	2100	2455000
	2-3x10	1420	1657000	1655	1934000	1890	2209000
	2-4x10	1010	1183000	1175	1380000	1345	1577000
	3-3x10	945	1105000	1100	1289000	1260	1473000
	1-6x10	1225	1392000	1430	1624000	1635	1855000
	1-8x10	900	1020000	1050	1190000	1200	1360000
	4-2x10	1180	1381000	1375	1612000	1575	1841000
	2-2x12	1600	1536000	1865	1792000	2130	2047000
	3-2x12	1065	1024000	1240	1195000	1420	1365000
	1-3x12	1920	1843000	2240	2151000	2560	2457000
	4-2x12	800	768000	935	896000	1065	1024000
	2-3x12	960	921000	1120	1075000	1280	1228000
16	2-3x10	1610	2011000	1880	2347000	2145	2681000
	2-4x10	1155	1438000	1350	1678000	1540	1917000
	3-3x10	1075	1341000	1255	1565000	1435	1787000
	1-6x10	1395	1687000	1625	1969000	1860	2249000
	1-8x10	1020	1237000	1190	1443000	1360	1649000
	4-2x10	1345	1675000	1570	1955000	1790	2233000
	2-2x12	1820	1861000	2120	2172000	2425	2481000
	3-2x12	1210	1242000	1410	1449000	1610	1655000
	4-2x12	910	931000	1060	1086000	1215	1241000
	5-2x12	730	745000	850	869000	975	993000
	2-3x12	1095	1117000	1280	1303000	1460	1489000
	3-3x12	730	737000	850	860000	975	982000

20	4-2x12	1280	1819000	1495	2123000	1705	2425000
	5-2x12	1135	1456000	1325	1699000	1515	1941000
	3-3x12	1135	1440000	1325	1680000	1515	1919000
	2-4x12	1220	1560000	1425	1820000	1625	2079000
	1-8x12	1085	1363000	1265	1591000	1445	1817000
	3-4x12	810	1039000	945	1212000	1080	1385000
	4-3x12	850	1092000	990	1274000	1135	1456000
	2-6x12	740	930000	865	1085000	985	1240000
	1-10x12	860	1075000	1005	1254000	1145	1433000
	2-3x14	1230	1336000	1435	1559000	1640	1781000
	1-6x14	1075	1149000	1255	1341000	1430	1532000
	2-4x14	845	903000	985	1054000	1125	1204000
21	5-2x12	1255	1686000	1465	1967000	1675	2247000
	3-3x12	1255	1666000	1465	1944000	1675	2221000
	2-4x12	1345	1806000	1570	2107000	1795	2407000
	1-8x12	1200	1578000	1400	1841000	1600	2103000
	3-4x12	895	1204000	1045	1405000	1195	1605000
	4-3x12	935	1264000	1090	1475000	1245	1685000
	2-6x12	820	1075000	955	1254000	1095	1433000
	1-10x12	950	1246000	1110	1454000	1265	1661000
	2-3x14	1355	1548000	1580	1806000	1805	2063000
	1-6x14	1190	1330000	1390	1552000	1585	1773000
	2-4x14	930	1045000	1085	1219000	1240	1393000
	3-3x14	905	1032000	1055	1204000	1205	1375000
22	5-2x12	1375	1938000	1605	2262000	1830	2583000
	3-3x12	1375	1915000	1605	2235000	1830	2553000
	3-4x12	985	1384000	1150	1615000	1315	1845000
	4-3x12	1030	1453000	1200	1695000	1375	1937000
	2-6x12	900	1237000	1050	1443000	1200	1649000
	1-10x12	1035	1432000	1205	1671000	1380	1909000
	2-3x14	1485	1780000	1730	2077000	1980	2373000
	1-6x14	1305	1530000	1525	1785000	1740	2039000
	2-4x14	1025	1201000	1195	1401000	1365	1600000
	3-3x14	995	1177000	1160	1373000	1325	1569000
	3-4x14	680	801000	795	935000	905	1068000
	1-8x14	955	1122000	1115	1309000	1275	1495000

References

Audel's Carpenter and Builder Guide, Theodore Audel & Co., Boston.
Grading Rules for Western Lumber, Western Wood Products Association, Portland, OR.
Guides to Improved Framed Walls for Houses, US, U.S. Department of Agriculture, Madison, WI.
Maguire, Byron W. *Carpentry for Residential Construction,* Craftsman Book Co., Carlsbad, CA.
Maguire, Byron W. *Carpentry in Commercial Construction,* Craftsman Book Co., Carlsbad, CA.
Manual for House Framing, National Forest Products Association, Washington, DC.
Maximum Spans for Joists and Rafters, Southern Forest Products Association, New Orleans, LA.
MOD 24 Building Guide, American Plywood Association, Tacoma, WA.
Plank and Beam Framing for Residential Building, National Forest Products Association, Washington, DC.
Plywood Residential Construction Guide, American Plywood Association, Tacoma, WA.
Southern Pine Grade Use Guide, Southern Forest Products Association, New Orleans, LA.
Span Tables for Joists and Rafters, National Forest Products Association, Washington, DC.
Thermax Sheathing System, Celatex Corp., Tampa, FL.
Western Woods Species Book, Vol. 1, Western Wood Products Association, Portland, OR.
Wood-Frame House Construction, U.S. Department of Agriculture, Handbook 73, Washington, DC.
Wood Decay in Homes, U.S. Department of Agriculture, Bulletin 73, Washington, DC.
Wood Preservation, Southern Forest Products Association, New Orleans, LA.
Working Stresses for Joists and Rafters, National Forest Products Association, Washington, DC.
Uniform Building Code, Whittier, CA.

Index

C

D

E

F

K

L

M

N

O

P

Q

R

S

T

U

V

W